MW01200113

Clandestine Marriage

CLANDESTINE MARRIAGE

Botany and Romantic Culture

THERESA M. KELLEY

The Johns Hopkins University Press
Baltimore

Printed in the United States of America on acid-free paper
2 4 6 8 9 7 5 3 1

The Johns Hopkins University Press
2715 North Charles Street
Baltimore, Maryland 21218-4363
www.press.jhu.edu

Library of Congress Cataloging-in-Publication Data

Kelley, Theresa M.
Clandestine marriage : botany and Romantic culture / Theresa M. Kelley.
p. cm.
Includes bibliographical references and index.
ISBN 978-1-4214-0517-9 (hdbk. : acid-free paper) —
ISBN 1-4214-0517-2 (hdbk. : acid-free paper)
1. Literature, Modern—19th century—History and criticism. 2. Botany in literature. 3. Plants in literature. 4. Literature and science. 5. Romanticism. I. Title.
PN56.B73K45 2012
809'.93364—dc23 2012002236

A catalog record for this book is available from the British Library.

The Johns Hopkins University Press uses environmentally friendly book materials, including recycled text paper that is composed of at least 30 percent post-consumer waste, whenever possible.

Contents

List of Illustrations vii
Acknowledgments ix
List of Abbreviations xi

1 Introduction 1

2 Botanical Matters 17

3 Botany's Publics and Privates 52

4 Botanizing Women 90

5 Clare's Commonable Plants 126

INTERLUDE ONE *Mala's Garden: A Caribbean Interlude* 159

6 Reading Matter and Paint 162

INTERLUDE TWO *A Romantic Garden: Shelley on Vitality and Decay* 210

7 Restless Romantic Plants and Philosophers 216

8 Conclusion 246

Notes 263
Bibliography 299
Index 325

Illustrations

Following page 66

1 *Wheat Stalk study*, Franz Bauer
2 *Stems and Petioles*, Franz Bauer
3 *Rafflesia arnoldi*, Franz Bauer
4 *The American Cowslip*, Peter Henderson
5 *Dragon Arum*, Peter Henderson
6 *Curious American Bog Plants*, Philip Reinagle, artist, D. Maddan, engraver
7 *Venus Fly-Trap*, William Bartram
8 *Bandana of the Everglades or Golden Canna*, William Bartram
9 *American Lotus, etc.*, William Bartram
10 *Exhibition Extraordinaire in the Horticultural Room*, G. Cruikshank
11 *Trapa bispinosa*, Roxburgh Indian artist

Following page 98

12 *Young Daughter of the Picts*, Jacques Le Moyne de Morgues
13 *Sensitive Flower*, J. J. Grandville
14 *Lichen pyxidalus*, Frances Beaufort Edgeworth
15 *Ophrys apifera. Bee Orchis or Beeflower*, Frances Beaufort Edgeworth
16 *Dryas octopetala*, Frances Beaufort Edgeworth
17 *Bigonia and Aristolochia*, Frances Beaufort Edgeworth
18 *Passiflora* (species not indicated), Frances Beaufort Edgeworth
19 *Magnolia grandiflora*, Mary Delany
20 *Passiflora laurifolia*, Mary Delany
21 *Ixia Crocata (Triandria Monogynia)*, Mary Delany
22 *Physalis, Winter Cherry*, Mary Delany
23 *Rara avis*, after Katherine Charteris Grey
24 *Cycnoches Loddigensii*, Katherine Charteris Grey
25 *Pterostylis obtusa*, Katherine Charteris Grey

26 *Erica pubescens*, Katherine Charteris Grey
27 *Orphrys muscifera, Fly Orchis*, James Sowerby

Following page 178

28 *Banian Tree*, Maria Graham
29 *The Indian Burr, or, Banian Tree*, James Forbes
30 *Convolvulus gangeticus R.*, Roxburgh Indian artist
31 *Pandanus odoratissimus* [fruit], Roxburgh artist
32 *Pandanus odoratissimus* [whole plant]
33 *Borassus flabelliformis*
34 *Trapa bispinosa* [detail], Roxburgh artist
35 *Nymphaea lotus alba*, Roxburgh Indian artist
36 *Nelumbium speciosum*, Roxburgh Indian artist
37 *Nymphea Lotos, Blue Water-Lily of Guzerat*, James Forbes
38 *Leea macrophylla*, Roxburgh Indian artist
39 *Bombax heptaphyllum*, Roxburgh Indian artist
40 *Saccharum exaltatum*, Roxburgh Indian Artist
41 *Calamus rotang*, Roxburgh Indian Artist
42 *Menyanthes cristata*, Roxburgh Indian artist
43 *Pistia stratiotes*, Roxburgh Indian artist
44 *Trichosanthes heteroclita*, Roxburgh Indian artist
45 [*Orchidaceae of Mexico and Guatemala*], G. Cruikshank
46 *Orchidaceae title page*, Bateman
47 *Vignette*, after *Schomburgkia Tibicinis* orchid plate
48 *Cycnoches Ventricosum*, Jane Edwards
49 *Orchidaceae tail-piece*, after Katherine Charteris Grey

Acknowledgments

I dedicate this book to Thomas Baylis, always my companion as I wrote this book in Austin, then Madison, on Lake Superior, across three continents, and over several years.

My work on the book has received generous support, beginning with fellowships from the University of Texas at Austin, the American Society for Eighteenth-Century Studies, and the Harry Ransom Humanities Research Center. At the University of Wisconsin–Madison, the Marjorie and Lorin Tiefenthaler Fund, the English Department, the Graduate School, and the Institute for Research in the Humanities have provided time to write. Residential fellowships at the Henry E. Huntington Library and the Yale Center for British Art supported the final stages of research. The National Endowment for the Humanities and the Simon D. Guggenheim Foundation awarded year-long fellowships as I completed this book. I am most grateful to all.

Numerous archives and their professional staffs have made this book possible, among them the Botany Library of the Natural History Museum, London, in particular Malcolm Beasley, the librarian, and Judith Magee, now curator for the Natural History Library. At the Library and Herbarium of the Royal Botanic Garden, Kew, James Kay, former assistant curator, Prints and Drawings, willingly made resources at Kew accessible. Gina Douglas, former librarian of the Linnean Society, London, introduced me to the resources of that library early on, and on one visit invited me to ride in a van taking Linnaeus's butterfly collection across London. Henry J. Noltie, taxonomist at the Royal Botanic Garden, Edinburgh, has advised this project with unstinting generosity, answered questions of all kinds, and introduced me to the Indian botanical drawings at Edinburgh and elsewhere. If there are mistakes in this book, he is decidedly not responsible. Peter Stevens and David Mabberley have answered nearly as many questions.

Jim Folsom, director of the Huntington Botanical Gardens, and staff members Kitty Connolly and Katrina White have welcomed inquiries and entertained notions. At the Henry E. Huntington Library, Alan Jutzi, Mary Robertson, Steve Tabor, and Jean-Robert Durbin offered archival direction. At the Yale Center for British Art, its Director Amy Meyers and her colleagues, Michael Hatt, Elizabeth Fairman, Gillian Forrester, Scott Wilcox, Lisa Ford, and Melissa Fournier Gold have given this project imaginative and archival support. Librarians at the University of Wisconsin–Madison, Robin Rider, Mary Rader, and Elsa E. Althen, have readily responded to inquiries and located documents. Donald C. Johnson, Director Emeritus of the Ames Library of South Asia, University of Minnesota, Minneapolis–St. Paul, guided my early study of Indian textile traditions.

For invitations to lecture on parts of this book, I thank Anne Secord and members of the Cabinet of Natural History Seminar at Cambridge University, including James Secord, Nicholas Jardine, and Martin J. S. Rudwick; Sonia Hofkosh and other participants in the Romantic Literature and Culture seminar at Harvard University; Karen Weisman and Alan Bewell at the University of Toronto; Paul Cirico at Cambridge University, Leah Marcus at Vanderbilt University, W. Roger Louis, Director of British Studies at the University of Texas at Austin, and Guillermina de Ferrari at the University of Wisconsin–Madison. I am thankful to colleagues have read chapters of this book and debated its claims: Elizabeth Hedrick, Lisa Moore, Lynn Festa, Aparna Dharwadker, David Clark, B. Venkat Mani, Robin Valenza, Julie Carlson, Daniel White, Deidre Lynch, Richard Sha, Rob Nixon, Anne McClintock, Kevin Gilmartin, and Eric Rothstein. Fernando Vidal, Gábor Zemplén, and seminar participants at the Max Planck Institute for the History of Science in Berlin commented on completed chapters.

In India, Deepak Kumar invited me to present my work on Indian botany and the British to other scholars and students at Jawaharlal Nehru University in Delhi. Druv Raina offered a wider historical lens for thinking about European and Indian intellectual exchange. Deepak and Neelam Kumar made us welcome in Delhi. Arun Bandopadhyay hosted our visit to Calcutta and my lecture to his colleagues and students at the University of Calcutta. Aditi Sarkar has long been an intellectual companion to this project, in Ahmedabad and at long distance. Abhay Mangaldas, Head of the House of Mangaldas Girdhardas, kindly introduced me to Mrs. Gina Sarabhai, Chairperson of the Calico Museum of Textiles. Dr. M. Sanjappa, Director of the Botanical Survey of India, and Mrs. Sanjappa welcomed us to Calcutta. Librarians at the Botanical Archives in the Calcutta Botanical Garden in Howrah provided generous access to those remarkable holdings.

For their hospitality to this traveler and project, I thank Sue McMahan, Sherrl Yanowitz, Neil Rogall, John Archer, and Brenda and Martin Adlam. Several graduate assistants have provided valuable archival assistance: Sharon Twigg, Jonathan Ewell, Mark Lounibos, Mary Mullen, Kate Vieira, Matthias Rudolf, Jessica Citti, Jason Cohen, Gwendolyn Blume, Victor Lenthe, Lenora Hanson, Joshua Taft, and finally Naomi Salmon. I thank Yvonne Schofer, librarian emerita indeed, for editing this book superbly as a gift of friendship. Dennis Marshall expertly copyedited the book.

Abbreviations

BG	Erasmus Darwin, *The Botanic Garden* (London: J. Johnson, 1791).
BL	British Library, London
BMNH	British Museum (Natural History), London
CJ	Immanuel Kant, *Critical Judgment*, trans. and ed. Paul Guyer (Cambridge: Cambridge University Press, 2000)
EP	John Clare, *The Early Poems of John Clare, 1804–1822*, ed. Eric Robinson and David Powell, 2 vols. (Oxford: Clarendon Press, 1989)
JC	John Clare, *John Clare by Himself*, ed. Eric Robinson and David Powell (New York: Routledge, 2002)
KU	Immanuel Kant, Vol. 3 of *Immanuel Kant: Schriften zur Ästhetik und Naturphilosophie*, ed. Manfred Frank and Véronique Zanetti (Frankfurt am Main: Deutscher Klassiker Verlag, 1996)
LA	Johann Wolfgang von Goethe, *Die Schriften zur Naturewissenschaft*, ed. Dorothea Kuhn, Wolf von Engelhardt, and Irmgard Müller, 23 vols. (Weimar: Hermann Böhlaus Nachfolger, 1947)
LP	John Clare, *The Later Poems of John Clare*, ed. Eric Robinson, 2 vols. (Oxford: Clarendon Press, 1984)
MP	John Clare, *Poems of the Middle Period, 1822–37*, ed. Eric Robinson, David Powell, and P. M. S. Dawson, 5 vols. (Oxford: Clarendon Press, 1996)
NHPW	John Clare, *The Natural History Prose Writings of John Clare*, ed. Margaret Grainger (Oxford: Clarendon Press, 1983)
SC	John Clare, *The Shepherd's Calendar*, ed. Eric Robinson, Geoffrey Summerfield, and David Powell, 2nd ed. (Oxford: Oxford University Press, 1993)
SPP	Percy Shelley, *Shelley's Poetry and Prose*, ed. Donald H. Reiman and Neil Fraistat, 2nd ed. (New York: W. W. Norton, 2002)
SPPW	William Jones, *Sir William Jones: Selected Poetical and Prose Works*, ed. Michael Franklin (Cardiff: University of Wales Press, 1995)

Clandestine Marriage

CHAPTER 1

Introduction

> This place affords legions of monstrous Plants, enough to confound all the methods of Botany ever hitherto thought upon.
>
> —*James Wallace, "Part of a Journal"*

CONFOUNDING PLANTS

Writing in 1700 about Darien, the site of an early and soon abandoned Scottish colony in Panama, the botanist James Wallace insisted on a confusion that narratives of imperial conquest and taxonomic mastery rarely specify, except as a prompt to greater efforts to create a taxonomic home and name for those plants that seem unwilling to recognize their place in a modern episteme in which mastery, not errancy, is the desired goal. I argue in this book that Wallace's official report to the Royal Society on the strange species he encountered in the New World belongs to an extended counternarrative that within a century made romantic era thinking about plants and nature anything but a settled project. For romantic writers and botanists, as for Wallace, strange plants invited an attraction to material and figurative differences that pushed against epistemic mastery.

It is an irony dictated by the physical form of a book that I must characterize the role of plant matter in romantic botany and its figurations without recourse to that matter. Imagine live plants and dried specimens crossing the globe, some sent in or with letters and across seas, sinking with ships that sink, arriving at destinations that have preserved them since, as best as local climates allow (and some certainly do not). Imagine now their current arrangement in cabinets of natural history museums, not as assorted curios that were arranged, more or less and at times whimsically, in earlier cabinets of natural history but as specimens associated with orders, genera, and species, along with whatever else might illustrate their traits: magazine clippings or early illustrations as engravings or drawings and notes about their taxonomic location and whatever disputes have arisen about the integrity of that location. Much has been preserved, and much lost, all of it matter, including plant matter.

Such arrangements are not innocent of a plan; nor are they instances of an innocent, unscientific seeing, for plants as things cannot be dislodged from the circuits through which they have passed. Bill Brown's hope that material presence might precede or survive cultural objectification as something more or "other" misses what the romantic history of plants insisted on.[1] It is precisely the embeddedness of plants as matter and thing in romantic concepts, particularly as those concepts roll round and back to take in new particulars, that stages one of modernity's most sustained arguments over whether things or their conceptual frameworks constitute a reliable index of what we know. Letters between botanists linger over plants, their habits, and locations. In the romantic era, plants did circulate, they were commodities and were treated as objects, yet they were also treated as particulars, their "curiosity" the delight of those who looked at them or at the illustrations that so many looked at in magazines and books. Against the pull of reification, for profit or pleasure, plants invited notice of their physicality and taxonomic pleasure in their particularity and difference. Here, I surmise, is an instance of a materiality kept in mind not by an unschooled mind but by minds trained to see particularly, even without the better microscopes that came into production in the early decades of the nineteenth century. Because even older microscopes allowed the same plant to appear very large or very small, looking at plants with or without magnification invited attention to things that could be so manipulated or, more precisely, things that could present a variation in structure and detail depending on whether they were or were not seen through a microscope. Brown quotes Theodor Adorno's critique of a man (arguably Hegel made near-sighted by the demands of Spirit) who "looks upon thingness as a radical evil . . . he tends to be hostile to otherness, to the alien thing that has lent its name to alienation, and not in vain."[2]

The romantic era attraction to botanical difference and Hegel's wariness of that difference together acknowledged nature, and in particular plant nature, as an unsettled domain for matter and figure. For despite efforts to bring such plants home by folding them into a systematic and giving them names to insist on their taxonomic location, romantic writing about botany was repeatedly drawn to the way that plant differences give systematic protocols the slip. We could call this admiration another moment in the history of the material sublime, but doing so would also miss the appeal of something edgier that has mostly been absent from aesthetic writing, then as now.[3]

Wallace's account of the "legions of monstrous Plants" he found in Panama was emphatic on this point. Like a modern Noah, he listed the species of animals and plants he collected to bring back, including a "Monkey . . . that chirps like a lark" yet "will never be larger than a rat"—a description that conveys the taxonomic bending of categories that plants also invite throughout the modern era. Looking back at the plants he had to leave behind in Darien, he noted that "if I should gather all, 'twould be enough to load the *St. Andrew*, for some of their Leaves exceed three Ells in length, and are very broad; besides these Monsters, reduceable to no Tribe, there are here a great many of the *European* kindred (but still something odd about them)."[4] Half familiar but still alien, in size and kind these plants registered differences that resist taxonomic order and invite figuration. This impulse arises, for Wallace and the generations of botanists that follow him, from their attraction to the strange, even monstrous, particularity of plants.

Why else would Wallace take home from Darien plants that were strange instead of those that were in some sense familiar? Botanists took on the unfamiliar, at home or abroad, in ways that recognized their material strangeness. Brown takes note of Walter Benjamin's hypothesis about the lingering presence or trace of emancipatory potential in material and visual forms that refuse to be altogether occluded or smoothed over by what Benjamin jeeringly called "homogeneous empty time." I make a related claim for the role of plants in the romantic era as the discovery of new ones or new examinations of domestic species disturb, complicate, or prompt reconfigurations of what Brown calls "relations between the particular and the general, the material and the conceptual . . . the local and the national."[5]

As I use it here, *material culture* refers to the way that human culture stubbornly refashions material objects and lived experience, a refashioning that nonetheless retains some residual sense of the material. I do not claim that culture is limited to or circumscribed by the material. My point is rather that the lived experience of materiality warrants notice. The writers, antiquarians, and wider reading public that were fascinated by fossils, stone circles, geological singularities of many kinds, bird species, habits and habitats as well as floral habitats, names, and tribes persistently speculated about what such evidence might mean and what view of human and world history this evidence might convey.

Botanists who traveled the globe in the early modern era often encountered plants that seemed either wildly unfamiliar, hence monstrous, or spectral evocations of something familiar that was nonetheless outside their ken. With the curious reflexivity that hovered between plants and botanists in the early modern and modern eras, plants mirrored the foreignness of colonial discoverers, a point muted or waylaid by efforts, from Linnaeus on, to domesticate exotic plants on European soil or transplant them around the globe, as if doing so would shake off their foreignness, a stubborn residue of other lands. Here indeed is another instance of the odd affinity that Lorraine Daston and Peter Galison have identified between scientific objects and their discoverers or makers. That affinity in turn questions the narrative rise of objectivity that Daston and Galison trace in the modern era.[6] For seeing in the strangeness of plants a reflection of their own exceptionalism oddly conjoins the object of scientific discovery to a constructed scientific persona. At such moments, the divide between objective and subjective categories is at best a porous membrane. Linnaeus's plant systematic suggests another such moment.

Contemporary thought about Linnaeus's classificatory project has emphasized its role in the desire for, and claim to, epistemological mastery as the work of modernity. Richard Drayton's *Nature's Government*, Richard Grove's *Green Imperialism*, Alan J. Bewell's work on colonial natures in romanticism, and other studies have emphasized the role of taxonomic inquiry in discovering plants and creating empires.[7] In Michel Foucault's early work, this project signals a coming to order that is both linguistic and taxonomic, as well as predictive of the contested biopolitical claims in his later writing.[8] In *The Order of Things*, Foucault describes the rupture between a sixteenth-century regime in which plants and all else were organized according to a doctrine of signatures that explained as well as declared hidden relations among all nature and Linnaeus's eighteenth-century invention of a systematic for plants and then all living beings that relies on visible traits, not hidden correspondences. For Foucault, this rupture is the birth of the modern episteme.

My argument is not that none of this happened. For it is evidently the case that more powerful nations still mine the herbal and pharmacological wealth of less powerful nations. Nor do I wish to debate whether Foucault was right or wrong to insist on an epistemic break between modern and premodern worldviews rather than a longer continuum in which signatures and visibility competed as explanatory models of nature.[9] What interests me is Foucault's account of epistemic visibility, which subtly registers the limiting and filtering that such visibility requires. From his rehearsal of Borges's putative rendition of a Chinese taxonomy that disrupts the ordered patterns modern thought has long used to "tame the wild profusion of existing things" to his lingering notice of plant–animals and monstrous forms that interrupt classifying schemes with "the endless murmur of nature," Foucault's analysis of modern systemic protocols veers repeatedly toward what they cannot control, even as they attempt to do so by identifying and then economizing difference as either the work of classification or an evolutionary project not yet fully understood but to be resolved in some taxonomic future.[10] As a material, literary, and cultural practice, romantic era botany foregrounds what Linnaeus and later taxonomists put aside and Foucault cannot: that plant nature occupies the disputed middle kingdom of nature, neither fully mineral nor fully animal but disturbingly in between.

Plants disrupt the kingdoms of nature not because they are receptive and available to viewers and poets in the sense that Anne-Lise François implies plants might be, but because they convey a resilient hiddenness that defies the Linnaean regime of visibility, even as that regime orders species according to the number and arrangement of reproductive parts.[11] As many readers will already know, Linnaeus's sexual system of plants is called that because he organized it according to the number and arrangement of male stamens and female pistils in each species. The plants he lumped together in the twenty-fourth and last class, which he named Cryptogamia—literally, "clandestine marriages"—were those whose reproductive organs and functionality he could not identify. The mysteriousness of the Cryptogamia (Linnaeus included ferns, mushrooms, lichens, and algae in this class) and their status as outliers in the sexual system of plants prompted even more interest after his death, as botanists, amateur and thoroughly professionalized, kept returning to these plants, some of which offered evidence that eventually prompted the decision to abandon the Linnaean systematic for some version of what came to be called the Natural System.[12]

Linnaeus's curious nomenclature, which emphasized "marriages" of fairly unusual and lurid kinds among various numbers of stamens and pistils, invited still further notice of the figuration his and earlier plant names—common as well as learned—made at once notorious and available, if barely hidden in Linnaeus's frequent use of Latinized or modified Greek versions in his binomial system of Latin names. My title, *Clandestine Marriage*, borrows (or purloins) Linnaeus's term *Cryptogamia*, or clandestine marriages—here the Latin and English plural deftly echo the figurative proliferation of the amazingly diverse "marriages" in his other twenty-three classes—to mark how Linnaean botany hides and harbors the very class of plants that undermines his claim to have created a global systematic based on visible criteria. Equally invisible and equally unclassifiable, this book argues, are the aesthetic pleasure and invitation to figure that move just beneath the surface of global botanizing as a commercial and imperial venture. To understand how plants came to be so understood in the romantic era, in the midst of concurrent claims about knowing plants, their names, and their taxonomic places, I briefly rehearse the early modern view of plants that Wallace echoes in 1700.

For Andrea Cesalpino in the sixteenth century, the task of identifying plants looked so daunting that he compared it to going to war against "legions of monstrous Plants." In *De Plantis* (1583), the first of early modern efforts to systematize the plant kingdom, Cesalpino declared that plants must be organized into categories "just like the battle line of an army," for otherwise "some plant might by chance escape our notice, and, in a way equivalent to those soldiers which at times move on to different groups, a plant can be placed in a category to which it does not belong."[13] Slightly clumsy yet strikingly resonant, Cesalpino's figure of plants that behave like undisciplined soldiers, marching off to join other units, conveys the hint of risk that early modern taxonomists identified as they tried to organize the world's plants, even as explorers were bringing new or newly discovered species back to Europe.

The global array of strange or strangely familiar plants suggested a nature so complex or so secret or both that some early or premodern botanists tried to decode nature's occult or hidden signs by invoking the medieval "doctrine of signatures," whereby certain plants might be identified as cures for illness. As Cesalpino and those that followed him increasingly relied on observation to make taxonomic claims, they still found themselves facing a "legion of monstrous Plants" that needed if not a home at least a set of nested taxonomic enclosures that would not fall asunder with every arrival of plant species Europeans had never seen before.[14] A residual and carefully reduced echo of this anxiety survived in Linnaeus's regime of visible and thus organizable taxonomic differences, inasmuch as the twenty-fourth class of plants—those that cannot be classified elsewhere in the system—was nonetheless named as though this class might still, if perversely and clandestinely, offer plant marriages that the other twenty-three classes conduct in the open.[15]

Romantic era frictions between the ambition to name and classify all plants and a strong suspicion that plants might "confound" any system devised to accomplish this goal, together with its middle position among the kingdoms of nature, made botany an epistemic minefield in an era when collecting and identifying plants (or not identifying them) was a popular field of inquiry among amateur naturalists. Long before Linnaeus published *Species Plantarum* in 1753, the book trade in botanical illustration that began in Holland in the seventeenth century migrated to France, reaching England at the end of the eighteenth century, helped to make botany part of illustrated book culture as well as an outdoor pastime.[16]

Early in the romantic era, British and North American fascination with botany coincided with wider public access to botanical illustration. As artists, engravers, and printers followed the illustrated book and magazine trade from Europe to England, they picked up recently invented printing techniques (aquatint, steel engraving, and color print methods) that made it possible to offer botanical illustrations in books and magazines that middle- as well as upper-class consumers could afford to buy or hand round, even as hand-colored watercolor continued in book and magazine publishing and the production of fine papers insured a still higher standard for botanical illustration. Aesthetic, potentially profitable, and alluring, plants became the leading indicator for public as well as learned inquiry as systematic debates, a burgeoning print culture in natural history books and magazines, and questions about the kingdoms of nature and the nature of species collided with different intensities. In romantic era debates about classification and nomenclature, botany was the disputed middle kingdom between inert forms (minerals) and animals; plant families included species that seemed at the time to include several that were neither all plant nor all animal.[17] For all these reasons, botany offered a conduit to some of romanticism's most persistent inquiries, beginning with the nature of nature and of life and including the debate about whether nature or spirit should dominate, the global market of commodity plants, the relation between scientific inquiry and aesthetic pleasure, and the epistemological value accorded concepts and particulars.

Imperial botany was thus only part, if the most visible part, of a disciplinary and cultural array of practices and commentaries in which botanical ideas operated across romantic culture, from taxonomic efforts to identify thousands of plants that Wallace, among many others, brought back from around the globe to philosophical questions about life and agency and poetic and aesthetic notice of plants, their assorted names, and processes. Wallace's bafflement conveys the epistemological difficulty that, a century later, would make botany something other and more than the work of imperialism. Hidden even in the folds of that imperial story is a highly nuanced and sustained recognition that plants invite notice of an excess that is by turns material and imaginative, natural and human.

Clandestine Marriage takes up the role of botany in romantic debates about life, nature, and knowledge and the work that botanical figures do to unsettle differences between plant and name, between names and ideas, and between scientific and aesthetic inquiries about plant matters. When romantics—poets, botanists, philosophers, and assorted others—thought about plants, they did so across several registers, from the material and the illustrative to the figurative and philosophical. What happened in the name of botany during the romantic era spoke to questions that proved difficult to negotiate in the ebb and flow of new plants, new geopolitical narratives, and debates about how concepts and categories bear a difficult, constantly renegotiable relation to things. These tensions, rather than a simple love of flowers, constituted the ground of what it meant to think about plants as a romantic writer, botanist, artist, or philosopher. Running through all of these was the aesthetic appeal of plants, an appeal that complicated commodity speculation in ways that conveyed more precisely why botany occupied a vexed position in romantic culture.

Romantic era skirmishes about plant nature registered just this turn and difficulty in romantic thinking about whether or how to keep these kingdoms separate. From Aristotle forward, it had been supposed that the kingdoms of nature were more or less distinct. When eighteenth-century experimentalists gathered evidence that some species had traits that resembled species that belonged to other kingdoms, it became more difficult to insist on this point: minerals, like plants, had organic patterns of growth; plants, like animals, seemed to have circulatory systems, and some appeared to be sentient or at least capable of motility and even respiration. In ways that exceeded Wallace's exuberant account of the taxonomic strangeness of Central American plants, it seemed possible that all plants were potentially unruly in kind as well as number. Some plants—the Venus flytrap (*Dionaea muscipula*), for example—were aggressive carnivores. Instead of being specimens that could be passively housed in a given species, genus, family, and kingdom, some plants seemed, from that perspective of allocating categories, restless, unplantlike.

This view of plants as spectral, cross-kingdom travelers was, if anything, a bonus for poetic figure, all the more so because plant names, most apparently common names, often imagined them as cross-kingdom beings, the most common of them bird-to-plant/plant-to-bird crosses like bird's foot trefoil or John Clare's "pouch lipd cuckoo bud." Erasmus Darwin made Linnaeus's marriages the scandalous vehicle for a much wider array of suppositional crossovers between plants, animals, and minerals in *The Loves of Plants*, part 2 of *The Botanic Garden*. Here and elsewhere in romantic era writing, being averse to fixed categories worked within but also against the ongoing taxonomic project of finding classificatory homes and arrangements for all plants.

Romantic era philosophical thought about plant nature took up two consequences of thinking about plants and classification: the first concerned the stability of a hierarchized nature, with plants safely positioned in the middle, below the animal and even further below the human. When Immanuel Kant suggested in his *Critique of Judgment* that plants were innerdirected in their organic development and for this reason seemed to be lifelike, he let loose a possibility that early *Naturphilosophie* writers would celebrate, but that the older Hegel came to regard as a dangerous analogy. One aspect of that danger followed from the relation, in botanical inquiry as well, between individual plants, species, and every other higher category in the taxa developed for ordering nature into kinds or species.

Plants with traits that appeared anomalous to species or kingdom designations dramatized the pull of singularity against supposed secure categories. For Hegel, this possibility conveyed the philosophical unruliness of plants and material nature, unruly because they might not accept spirit's dominion and, even worse, because their contingent, changing status as singularities evaded Hegel's vision of history, in which the world spirit must ultimately speak for the nature from which it arises. Goethe's complex brief for plant materiality and vitality in his essay *The Metamorphosis of Plants* (1790) offered a trenchant challenge to Hegel's mature philosophy of nature. The modern claim that mechanism stands on one side of romantic arguments about nature and vitalism on the other mistakes the fundamental difference exposed by Goethe and Hegel's real disagreement, which concerned not whether plant nature involved mechanical processes, but precisely how those processes operated in tandem with others whose invisible and apparently vital character could not be located outside plants nor seen in their developmental operations.

This romantic debate expanded a much longer and older inquiry about plants, language, and classification. Echoing principles that derived from Aristotle, Cesalpino put it this way: "All science consists in the gathering together of things that are alike, and in the distinguishing and separating of the unlike, that is, distributing them into genera and species, as into companies, according to the difference that nature points out."[18] The crux, as every taxonomist since has acknowledged, is difference, usually and accurately presented by the plural *differentiae* to refer to the traits that are said to distinguish one species from another and finally many others.[19] The seventeenth-century English taxonomist John Ray remarked that this project was "not the task of one man or of one age," and perhaps not any age:

> But I would not have my readers expect something perfect or complete; something which would divide all plants so exactly as to include every species without leaving any in positions anomalous or unique [*sui generis*]; something which would so define [*circumscribat*] each genus by its own characteristics that no species be left, so to speak, homeless or be found common to many genera. . . . I dare not promise even so perfect a Method as Nature permits—that is not the task of one man or of one age.[20]

Ray's distinction between "Nature," which will permit no "homeless" species, and human knowledge, which has to content itself with a partial record that includes homeless species or species that might belong to various genera, at once conveyed and deferred the goal of modern taxonomy: to find a habitation and a name, in taxonomy and in nature, for all species. His Latin diction conveyed the taxonomic goal: to circumscribe and thereby limit membership in each genus and species. Increasingly burdened with more information about newly discovered species, later botanists found it difficult, if not logically untenable, to claim that species are the essential and unchanging taxonomic units of classification.

The problem of botanical names foregrounded what was at stake in Linnaeus's invention of the Latin binomial as a universal, permanent name assignable to all the world's species. Put succinctly, pre-Linnaean botanical names illustrated the unsettled and multiform relation between words and things. If this concern was specifically John Locke's in the *Essay concerning Human Understanding* (1690), it echoed the epistemological project of the newly established Royal Society in 1660, whose members undertook to save knowledge from excessive, wild metaphor that had taken a particularly virulent form in the political rhetoric of the Civil War and interregnum eras. To protect language from the vagaries of metaphor and wandering signification, John Wilkins proposed a lexicon and a series of descriptive tables that would, he believed, establish the "real universal character" of things.

To demonstrate the principle that words lack authority unless carefully embedded in language and grammar that minimize the insufficiencies of words by giving them fixed roles, he presented his friend John Ray's taxonomy of "herbs" or plants as just such a system, as plainly offered as God's visible world. Wilkins characteristically read Ray with a more fixed view of species and their relations than Ray himself had proposed or that his use of local names, sometimes more than one, asserted. Noting twice that when species are repropagated "subordinate species" are continually introduced, Wilkins retreated to higher ground: "I design in the following tables to take notice only of the *chief* families of plants, to which others are to be reduced," and further: "in the description of those Plants which are heads of *numerous families*, I take notice only of that *Communis ratio*, which belongs to all the subordinate varieties of them." At the back of Wilkins's *Essay*, a summary of the parts of plants presented them as a nicely boxed set, one among the many this work advanced to insure that words could not, would not, lead knowledge astray.[21]

Linnaeus hoped to do the same, first for plants and then for all animals, by giving them names and using those names to fix them in a hierarchy of being. Although the Linnaean system seemed to work better for plants than for animals (Buffon's famous attack on Linnaeus's classification of animals marked the most troubled moment in Linnaeus's eighteenth-century reputation), it nonetheless remained the template for later zoological inquiry.[22] When the Linnaean system proved unequal to the taxonomic project, the English turned, long after their European counterparts, to the Natural Orders or Natural System and then, as the evolutionary thesis of Darwin's *Origin of Species* gained authority, to successive phylogenetic taxonomic models that further challenged the status of species as coherent groups over time.

This is not of course to suggest that taxonomists named no species, established no systematic principles, no taxonomic names. To the contrary, many of their early names, including many Linnaeus invented, are still in use. My point is rather that the taxonomic project Alexander von Humboldt declared privately in 1833, the year he published *Cosmos*, is a particularly ambitious expression of a desire for complete systematic knowledge that hovers just below the surface of taxonomic inquiry from the early modern period on: "I have the mad idea to portray the whole material world, all that we know now of the phenomena of the universe and the Earth, from the nebulae of stars to the geography of mosses on granite rocks, all in one work."[23] He had begun that project during the first years of the nineteenth century, while traveling in South America, Mexico, and the United States with Aimé Bonpland. In 1806, Humboldt created the first map and essay on plant geography, *Essai sur la géographie des plantes*. Reissued, it included a folded, folio size, hand-colored chart that depicted the geographic zones of Andean plants from the summit of the Andean volcano Chimborazo to the sea.[24]

Among these Andean plants Bonpland and Humboldt found was a strawberry that looked nothing like its European counterpart.[25] Now even those species that had seemed to Wallace oddly familiar in 1700 had to be reexamined for differences. Put succinctly, the discovery of exotic flora helped to make the exfoliating, defamiliarizing project of botanical difference equally domestic and exotic. By the end of the eighteenth century, the scandal of Linnaeus's "loves of plants," as Erasmus Darwin titled part 2 of his scientific-poetic fantasy *The Botanic Garden* (1791), was hardly covert. Botanical figures emphasized the implicit contradiction between an exultant yet naive celebration of the capacity to gather and organize material nature and the persistent difficulty that beset efforts to do so.

Because plant nature seemed to produce material particulars endlessly (or so it seemed to Hegel and his contemporaries), it threatened Spirit's conceptual hold on particulars, even as new species pushed against or required shifting the terms of systems and species as concepts. Jacques Derrida emphasizes the systemic or inherent antagonism between singularity and abstraction in both empiricism and Hegelian spirit; Adorno similarly challenges Hegelian dialectics for its subtle but inevitable incorporation of contingency and particularity into the realm of Spirit. A cognate difficulty was lodged in the deep time of romanticism, at issue in its revolutionary hopes for individuals and societies, echoed in the darkened mirror of its temporizing defense of human slavery, and in its most material grasp of nature's particulars. As the most aesthetically commanding of those particulars, plants embodied the passage from matter to culture, literature and philosophy, and back.[26]

I argue in this book that botany is the cultural imaginary of romantic nature and, as such, is at issue there wherever nature matters, including nature as matter. Because doing botany meant then as now assigning plants to a tribe or family, a genus, and a species, it had powerfully to do with romantic considerations about where the interests of individuals and collective identities collided. Debates about botanical names foregrounded their invented nature and often their relation to common and local speech. Because plants seemed to be restlessly situated between inanimate and animate kingdoms of nature, their categorical fluidity troubled efforts to keep human and animal species distinct within the animal category, as well as efforts to make distinctions among the human races that would have relegated nonwhite races to animal or lower human species. As the most popular and to a significant degree the most visible of romantic natural histories, botany was the site romantic writers used to stage practical, figurative, and philosophical claims about nature.

SPECIES AND FIGURES

When William Wordsworth in book 7 of *The Prelude* uses the language of classification to survey the human types, inanimate forms, and individuals he sees in the "hubbub" of London, he conveys the paradoxical mix of risk and safety offered by classificatory protocols:

> See—among less distinguishable shapes—
> The Italian, with his frame of images
> Upon his head; with basket at his waist,
> The Jew; the stately and slow-moving Turk,
> With freight of slippers piled beneath his arm.
> Briefly, we find (if tired of random sights,
> And haply to that search our thoughts should turn)
> Among the crowd, conspicuous less or more
> As we proceed, all specimens of man
> Through all the colours which the sun bestows,
> And every character of form and face.[27]

Like its poetic counterparts—personification and allegory—the abstracting language of specimens and types allows a modicum of crowd control and a semblance of distance that is both physical and conceptual. Standing above the crowd, the poet expects to be relieved of "random sights" and unclassifiable individuals. No wonder the blind beggar who turns up later in the book stops the poet in his tracks. His blindness marks the limits of a taxonomic and visual array as schematic as any Wilkins devised to limit figurative malfeasance. For although the blind beggar's written paper explains "the story of the man, and who he was," for the poet these words are "a "label," a "type / Or emblem" that resist the project of identifying and locating individuals within a visible schema. Instead, the poet is compelled by what he cannot see. "As if admonished from another world"; 1805 *Prelude* 7, ll. 610–23, 260). Here what can neither be seen nor be categorized in terms of visible traits trumps the much stagier invisibility of Jack the Giant Killer's coat, which the poet had earlier recalled among the sights of London. Whereas Jack wears a coat with "invisible" written across the back in flames as a device for prompting the viewing audience to imagine he is invisible when he is anything but, here invisibility haunts the poet's effort to create an abstracting taxonomy of London humanity and stops that project cold.

The vantage point Wordsworth's poet wants for himself and for his readers—a showman's platform above London where it might be possible to see as a whole the spectacle that is at close range overwhelming in its parts—echoes an allied desire among romantic botanists. Imagine Wordsworth as a botanist, almost certainly Scottish, traveling up and down the vast Indian subcontinent, trying to order (let alone preserve) specimens of its exotic flora so that they might be assigned positions in a global taxonomy in the name of India and perhaps the British Empire.

Although botanical figures operate in different registers and occasions across romantic poetry, figures and plants that cross kingdoms or the distinction between human and plant species are commanding evidence of the intersections and negotiations of my argument.[28] For recent as well as nineteenth-century writers, plants that appear to mimic other forms of life constitute a material invitation to figure. Writing about such plants in *The Botany of Desire*, Michael Pollan argues that plants are agents of our desire for them, be they apples or orchids. Gilles Deleuze and Félix Guattari suggest in A *Thousand Plateaus* that "wasp and orchid, as heterogeneous elements, form a rhizome. It could be said that the orchid imitates the wasp, reproducing its image in a signifying fashion."[29] Plant mimicry, especially among the English orchises, is the romantic standard-bearer for plant to animal or human crosses that invite poets to perform their own figural mimicry.

What is engaging about the way romantic poets take up this invitation is their effort to sustain the game and its argument. Catherine de Zegher, echoing Pollan as well as Deleuze and Guattari, suggests that botanical illustrations are rhizomatically allied to plants such that the mimetic faculty traduces the difference between nature and art.[30] Yet this claim is not how Deleuze and Guattari talk about the orchid and the wasp; nor does it permit what romantic poets and botanists imagine: a shadow dance in which plants, art and figure sustain identities, with something like the force of a Kantian *as if* that keeps the difference between matter and figure in play. Collapsed into sameness, the rhythms of figuration and mimesis get foreshortened. Attributing them to nature muffles the difference between botanical form, name, and figure that is so compelling in romantic culture and our own. Contending that the agency of this exfoliation is the leaf or the orchid occludes a trace or network in which the mind plays off and on filiations between the animate, inanimate, sentient, and nonsentient. It is rather the interplay between the materiality of botanical forms and their figurative "residuals" that command notice. Among those residuals is the exotic and at times excessive relation between systematic identification and the discovery of new species. This, too, attracts mimesis and extension—in short, the work of figurality.

One argument of this book concerns what happens when botanical particulars resist or exceed conceptual location, either because they are hybrid, morphologically resistant, or because their names invite imagining them beyond the categories they putatively occupy and instantiate. These alien and playful remainders, schematic questions, and plants as particulars collectively represent the pressure of contingency and matter within and against conceptual systems. Operating within the horizon of imperial gain, which hoped that plants might create economic stability and surplus value, romantic botany speaks to the possibility of a material and aesthetic pleasure that cannot be economized.

I pursue this argument in seven more chapters. Chapter 2 characterizes debates about taxonomy and morphology in England and on the Continent between 1750 and 1850, with particular attention to discussions of the adequacy of Linnaean systematics and nomenclature, debates about the species category, the development of the Natural System, and evolutionary possibilities that Charles Darwin formulated for a scientific and wider public in *Origin of Species*.

Chapter 3 assesses the public and private reception of botany in the romantic era, in private letters, parodies, and the periodical press, its circulation among Quaker botanists, its prominence in illustrated print culture, and the poetic botany of Erasmus Darwin's "Loves of the Plants." Chapter 4, "Botanizing Women," tracks passive and not so passive versions of this category, from women as botanical figures to those who themselves botanized within a public sphere that was increasingly suspicious of women who worked and thought and, above all, published what they wrote. I consider some edgier moments in the way women botanize, among them an argument about women and botany between two radical writers, Mary Wollstonecraft and Anna Barbauld; women characters whose absorption in natural history means social or marital trouble; Maria Jackson's ambitious books that issue suppressed challenges to Linnaean principles; the botanical caricaturist of pressed flowers identified as the Lady Jane Grey of Groby, whose public name is half true, half disguise; the paper mosaic botanicals of Mrs. Delany; the botanical drawings of Frances Beaufort Edgeworth; and the poet and novelist Charlotte Smith (1749–1807), who wrote about natural history across several genres, maneuvering with some care in the highly negotiated cultural space available to women writers who depended on sales to live.

In chapter 5, I argue that John Clare is less a poet whose poetic career was in some sense curtailed by the enclosure of his native village, as many have argued, than a poet whose poetics of place, and especially of plants, would have had to posit enclosure had it not already happened. My claim is that legal enclosure and its affront to the traditions of commonable right triggered Clare's poetic invention of a system or network of "commonable" plants that hovers, imaginatively, holding together a network of rights and practices that enclosure was designed to erase. The romantic poet who wrote the most about botany and natural history, and arguably the one who knew as much about plants as any of his contemporaries, Clare used botanical figures to specify a resistance to nomenclature that works in tandem with his critique of enclosure and grammar. In this doubled lexicon of botanical figure and poetic resistance, the names of plants became for Clare the ground for his challenges to patrons and publishers and, behind them, the long-dead Linnaeus, whose designs on flora and fauna, Clare instinctively recognized, were similarly authoritarian. To these authorities and the history of enclosure in which the sway of political and local authorities is sharply inscribed, Clare responded with a retrospective and imagined topography in which the natural history of plants constitutes a moving, relational, and invented map that suggests how its human and plant inhabitants might yet cohere.

Interlude 1, which precedes chapter 6, introduces the imperial section of the book by considering the Caribbean side of Abbé Guillaume-Thomas Raynal's famous analysis of global and colonial commerce, *L'Histoire des deux Indes.*[31] I read Shani Mootoo's novel *Cereus Blooms at Night* as it looks back in anger on colonial botanical practices and documents that say little about Caribbean plant knowledge despite ample evidence that slaves and Amer-Indians knew perfectly well how to use plants to poison their masters and ruin crops. Chapter 6 engages the other pole of Raynal's treatise, the Indian subcontinent, by offering a double analysis of Indian botanical art and the British who commissioned that art and studied Indian plants. Using archival and pictorial evidence, I consider the mismatch between openly imperial British rhetoric about Indian botany and actual British and Indian practices. For William Jones, William Roxburgh, and other East India Company officials who hired Indian artists to paint plant specimens, Indian botany and illustration constitute a telling microcosm of the dynamics and instabilities that mark romanticism's engagements with this "other."

Interlude 2 reads Percy Shelley's poetics of plant nature, growth, and decay to introduce the philosophical argument that follows in chapter 7. The interlude begins with Shelley's *The Sensitive-Plant,* which chronicles the life and death of the *Mimosa*, a plant that is often invoked in romantic debates about the sentience or agency of plants, and then revisits claims about such plants and figures by reading the role of decay and death in the life of nature in Shelley's last poem, *The Triumph of Life.* Chapter 7 then assesses readings of plant nature that arise from German *Naturphilosophie* but assume their definitive, antithetical character in the writings of Goethe and Hegel. Hegel's more focused and wary critique of Goethe's 1790 essay on plant metamorphosis took several turns as the younger philosopher moved more decisively away from his early German colleagues. Read together, Goethe and Hegel remake the putatively tidy division between vitalist and mechanical forms of life that some have claimed as fundamental to the romantic idea of life.

Chapter 8, the conclusion, begins with a reading of James Bateman's evocative 1838–43 monograph *The Orchidaceae of Mexico and Guatemala,* a work that capitalized on the increasing attention given to wild and exotic orchids in romantic print culture from the 1790s to the 1840s. Bateman's elaborately emblematic structure and text include satiric caricatures that depict orchids mutating into animals and people. I argue here that romantic orchids embody the slippage between the scientific, the literary, and the aesthetic that botany presents across many species and genera, but nowhere so provocatively as it does with orchids. As the number of known species and genera of orchids spiraled, the invitation to imagine orchids as people, insects, and animals intensified, as if in admiring reply. I conclude the book with thoughts on plant mimicry, especially that of orchids, as a romantic and contemporary invitation to think about where plants belong and what they are.

CHAPTER 2

Botanical Matters

DIFFERENTIAE

> Je bornerai là ces exemples. Ils montrent que tous nos efforts sont impuissants, en présence des relations multiples qu'affectent de toutes parts les êtres qui nous entourent. C'est la lutte, dont parle le grand botaniste Goethe, de l'homme contre la nature infinie. On est assuré toujours de trouver l'homme surpassé.
>
> (I will conclude these examples here. They show that all of our efforts are powerless in the presence of multiple relations that affect all parts of the beings that surround us. It is the struggle, as the great botanist Goethe says, against infinite nature. We are always certain to find humankind surpassed.)[1]

Published after three centuries of European exploration and taxonomic activity around the globe, Henri Baillon's exhaustive classification of the Euphorbia, a large plant family with many exotic species and genera, offers the extraordinary admission, quoted above, that "it is the struggle . . . against infinite nature." Precisely because such monographs describe all known species and genera of a given plant family and analyze relevant taxonomical debates, they tend to offer confident hypotheses and learned conclusions. Baillon breaks with this convention only in the sentences quoted above. The rest of the monograph runs true to type, as it were. Three hundred or even one hundred years earlier, a monograph of several hundred pages and numerous engraved plates devoted to a single plant family would have been inconceivable. Yet Baillon appears to be nearly as uncertain about nature's things as his predecessors had been—among them Andrea Cesalpino in 1583 and James Wallace in 1700.

Some botanists practiced what Sir J. E. Smith characterized as "philosophical" or speculative botany rather than the analysis of plant groups that Baillon offers. However, the two practices were typically distinct. The theoretical and philosophical implications of taxonomic evidence might be extraordinary, but taxonomical papers were and still are characteristically circumspect in presenting those implications. Robert Brown, the most brilliant taxonomist and botanist of the first half of the nineteenth century in England, is a notorious case in point, reticent nearly to a fault about the larger implications of his botanical investigations.[2] His practice mirrors a widespread scientific and cultural disposition that supposes minute, particular observation and analysis constitute the work at hand.

Baillon's departure from this protocol is almost certainly not unique, but it is sufficiently unusual to warrant asking what it means. That it occurs not at the end of the work but at the end of its penultimate section suggests that its author recognized its anomalous relation to the conventions of the taxonomic monograph. He concludes the work without further reflection about who or what might win or lose in the struggle between infinite nature and taxonomic man. I argue in this chapter that the special interest of botany in and for romanticism derives from its complex engagement with the internal conflicts that Baillon's 1858 monograph reprises: tenacious investigation of plants coupled with closely reasoned taxonomical debate, on the one hand, and, on the other, doubts about the possibility of mapping nature's kinds. Committed to a level of particularity about nature that nonnaturalists and nontaxonomists would find numbing, botanists from the time of Linnaeus forward have experimented with successive protocols for mapping plant relationships: reproduction, natural affinities, evolutionary histories, and finally genetic mapping, called cladistics, which maps complex interrelations among plants and taxa. The history of this taxonomic effort witnesses what romantic era botanists began to recognize: that botanical nature persistently defied their best efforts.

Aristotle introduced the term *differentiae* to specify key differences between things. Centuries later Cesalpino used the same term to refer to features that distinguish one plant from others in a species or species within a genus. As this long continuum suggests, differentiae became the unending order of business in botanical thought. As Cesalpino presented it, the Latin term refers to the specific traits or "characters" of plants and animals, not in and of themselves but as markers of contrastive distinction.[3] In the modern translation of Linnaeus's *Philosophia Botanica*, the *differentiae* are called Definitions, in order to signal that species and genus definition is what differentiating characters make possible.[4] Historians of botany have mostly agreed that for Aristotle and his student Theophrastus, the notion of differentiating characters invokes Plato's theory of *eidos*, or universal ideas, insofar as the natural things we see, such as trees, flowers, birds, and animals, express, but are not the same as, universal forms or ideas.[5]

In the taxonomic debates that intensified with Linnaeus's publication of his *Species Plantarum* in 1753, the work of differentiation became more complex and also more fraught as naturalists worked with, debated, and then rejected the Linnaean systematic in favor of, at least in the short term, the Natural System. After the publication of Darwin's *Origin of Species* in 1859, his and subsequent accounts of evolutionary relationships in their turn challenged the stable familial relationships hypothesized by the Natural System. Whether between minute particulars and larger claims (including those that might unravel a tentative hierarchy or organization of those particulars); between one systematic for classifying plants and another; or between different versions of the same systematic; between species and individual plants; between species and genera; or between genera and families, differentiation is at issue in all the levels of systematic inquiry. The radical exfoliation of differentiation in this period became the disputed signature of botanical inquiry, and it has remained so since. The implications of this recognition extend from thinking about nature itself, that watchword of romanticism, to thinking about how botany, as a practice and idea, authorizes difference and permutation.

To begin mapping the roles of differentiation as the practical and philosophical work of botany during the romantic era, I open with a detailed overview of the major taxonomic debates and players to show what is at stake as naturalists and philosophers think about taxonomic inquiry. My argument then widens to engage philosophical and logical questions that reprise Aristotle's and Theophrastus's arguments for a logical schema of nature's kinds. Botany and natural history at large turn the tables on this ancient and modern synthesis of logic and matter.

I then consider species and individuals as the place where such differences begin as they are made to count or not, depending on whether one values individuals and singularities or their supervising categories or, alternatively, on whether such categories do indeed supervise individuals. My third and longest section here describes the Linnaean and the Natural Systems that jostled for preeminence from about 1750 to 1850. Charles Darwin's evolutionary model of species change and development radically undermined both, although Linnaeus's binomial Latinate nomenclature has survived more or less intact. Since the early twentieth century, phylogenetic and molecular systems have continued to refashion taxonomic understanding of species and the higher taxa, a generic term for taxonomic categories above the species category. A fourth section introduces the linguistic and figurative opportunities provided by botanical nomenclature on various systematic hooks. The last section sketches the philosophical implications of these botanical matters.

Linnaean systematics dominated botanical practice from about 1750 to 1810 and, among amateurs and field naturalists, long after that.[6] Based on an enumeration of the reproductive parts of flowers and their arrangements, the Linnaean system was easy to learn and use. It was also and significantly the first systematic that claimed to account for all plants and animals. Linnaeus divided the plant kingdom into twenty-four classes and assigned species to those classes, or, more precisely, to the first twenty-three classes, depending on the number of stamens or stamina. Within each class, the number and arrangement of pistils determined the subclass or "order" of the plant. The twenty-fourth class, the Cryptogamia, included all plants that at this period were thought to "hide" or disguise their reproductive parts and procedures. Readily admitting that this system is artificial in the sense that it imposed an order based on a single, if highly significant, feature of most plants, Linnaeus also identified several protocols for establishing a "natural" system of classification.

The so-called Natural System gained prominence in Europe in 1789, when Antoine-Laurent de Jussieu published *Genera Plantarum,* a work that lists one hundred natural plant families, a number that well exceeds those identified by Linnaeus and his predecessors. Jussieu determined plant membership in families by tracking the traits shared by species within each genus in that family and, within each genus, a smaller set of traits said to be common to all plants in a given species. These traits were to be identified, not imposed, by tracking leading familial resemblances that might be quite visible, as in the Umbelliferae, a plant family known for its umbrella-like stem and leaf structure, or minute, as in the triggering anatomy of many orchis species that were assigned to a genus of the large orchid family. Microscopic investigation and botanical analyses by Jussieu, Michel Adanson, Robert Brown, and Augustin Pyramus de Candolle, among others, established the Natural System over the Linnaean before the end of the eighteenth century on the Continent and by 1810 in England, at least among professional botanists.

The differences between Linnaean systematics and the Natural System are neither simply nor only that the first is an "artificial" method and the second "natural," despite the propaganda to that effect in the ground-clearing prefaces of advocates of the Natural System. Briefly put, an "artificial" systematic relies on a single trait (such as Linnaeus's pistils and stamens), whereas a "natural" one proposes to find resemblances in nature that direct the construction of plant families and higher taxa.[7] Taxonomists argue that the latter method is more robust because it yields taxonomic results that are likely to survive even if subsequent inquiry challenges the analysis of one or more traits. Linnaeus acknowledged that his system was artificial; he also acknowledged a number of natural families and apparently believed that his genera were natural rather than artificially constructed (Jussieu would later dispute the "natural" character of Linnaean genera). Linneaus used the term *affinity* to characterize the fundamental taxonomic protocol for the development of a natural system long before Jussieu did.[8] Larson suggests that Linnaeus probably believed that the discovery of a system of natural affinities among plants would be roughly akin to finding "the philosopher's stone."[9]

There are nonetheless compelling differences between Linnaeus and his challengers. The first leads the rest: Linnaeus chose to emphasize only one aspect of plants—their reproductive parts. The botanists who developed the Natural System urged attention to several morphological features that could show affinities between species that might otherwise be identified with different genera. As any modern reader of Goethe's *Elective Affinities* knows, extrafamilial affinities complicate and may even undo the stability of family relationships when wives, lovers, and babies resemble each other against the grain of ratified social and familial alliances. The history of the Natural System from Jussieu to Charles Darwin tells a similar tale of differences that disturb the systematic that gave those differences their taxonomic identity.

Although romantic era debates about whether to support or abandon the Linnaean system in favor of another rarely took the middle ground (as in "let's use them all"), the length and complexity of these debates created a space for thinking about competing models of relationship as the disputed and disputable ground of nature. In terms of political economy, the Linnaean system, as so many have remarked, imagines a closed and well-defined taxonomic economy in which species and genera are distinct and the sexual hierarchy is evidently that of male determinations first, female ones second, in a distinctly secondary role. The Natural System recognized diverse hierarchies. It also marked relationships that cross and complicate those hierarchies and, within the range of traits for species and genera, it noted variations that muddy sharp distinctions between categories. Linnaeus and many of his successors were keen about the political and sexual economy of Linnaean nature, although Erasmus Darwin's account of that system would make clear that it also makes available transgressive sexual and political valences.

As advocates of the Natural System emphasized, the attention to affinities across established categories invited a proliferating array of figurative economies akin to those suggested by Goethe's novel. For what looks "natural" in the Natural System may well require artificial and in effect figurative adjustments, akin to the work of metaphor as it turns away from literal to figurative. Neither all natural nor all constructed, the nature represented in romantic botany is a contested and various medium, its parts by turns material and conceptual, neither wholly natural nor wholly artificial.

For Linnaeus until late in life and for others as late as the mid-nineteenth century, the ancient conviction that species are fixed, universal, and eternal[10] jostled with an accumulating record of newly discovered species that demanded both the practical work of differentiation and, over time, more categories and more reassignments of plants to different genera and even species. For if on the one hand the material plenitude of newly discovered species demanded, or was understood to demand, a classificatory schema or taxonomy that could include and order every species, taxonomic inquiry from Cesalpino forward persistently urged the adoption of new criteria, further differentiation of categories, and not infrequently the removal or reassignment of plant groups, sometimes even to locations outside the perimeters of the plant kingdom.

The drive toward differentiation as much as toward similitude, so basic to the history of taxonomy, insists, then, on the very logical principle that would thereafter make it difficult to settle on one systematic over another. The world's botanical differences may, as categories are braided and unbraided, challenge the viability of categories, including that of species. This point becomes obvious if we compare Linnaeus's original set of categories (kingdom, class, order, genus, and species) with the expanded Linnaean model biologists developed and used until 1970 (kingdom, phylum, subphylum, superclass, class, subclass, infraclass, cohort, superorder, order, suborder, infraorder, superfamily, family, subfamily, tribe, subtribe, genus, subgenus, species, subspecies).[11] When the English naturalist J. E. Bicheno complained that French systematicists were forever subdividing Linnaean categories, he recognized, albeit negatively, the inconstant constant in modern taxonomy: its unremitting genius for subdivision, renaming, and reassignment of species, genera, families, orders, even membership in a kingdom.[12] The problem as Bicheno understood it is that the French muddle the great task of classification—to find the unity of things—by pursuing too many details and creating too many subcategories to organize those details.

But the problem goes deeper than the Anglo-French tendency to indulge in mutual caricature. The dispute about whether taxonomy is possible has a parallel history that begins with Buffon's mockery of the Linnaean system and continues to the present. John Dupré argues that "the disunity of science is not merely an unfortunate consequence of our limited computational or other cognitive capacities, but rather reflects accurately the underlying ontological complexity of the world, the disorder of things."[13] What contemporary debates about systematics offer instead is a field of taxonomic difference that can be submitted to different kinds of protocols, none of which is wholly congruent with all others and some of which work better for some kingdoms of natural history than for others.

Bicheno's critique of the French points to the centrifugal pressure of taxonomic differentiation that, insofar as it subdivides more than it unifies, disables the taxonomic project. For Bicheno, the history of taxonomy from Linnaeus to the Natural System marked a critical disabling of that project. Whereas Linnaeus was concerned to identify species and worked out a descending series of artificial categories down to the level of species, where the number and proportion of fructifying parts was "most obvious and least liable to vary," the Natural System works in the opposite direction, from "genera . . . upwards to orders, and orders we combine in classes." The effect of this shift in emphasis, Bicheno argued, was to "fritter away characters which are essential to the use of a genus, and destroy our power over it when we proceed to generalise."[14] The other side of this story is, however, the impediment to generalization that such particulars presented to systematicists from Jussieu on as they both identified distinctive characters of species and genera and admitted that members of these groups might not fully accord with the list of defining traits. As the taxonomic value of natural families lessened and that of the higher orders increased, Scott Atran observes, interfamilial as well as intrafamilial similarities and differences attracted more notice.[15] Scholars have implied a similarly retrospective effect in Linnaeus's emphasis on species identity over against intraspecies difference, an emphasis that devalued varieties within species. Critics of Linnaeus, among them botanists who relied on the Linnaean system, noticed intraspecies variations or recognized different species where Linnaeus identified none or one.[16] In an 1867 English translation of *The Vegetable World*, Louis Figuier remarked that Linnaeus's emphasis on distinguishing one species from another "insists much on their differences, little upon their resemblances."[17]

Modern taxonomic inquiry is at times skeptical about nearly every claim it makes about categories: whether there are species at all; whether higher taxa exist and can be identified; whether, in short, the work of taxonomy can ever be complete or completely right.[18] There are perhaps two ways to understand the problem: one is the pragmatic recognition among taxonomists that the work of taxonomic distinction is ongoing—that recalibrations are necessary adjustments to the complexity of things and new discoveries about how things might be related. The other is a more radical skepticism that supposes taxonomic work is never done because it cannot be. Whatever position one takes about the difficulty of ascertaining taxonomic order, that difficulty invites, has long invited, these questions. What happens if we cannot encompass the immense variety of the world and its peoples with a theory and organization of our own? What if all these—plants, animals, peoples—evade our desire to manage and profit from such complexity?

A more pointedly philosophical version of these questions operates in tandem with but also in opposition to imperial ambition. What if the effort to mark singularities—those individual plants—fails? If, as Michel Foucault insisted, the ordering of things and words was the great Enlightenment project, what happens if nature itself cannot be so ordered?[19] What will knowledge look like if it is not hierarchical and stable? Where do individuals fit if they are only individuals without categories? These questions are discernible in what comes before, as far back as Aristotle's theory of essential kinds, but more insistently in the taxonomic debates that Linnaeus's sexual system helped to inaugurate. Written all over botanical and extrabotanical debates in this period are problems of borders and definitions that register a highly metafigurative interest in the security of territories and bodies, whether human or animal or vegetable, native or alien. Those problems begin with asking what the term *species* means, a question at once radical and foundational to taxonomic knowledge.

SINGULARITY AND SPECIES

> Some thing I think was Due to Mee from the Common Wealth of Botany for the great number of plants & Seeds I have annually procur'd from Abroad, and you have been so good as to pay It, by giving Mee a species of Eternity (Botanically speaking), That is, a name as long as Men and Books Endure. This layes me under Great Obligations, which I shall never Forget.[20]

The eighteenth-century London physician and naturalist Peter Collinson, who wrote these lines to Carl Linnaeus in 1739, corresponded with naturalists in America and Europe as well as England, requesting or sending plants and commenting on a wide variety of topics in natural history. In this letter to Linnaeus, Collinson thanks him for naming the North American horsemint plant *Collinsonia Canadensis*. This "species of Eternity"—hovering somewhere between botany and figurative coin—began as a plant specimen that Collinson received from his fellow Quaker, the Philadelphia botanist John Bartram, and then sent on to Linnaeus.[21]

Collinson's rhetoric illustrates the value attached to having a species named after oneself, a distinction he urged that Linnaeus extend to Bartram as the naturalist who had supplied them both with so many new plants.[22] As late as the 1840s, when few still argued that species were fixed and eternal, botanists still supposed that naming plants and an occasional genus honored plants as well as persons.[23]

Yet the term *species*, quite apart from whether it could be said to invoke "Eternity (Botanically speaking)," as Collinson believed, has a troubled history in philosophical definition and taxonomic practice, where the proclaimed stability of a species matters a great deal to the task of classifying matter. Taken from the Greek *eidos*, *species* means "appearance"; in the singular, *specie*, it can also mean money, as in coin or kinds of money.[24] Dr. Johnson's eighteenth-century *Dictionary of the English Language* canvassed the range of meanings available in modern English: *species* can mean "a sort; a subdivision of a general term"; "a class of nature" or "single order of beings"; "appearance to the senses"; a "representation in the mind" or "a show or exhibition"; "circulating money"; or "simples" in a pharmaceutical compound.[25] Biologists have long claimed that Aristotle believed natural kinds or species were eternal, fixed, platonic categories.[26] Atran argued against this engrained consensus, suggesting that whereas Aristotle supposed the array of geometrical forms to be observed fact, he did not make the same claim about species or natural kinds. These he instead treated in terms of their "*materially* adapted nature."[27] The problem of definition turns, then, on how one understands the term *eidos*. For Plato, it evidently signaled an idea in the mind to which things in the world are approximations. Aristotle presents a more generally pragmatic insistence on appearance and phenomena as the grounds of evidence. Yet early modern biologists followed Cesalpino in thinking that Aristotle's view of the species was in some sense platonic.

Like any coin of the realm, a species is money in the bank: it organizes individual plants and thereby gives them taxonomic interest. As such the term *species* remained for centuries, as it had been for Aristotle, linked to appearance: one determined to which species a plant belonged by looking at it. Some of that looking might be under a microscope, but until the nineteenth century most taxonomic gazing began with what the naked eye could see more or less, hence the mystery of Linnaeus's Cryptogamia, those plants whose reproductive apparatuses were neither apparent nor schematically recognizable when compared with those of the flowering plants. Curiously, Aristotle did not discuss the terms *species*, *differentiae*, and *genus* in his biological works, yet from Cesalpino forward these terms are critical to the classificatory project.[28]

The story of modern doubts about whether species are fixed or even natural begins, as do so many modern stories about language and knowledge, with John Locke. The Lockean definition that turns up in etymological dictionaries looks unremarkable: "the individuals that are ranked into one Sort, called by one common Name, and so received as being of one Species" (Locke, *Essay*, 3.6.8, p. 443).[29] But lexical appearances can be deceiving. In the paragraph from which this definition is taken, he assails the ontological status of species as the log-splitter of taxonomic inquiry, asserting that "Species of Things" are given names that correspond to "the complex Ideas in us; and not according to precise, distinct, real *Essences*" (*Essay* 3.6.8, p. 443).

Returning to this point by different routes throughout the *Essay*, Locke insisted that what species mark out are "nominal" rather than "real" essences and, further, that species mark out "properties" humans identify as belonging to a group, not real individuals that they assign to that group (*Essay* 3.6.6, p. 442; 3.6.20, p. 449). Nor was he persuaded that propagation constitutes a basis for identifying a "real *Species* distinct and entire," inasmuch as mating across nominal species occurs with some persistence, yielding the odd admixture (as in the issue of a cat and a rat) or "monstrous Productions" that defy efforts to assign them to a species (*Essay* 3. 6. 23, pp. 451–52). With the rhetorical flourish of a peroration—which this would be if he had not also returned to the topic as yet another instance of "the Abuse of Words" later in the *Essay*—Locke asked: "Wherein then, would I gladly know, consists the precise and *unmovable Boundaries of* that *Species* [Man]? 'Tis plain, if we examine, there is *no* such thing *made by Nature*, and established by Her amongst Men" (*Essay* 3.6.27, p. 454). Returning to this theme with variations in his discussion of how words and language may be abused, Locke rounded back to Aristotle's definition, but then flips the difference: our understanding of species is grounded in their appearance and thus founded in the knowledge we gain from our senses and not in an unverifiable notion of biological essence. He argues the point by invoking the more typical English translation of *eidos* as "idea," which Locke understood to mean visual appearance, not platonic idea (*Essay* 3.11.19–24, pp. 518–21). Atran might say that Locke here reclaimed Aristotle's pragmatic rather than platonic view of species.

At the edges of otherwise predictable deployments of zoological and botanical categories, the idea of a species is similarly entangled. One definition understands humankind to be "the species," suggesting thereby that all other species are subordinate to the human, as in this statement from the first number of Joseph Addison's *Spectator*: "I live in the World, rather as a Spectator of Mankind, than as one of the Species" (*Spectator*, 1: I, 4). Dr. Johnson's use of the term *species* to describe Shakespeare's characters flattens what is elsewhere a key distinction between individuals and species: "in the writing of other poets a character is too often an individual; in those of Shakespeare it is commonly a species." Johnson's early enunciation of the now familiar claim about the universality of Shakespeare's characters works against the modern, realist preference for characters who are individuals. Johnson defends instead the collective type as a guarantor of universal recognizability, a definitional move that is close kin to Addison's universalist rendering of humankind as "the species."

The practice of botany during most of the romantic era is resolutely incurious about these definitional and philosophical wobbles. In part, the reason for this derives from the powerful use value of the role of species as a consolidating category that handily supervises individual plants. Perhaps for allied reasons, some botanists during this period were less attentive to varieties or subvarieties that would dilute the force of a species as a collective and, within the rubric of its specified type, undifferentiated group. Yet if species were the taxonomic rule of practicing botanists for much of this period, several reiterated taxonomic events or questions disturbed that rule. These included moments in which taxonomists or commentators speculated about whether a class or species might be related to animals taxonomically or in terms of reproductive mechanisms or, as they were typically called, *contrivances*—a word that nicely points up the notion that plants may appear to contrive their mode of reproduction; whether an existing species might be undone by the discovery of a new plant; the taxonomic problem of monsters or aberrant, unclassifiable forms; and the status of singular plants within a species.

Romantic philosophical accounts of species in terms of part and whole distinctions registered a problem in usage that went back to Aristotle. In strictly logical terms, the part-to-whole distinction assumes precisely what the history of thinking about species cannot: that parts maintain a clean logical relationship to the wholes of which they are part. Biological diversity among individual plants in species materially resists the formal and logical assignment Immanuel Kant asserts in arguing that intuitive understanding "goes from the synthetically universal (of the intuition of the whole as such) to the particular, i.e., from the whole to the parts."[30]

The philosophical confidence proclaimed in Samuel Coleridge's definition of the symbol conveys the double edge of hope and doubt in thinking about biological systems: "a translucence of the Special in the Individual or of the General in the Especial or of the Universal in the General. Above all by the translucence of the Eternal through and in the Temporal. It always partakes of the Reality which it renders intelligible; and while it enunciates the whole, abides itself as a living part in that Unity, of which it is the representative."[31] The uninterrupted, nested set of translucent relations that, Coleridge imagines, link the individual to the "special" or "especial" and these to the "General," to the "Universal," and finally to the "Eternal" might be said to be the best, most essentialist, hope for the idea of species, for which one definition, common in the seventeenth century but obsolete now, was "the outward appearance or aspect, the visible form or image, *of* something, as constituting the immediate object of vision."[32] Coleridge's figure of translucence corresponds to the reliance on visibility for making taxonomic distinctions, both before and after the development of microscopes sufficiently refined to allow one to see minute plant differences.

In *Naïve and Sentimental Poetry*, Friedrich Schiller mapped the relation between species and genus to show how naïve poetic genius fares without the assistance it requires "from without." He (this poetic genius is "he" in Schiller) may either become wholly identified with his genus, in which case he abandons his species as a being in nature; or he may become wholly identified with his species. In this latter instance, he "abandons his genus" to become wholly part of nature.[33] *Genus* here means something like the category of recognizably human belonging, whereas *species* designates a belonging to nature that operates outside the organizing principle of the genus. It is as though, for Schiller, to rise to the level of genus one must be human; nature can only aspire to offer itself as species that look more like unorganized monads in patent need of a higher, organizing principle.

Hegel presented an allied antinomy between nature and the human in his critique of German romantic *Naturphilosophie* and Goethe's theory of plant metamorphosis that moves one step further down the scale of nature. Here Hegel's philosophical project on behalf of the world Spirit requires that individuals or particulars, including plants, be understood as instances of a species rather than as particulars with a separate, and potentially different identity.[34] If the imagined or desired whole in question is the world of plants, a single plant individual may be, categorically speaking, in some danger. Hegel's subordination of nature to spirit or mind conveys the risk entailed by being assigned to a category.

Romantic *philosophical* notice of the species category suggests that the status of individuals is the most contested and perhaps the most romantic aspect of the classificatory task. Perhaps we should not be surprised on this point, for as the epoch that professed itself concerned as never before with what is singular or individual, romanticism had a vested interest in how categories or hierarchies or societies would organize singularity. Romantic touchstones for this sense of the individual as a singular, remarkable entity are many, among them William Wordsworth's phrase, "there is a tree of many one" in the *Intimations Ode* or Barbara Stafford's discussions of the role of geological singularity in travel literature of the period.[35] Romantic notice of singularities in nature evidently blends into a parallel notice of singular beings in landscape in romantic poetics and visuality, among them Wordsworth's solitary reaper, J. M. W. Turner's tiny Hannibal in a snowstorm and a confused melee of troops or, conversely, his painting of the giant Polyphemus hurling boulders at a jeering Ulysses. In discussions of classification, romantic singularity becomes as endangered as a species as Turner's Hannibal is somewhere in the Alps.

Thinking about justice before the law, Jacques Derrida argued that the very category of the singular is both compromised by its placement within a system and sustained by the very existence of that system.[36] To put this point in its most romantic form: to be a citizen means in principle to have one's singularity protected, under the cover, as it were, of the laws that govern and give a place to the very category of citizen. And yet, philosophically speaking, being singular means just that: being without parallel, individual, named only for oneself. One could perhaps argue that this is a problem only if one imagines the relation of the singular individual to the society of citizens as one of a part to a whole. Yet this is precisely what we do. It is even more precisely what taxonomic inquiry does.

Some modern biologists and philosophers of science have turned this hypothesis on its head, suggesting that species are themselves individuals and as such not categories to which individuals can be appropriately assigned. If, as David Hull has argued, an "individual" is "a particular, a thing denoted by its name and nothing else,"[37] it becomes something else (and/or something more) once it is identified as a member of a species or kind. If the relation between individuals and species is one of nature's "joints," and Hull for one doesn't believe this to be the case,[38] this joint would be the first place in a classificatory system where singularity is overseen by a category that introduces another singularity that, logically speaking, suppresses the first. For if the markers of a given species are what distinguish it and its members from other plants and species, these markers constitute another, now collective, "singularity" that puts aside that of the individual plant, which now functions as an exemplar of the species,[39] even if it is not technically what modern taxonomists identify as the ur-species or *type*—a plant that would so accurately represent the traits of a species that it would be preserved in herbaria so that new specimens could be compared to it and to other data about the species.[40]

It is useful to emphasize what follows from the claim that species are individuals. If we suppose that species are not classes, but individuals, the class or whole has little or no function. That is to say, a whole requires parts. Individuals require a species only if they are said to belong to one. This chiastic relation is unequal: species need individuals less than individuals need species. For individuals only need species if there are species. In romantic thinking about plants and other natural kingdoms, this distinction is less clear or more malleable. It may also be more opaque about its implications, much as Coleridge's account of the symbol's relation of part to whole in the symbol is fairly opaque about the status of the individual so slipped into the category of the eternal.

Contemporary debates about species have, then, a latent history in earlier taxonomic inquiry. Charles Darwin's evolutionary thesis, together with phylogenetic extension of Gregor Mendel's experiments, suggested, along with the origin of species, their end as a concept. Once biologists recognized the role of evolutionary modification in species, it was only another step in the chain of reasoning and experiment to recognize that evolutionary change could and did extinguish species and, further, that the very process of evolutionary change depended not on the species traits but on individuals that reproduce at some spatial or genetic edge of the species where change occurs. From this perspective, speculations about biological monsters from the eighteenth century forward had it exactly right: it is individuals, including "aberrant" ones, that direct species change and thereby weaken the reach of the species category. Whatever we have to say about species, so this argument goes, turns out to depend on their status as individuals. Just as an individual is born, lives, and dies, so do species; just as an individual is organized such that we recognize it as distinct from its surroundings, so is a species.[41] But once we recognize these similarities, we must also grant that species as a category is also potentially or actually erasable. So, proponents of this view conclude, species are in fact individuals, not species.

Among biologists who have disagreed with the "species-as–individuals" hypothesis, Ernst Mayr proposes that whereas the species concept may be in trouble, the species category works fine because it recognizes "the integrity of species-specific genotypes," by which Mayr means that "the individuals of a species form a reproductive community" (Locke had not been convinced by a similar definitional strategy).[42] Mayr, to be sure, recognizes that plants of one species may hybridize with plants of another. He suggests that we can understand hybridization only if we have already recognized that plants are hybrids because they diverge from the set of traits identified with a given species. Mayr and others (but not Atran or Hull) suppose that this view of what constitutes a species marks a key shift in the idea of species that biologists have long attributed to Aristotle—the claim that a set of traits, whether morphological or having to do with reproductive mechanisms, marked membership in a species.[43]

Philosophers of science who argue that species are individuals suggest further that the record of hybridity or variation is the real story of biology, one at best arrested only momentarily by claiming that species exist as groups of individuals. Hull is emphatic on this point: "everything involved in selection processes and everything that results from selection are spatiotemporal particulars—individuals."[44] Whether the biological entity, be it plant or animal, simply replicates itself (whether asexually or sexually) or interacts with an existing gene pool by changing it, it does so as an individual that acts on and out the evolutionary process in time and in space. Mapping how individuals do this may also require a radically pluralist adoption of several, distinct classificatory systems, rather than plumping for one system over others, as scientists have continued to do before and since Linnaeus.

Marc Ereshefsky has argued that although each of the three available classificatory protocols—ecological, interbreeding, and phylogenetic—contribute to taxonomic understanding, they are also in conflict. Further, no single protocol would suffice to account for the diversity of nature's things. That is to say, whereas the ecological approach argues that the environment rather than reproductive factors controls the stability of species, the interbreeding approach argues to the contrary that reproduction controls species stability and change. The phylogenetic or cladistic approach seeks to identify a common ancestor as the guarantor of subsequent evolutionary patterns for its descendants.[45] Together they present what might be described as a many-worlds map of the classificatory complexity of the biological world. The analogy works as long as one understands that this resulting map displays slippages and border conflicts.

Enmeshed in the coils of these questions and biological problems, romantic botanical inquiry pointed toward taxonomic difficulties that gradually attracted wider and deeper attention into our present. Biologists and specifically botanists who in this period assigned plants to species, established or disputed classificatory protocols, and named plants did not argue for abandoning the species category. But their ruminations about species membership, aberrant or monstrous plants, and the relationship between species and taxonomic systems prompted some surprising declarations. Essays and letters to editors in scientific journals of the mid-nineteenth century quarreled about whether names were descriptive or arbitrary, which plant systematics were useful (or not), and, finally, what a species was. On this last point, there appears to have been little agreement.[46]

Romantic botanists debated the idea of species well past the mid-nineteenth century. They argued either that species were essential kinds, more or less the same as those that existed at creation (more likely less as the years went by), that species marked resemblances among individual plants in the present, or that they were, like genera, the next step up in taxonomy, simple useful abstractions for organizing individuals. Although Charles Darwin might have been expected to argue in *The Origin of Species* (1859) that genealogy or descent, rather than creation or resemblance, is the marker of a species, he chose instead to list the various claimants and then declare that "in the following pages I mean by species, those collections of individuals, which have commonly been so designated by naturalists."[47]

Darwin's consensualist gesture toward his predecessors registers a deeply embedded taxonomic impulse to preserve order, even an old order, as long as possible. For without it plethora and disorder might otherwise rule; names would be changed, distinctions lost, and confusion would replace the jerry-rigged but adequate series of conventions for classifying and naming plants. And yet, confronted by the world's burgeoning plant record, as any modern botanist will readily admit, there is no prospect of an end or even of a middle in which all might agree on the notion of a species or the higher taxa.[48] For botanists of the romantic era, the risk of disorder rather than order was, as it were, a new reality—one that required a counterargument for ordering species, even as it became apparent that no principle for doing so would be likely to suffice.

SYSTEMATICS

Building on Joseph Pitton de Tournefort's earlier notice of the differences in fructification structures among plants and, closer at hand, Sébastien Vaillant's early eighteenth-century observations about the sexuality of plants, Linnaeus's "Sexual System" was the first systematic that aspired to classify all plants and animals. Linnaeus is credited with two powerful innovations, one more or less permanent, the other temporary. Neither was in fact his invention, but by combining them he created the first globally applicable systematic for classifying plants and animals. The first of these was binomial nomenclature—the adoption of a two-word Latin name (sometimes derived from Greek and other languages) that would mark the genus and species of a plant or animal, which has so far survived modern complaints that it ought to be jettisoned along with other conceptually indefensible leftovers of the Linnaean system.[49] Prior to Linnaeus, others had experimented with ways to manage the rampant biodiversity of different common names and extensive plant descriptions, either by substituting a concise name or by offering all the names as synonyms. Linnaeus made the binomial name axiomatic and put it into a classificatory system organized to reflect physical evidence of the reproductive apparatus of plants.[50]

The twenty-fourth class was anomalous because plants assigned to it "hid" their reproductive parts; hence, Linnaeus's name for this class, Cryptogamia (hidden gametes); that is, plants that lacked flowers and seeds, which included ferns, mosses, algae, fungi, and lichens.[51] Although membership in the cryptogamic plant group has changed and the term itself is sometimes omitted in modern biological summaries,[52] this group includes plants and plant systems whose complexity and significance for the phylogenetic record Linnaeus could not have known. The flowering plants, or Angiosperms, now identified as higher-order plants because they evolved later than other plant groups, were the focus of both the Linnaean system (occupying its first twenty-three classes and their orders), and the Natural System. Together with nonflowering plant families (cycads, conifers, ferns, and fern allies), the Angiosperms constitute the vascular plants.[53]

Linnaeus assigned the classes and orders names with Latin and Greek derivations that accord with his systematic protocols. The first class includes plants that have one stamen and is named Monandria, literally one male; the second class is Diandria, literally two males; and so on up through the eleventh class, for a variable number of stamens between twelve and nineteen, and the twelfth, for twenty or more stamens. Thereafter, the classes are identified by the position of the stamens: class nineteen, Syngenesia, so named because the male organs are attached to the female; class twenty, Gynandria, with stamens and pistils in different flowers; class twenty-one, Monoecia, male and female flowers on the same plant; twenty-two, Dioecia, male and female flowers on different plants; twenty-three, Polygamia, male and female flowers mixed with hermaphroditic flowers, occurring on the same or different plants; and finally, the Cryptogamia.

The orders within each class were similarly coded: the order for plants with one pistil is Monogynia, literally one woman; for plants with two pistils, Digynia, or two women, and so on. A recent University of Toronto website for a botany course explained the Linnaean system by using the American cowslip, or *Dodecatheon meadia L.*, as its example. Because Linnaeus named the species, *L.* appears after the species name; it is identified as Pentandria Monogynia, literally five stamens, one pistil, the site explained, so that one "needed merely to be able to understand the construction of a flower, and be able to count."

Erasmus Darwin's presentation of the American cowslip as *Dodecatheon meadia* L. in *The Botanic Garden* suggests some of the possibilities the Linnaean sexual system made available, both for the presentation of a flamboyant array of sexual differences and the insistence on a highly patriarchal structure. Both features had a sustained impact on the way botany could be mined for figurative as well as sociopolitical purposes in romantic culture.[54] Here is Darwin:

> MEADIA's soft chains five suppliant beaux confess,
> And hand in hand the laughing belle address;
> Alike to all, she bows with wanton air,
> Rolls her dark eye, and waves her golden hair.[55]

Linnaeus's own table of the classes is still more suggestive. Published in Latin in his *Systema Naturae* in 1735 and eventually translated into unequivocal English by Darwin and two other fellow members of "a Botanical Society, at Lichfield," it declared that all but the Cryptogamia were involved in "public marriages," in one or more "beds" and "houses" with varying familial relations, called "brotherhoods," among the male stamens, or no such relation, occasional "confederate males" and "feminine males," some "hermaphrodites" identified as "husband" and "wife" living in the "same bed," and females whose unlawful status shifts from the 1759 to 1765 editions from "*meretrices*" to "*concubinae*."[56] When John Hill introduced the Linnaean system to English readers in 1751, he made no mention of marriages, concubines, beds, or houses, let alone hermaphrodites, perhaps because the initial response to Linnaeus had included rumblings (in Latin) about Linnaeus's "lewd method" and the "loathsome harlotry" on display in "so licentious a system."[57]

Among later naturalists, objections tended to be more practically directed. Some, like John Bartram and the poet naturalist John Clare, were intimidated by Linnaeus's use of Latin, a scientific protocol that he inherited and continued but one that may contradict the widespread claim that his system could be easily taught and used.[58] Writing Collinson in 1757, Bartram complained that Linnaeus "crowded too many species into one genus." Philip Miller, director of the Chelsea Physic Garden, London, and the author of the popular *Gardeners Dictionary* (1752), repeatedly complained to Bartram about Linnaeus's genera: "*Doctor Linneaus*, has joyned so many genera together as occasions confusion. The Apple and Pear are undoubtedly different genera: they will not take upon each other either by buding or grafting; and it is well known from experience, that all trees and shrubs of the same genus will grow upon each other."[59] Bartram, ever the discriminating naturalist, challenged Miller on this point, apparently using as evidence fruit produced by grafting Miller's own pears in Bartram's orchard: "ye pears I have grafted in aples thorns & our crab stocks & growed well for a year or two I have seen an aple tree bear both aples & pears at once." Bartram concludes diplomatically, if evenhandedly, "there is no general rules without some exceptions so it is with Lineus system."[60]

Bartram's observation registered a crucial difficulty. In his effort to secure the species category, Linnaeus for most of his career argued that apparent variety within a species could be explained by external fluctuations in climate.[61] Plants that seemed to be hybrids or even "monsters" were thus by definition not true or reliable members of a species. But for later taxonomists Linnaeus's linear hierarchy invited several questions: How might relationships among plants and their higher taxa be represented? If not in the linear fashion of Linnaeus's twenty-four classes of plants, would the structure be tree-like, web-like, net-like, or map-like?[62] If there are, as Gottfried Leibniz, John Locke, Linnaeus, and even Jussieu (but not A. P. de Candolle) insisted, no gaps in nature, it was equally clear that human taxonomists could not fill in all the gaps in a classificatory record that had not yet or could not expand rapidly enough to include newly discovered plants. Beginning in the nineteenth century, the identification of botanical *types* worked on the assumption that species are fixed, yet even this conviction had become tenuous well before Charles Darwin published his evolutionary hypothesis. At its most macroscopic level, systematics argued for distinct kingdoms in nature: mineral, plant, animal. But from the early eighteenth century on, the list of plants that behaved as though they were animate beings (the mimosa or sensitive plant, for example) or whose mechanisms seemed curiously contrived to look like they might be agents in their own reproduction, disturbed this patently hierarchical and tidy ordering of kingdoms. Linnaeus's Cryptogamia produced notable offenders in this regard, inasmuch as the ferns, mosses, and lichens Linnaeus grouped in this class have, as it were, gone on to inhabit different places in the current Five Kingdoms schema of nature.

Georges L. Leclerc, known by his aristocratic name Buffon, offered objections to the Linnaean system that were neither prudish nor exclusively local and practical, although he certainly illustrates those objections with local examples. Beginning in his *Histoire naturelle, générale et particulière* (1749), Buffon challenged in principle and by name the ambition of the Linnaean system on the grounds that without a knowledge of the whole of nature, it is impossible to order plants or animals in a system, let alone determine which belong to which category in that system. Buffon famously mocked Linnaeus's zoological categories, arguing that they present the India pig as a species of rabbit, the rhinoceros a species of elephant, and so on. Buffon's criticism of Linnaean botany is fundamental: if one cannot see the stamina of a plant, one "knows nothing" about it. Inasmuch as the cryptogamic class of plants was by definition without visible reproductive organs, Buffon here indicted an entire Linnaean class, as well as plants whose reproductive parts are "*presqu' infiniment petites*" (almost infinitely small), such that to identify a tree or a plant one needs to carry a microscope.

Buffon's preference for doing botany without a microscope is only partly explained by the relative insufficiency of such instruments in the mid- to late eighteenth century. At issue for Buffon, as for Linnaeus and earlier naturalists, was the ancient claim that species is both kind and appearance. If you can't see it, you can't know what species it is.[63] In his *Encyclopédie* article on botany, Buffon imagined a future when botanists would "renounce their chimerical pretension to follow in their systematics the unintelligible order of nature, which can only be conceived by the Creator." To do this, botanists must put aside invented taxonomic hierarchies or systems of nomenclature and focus instead on knowing the "properties" of plants and "how to cultivate useful ones and destroy those that are not useful, to observe their structures and all the parts that have to do with the vegetable economy."[64]

Although objections to Linnaeus argued that the classificatory rubric of his sexual system did not recognize natural plant families, in *Philosophia Botanica* Linnaeus identified a number of plant families that survive in subsequent taxonomies, among them Cucurbitaceae, Papilionaceae, Palmae, and Compositae. Linnaeus's Gramina became Gramineae [grasses]; his Calamariae (no relation to the crustacean) became the family Cyperaceae.[65] Even after the Linnaean system had been largely discarded, it survived in odd places. In a folio chart of *The Linnaean, artificial or sexual system* published sometime before 1847 in Calcutta, where so many Europeans had discovered so much exotic flora, William Ondaatje explained in a note that the sixth Order in Linnaeus's Class 19, Syngenesia Monogynia, had been "abolished as being unnatural and uncertain" because its members, simple flowers with leafy calyx, widely differed "from the natural assemblage of this class, except in the point of having united anthers." This criticism rehearses a by now familiar pair of objections to the Linnaean system: first that it is more valuable to recognize natural groups of plants—so designated according to whatever morphological principle seemed to predominate in their grouping—than it is to privilege, taxonomically speaking, sexual organs like the anthers, part of the structure of the stamen or stamina. Ondaatje also noted that Linnaeus's followers had also abolished his twenty-third class, Polygamia, "and very properly so," presumably because this class features plants that are, except for the presence of bisexual (= hermaphrodite) and unisexual (male or female) flowers on the same plant, highly dissimilar. The class included bananas, the sycamore tree, and many grasses.[66]

In 1810, Robert Brown published experimental evidence that the Linnaean system was inadequate because it ignored morphological evidence that the Natural System was conceptually prepared to recognize and interpret.[67] Although he presented this evidence to fellow members of the Linnean Society of London, then published it in the *Transactions* of that body, until at least the late 1820s fellows of the same society were still writing essays defending the Linnaean system. In private and public communications (the latter were more diplomatic), Sir J. E. Smith was loathe to abandon Linnaeus, whose herbarium the English botanist had bought (it thereafter become the material base for Smith's career). In the same *Transactions* that featured Brown's remarkable demonstration of the explanatory power of the Natural System, Bicheno's putative comparison of the "French" (that is, Natural) and Linnaean systems concluded that the Linnaean remained more useful in the field because easier to learn.[68]

Bicheno was not the only British botanist to so argue. A random sampling of works about botany published in England after 1810 turned up many that are thoroughly Linnaean: Stephen Clark's 1822 *Hortus Anglicus*; the first edition (1830) of J. C. Loudon's *Hortus Britannicus* (its 1839 edition added an account of the Natural System); and an herbarium collection in 219 volumes of plants that Emma Radcliffe collected between 1837 and 1840 was also arranged in Linnaean classes. The Natural System did not gain popular acceptance until the 1830s and even 1840s, probably for several reasons: its differences from the Linnaean system were not as yet widely acknowledged; the Linnaean systematic was indeed easy to learn and use as long as one did not take note of its anomalies; continued use of Linnaean names masked those differences; and taxonomic variability was perhaps not a principle that advocates of the Natural System were eager to publicize at a time when the project of replacing one systematic with another required, at the very least, a rhetoric of systematic certainty and reliability.

THE NATURAL SYSTEM AND BEYOND

The development of the Natural System was strikingly familial. Two families—the Jussieus before and after the French Revolution, then the father and son Swiss team Augustin Pyramus and Alphonse de Candolle—experimented, codified, and elucidated taxonomic principles first developed by Adanson, who credited Buffon with the suggestion that one might pursue a natural method by scrutinizing an ensemble of parts.[69] Soon afterward, the Jussieu brothers Antoine, Bernard, and Joseph, one of their nephews Antoine-Laurent, and finally his son Adrien de Jussieu, expanded the system as they accumulated experimental knowledge about plant characters and additional plant families.[70] The result was Antoine-Laurent de Jussieu's 1789 *Genera Plantarum*. The one hundred families listed in this work included eleven Linnaean family names, forty-six identified by Bernard de Jussieu, six identified by Adanson, and thirty-four that were new. Seventy-six of the angiosperm families that A.-L. de Jussieu reported have survived in the international code of botanical nomenclature.[71] Early in 1793, the Jussieus were asked to replant a plot of ground in central Paris that had belonged before the Revolution to the king to illustrate the Natural System. Now known as the garden of the Musée d'histoire naturelle, the plot would be replanted in later years to reflect successive versions of the new system. Investigations and microscopic discoveries by Brown in England, Augustin Pyramus de Candolle (usually identified as A. P. de Candolle) and his son Alphonse in France and Geneva, and many others gradually established the Natural System over the Linnaean.

John Ray's earlier proposition that a system based on natural affinities among groups of plants would require careful attention to several characters became the guiding principle of the Natural System, which examined plant morphology structure by structure in order to identify the greatest number of affinities between plants. Instead of relying on one feature of the plant, such as the shape and number of flower petals (Tournefort)[72] or reproductive organs (Linnaeus), the Natural System required that botanists look for as many morphological similarities as they could identify. These similarities could be used either to assign plants to appropriate genera, families, and orders or to acknowledge affinities that cut across recognized families and orders.[73]

Plant affinities had long interested taxonomists. Some of Linnaeus's early critics had objected that his sexual system did not emphasize such relations. What "affinity," asked one of his contemporaries, marked several kinds of plants gathered into a single class except the number of stamens?[74] Until Darwin's *Origin of Species*, systematicists chose not to speculate overmuch about the reasons that members of specific groups were so similar, either because, Julius von Sachs argued, they chose to believe in the constancy of species since Creation[75] or because they found it easier not to comment on a by now fraught claim.

The debate about the rules or method for identifying families, which was continued for most of the nineteenth century, conveys the taxonomic and evaluative complexity of the Natural System. A.-L. de Jussieu argued in 1789 that one needed first to draw on [*puiser*] the principal rules for the formation of other genera by distinguishing, among the common characters that draw their species to each other [*rapprochent*], those that are most from those that are least constant. In doing so, one must also note when a character that is uniform in one genus is not so in another. Finally, in listing these characters, a single constant has a value equal or superior to that of many variable characters.[76]

A. P. de Candolle was more explicit in 1819 about the role of comparison in taxonomic judgments. Admitting that the fundamental desire of a taxonomic system—to determine the rank that plant organs must occupy in a natural hierarchy—is difficult to satisfy, he advised that the "degree of importance of each organ cannot be calculated exactly except relative to those organs that belong to the same class of functions."[77] This protocol is clear enough about how to compare organs with the same functions, but it does not explain how taxonomists would decide which organs or characters are more or the most important for determining membership in a family. Candolle hinted at something like an answer when he discussed how to proceed if one lacked the information needed to "establish a complete and rigorous classification of functions": "one must choose the function that allows us to go forward with the most certainty." He added that in the current state of botanical knowledge, reproduction was the function best understood and thus the one that should for the moment lead in classifying plants.[78]

Whereas Ray had insisted that nature makes no leaps—a claim reiterated, with distinct emphases, by Leibniz, Linnaeus, and Charles Darwin—A. P. de Candolle, the most influential taxonomist of the nineteenth century, perhaps the most significant since Linnaeus, pointedly suggested that gaps in the "series of beings," not the limitations of human understanding, require that taxonomists make leaps.[79] Arguing that Linnaeus made much grander predictions about the reach of the natural system and the putative "naturalness" of genera that neither de Candolle nor his contemporaries believed were natural, de Candolle then attacked the consensus that Linnaeus had invoked to argue that nature does not make leaps. Linnaeus, de Candolle insisted, was simply wrong: "the most zealous partisans of the natural orders recognize today that there are leaps or gaps in the series of things."[80] Why there are leaps Candolle does not say, so it is difficult to know whether he also rejected the claim that no new species could be created.

On the one side of this dispute is the intuitive grasp of natural families like the Umbelliferae. Sachs suggested that some degree of Platonism may be involved or residually at issue in such judgments, inasmuch as botanists were still not ready to argue that species and therefore families might emerge or die out.[81] Atran has argued to the contrary that without intuitive judgments the complexity of living things would render taxonomic judgments about their membership in specific families nearly impossible. That is to say, if one did not begin somewhere with some sense of likely families in the process of deciding which battery of plant similarities ought to be leading indicators, one would never begin.

De Candolle's insistence that there are gaps in nature forecasts the work of botanists from Charles Darwin to the present, twenty-first-century generation. Precisely because radical adjustments in taxonomic method and conclusions since Linnaeus have continually altered the systematic landscape and most of its protocols, later botanists have been more acutely aware of gaps in the taxonomic record than were their predecessors. For Nicholas Jardine and many taxonomists, these gaps may register the continuing challenge to systematic inquiry and the deferred promise of absolute scientific truth.[82] What Jardine describes more broadly as "chronological hybridity," referring to shifts in meaning that accrue in the history of science, also litters the history of systematics with discarded or compromised hierarchies.

A. P. de Candolle's son Alphonse, who published the last account of the Natural System in 1835, there acknowledged that earlier botanists had also believed in the importance of developing a systematic of natural groups or families, even as they developed systematics that were artificial. The younger de Candolle reminded his readers that even Linnaeus had written that the goal of botany was to discover such a system. Between 1789 and 1835, several generations of French and Swiss taxonomists and their colleagues in England and on the Continent pursued the question of natural families and forged a systematic that aimed to offer a complete and ordered account of those families, their higher orders, and the guiding principles used to make such determinations.[83]

There was, however, no consensus about how to use these guiding principles. Even de Candolle father and son did not agree on the question of whether taxonomic judgments are wholly experimental—directed solely by observing plants—or at least partly a priori; and, if a priori, whether they are shaped by Platonic or Aristotelian notions of essential kinds or groups, or involve an intuitive grasp of familial resemblances. Looking back on the development of the Natural System from 1835, Alphonse de Candolle argued that beginning with Magnol in the seventeenth century until Adanson and Bernard de Jussieu in the late eighteenth, the rule for deciding membership in families was no rule at all but *tâtonnement*, a process that involved a "*sens intime*" that, because it is intuitive, cannot be codified to yield a general rule or method. And yet de Candolle the elder also claimed that *tâtonnement* was the first step to be taken to discern natural arrangement, the next two being general comparison and the careful subordination of less significant characters.[84] For de Candolle the younger, *tâtonnement* had to go because it implied that taxonomy was less than an absolutely experimental science. He insisted rather that clear rules or guidelines, not intuitive judgments, constituted the true work of discerning the natural families.

As it developed from the late eighteenth century into the early nineteenth, the Natural System consolidated powerful shifts and as many irresolutions in botanical culture. As a taxonomic approach that was no longer willing to pursue single characters—it is instructive that de Candolle the younger repeatedly refers to "*classifications naturelles*" in the plural[85]—the natural system elaborated a paradox: to identify communities of plants (or quasi plants, as lichens and fungi were increasingly suspected to be), taxonomists had both to recognize a plurality of differentiating functions and parts and insist on their common cellular basis.

The younger de Candolle also insisted that the task before botanists was to unify the investigation of plant groups. The disciplinary comparison he uses to make this point supposes that whereas the study of physics involves successive new disciplines, the study of plants should not: "instead of subdividing itself more each year into distinct sciences, as does physics, which divides itself today into optics, electromagnetics and so on, one senses more than ever the necessity of tying into a compact bundle the branches, formerly separated, of the study of vegetables."[86] The slightly awkward literalness of my translation preserves the figure of a tied bundle, a deliberately crafted gathering of the strands of botanical inquiry that have been separated. Much as Derrida argues that such bundles convey the shimmer of difference in the guise of resemblance, so is the history of botanical taxonomy riddled with allied moments when the tug of difference unravels a carefully crafted claim about resemblance.[87] What de Candolle the younger referred to as the separation, marked from Linnaeus on, between the cryptogams and the flowering plants, is no longer necessary because the cellular basis of plant life is common to both.

Discoveries at the cellular and subcellular level, materially advanced by the invention of finer lenses and microscopes, revolutionized earlier botanical theory and speculation.[88] In one sense, these discoveries resolved, or seemed to resolve, older questions about the vitalism of plants, a notion advanced by German *Naturphilosophie* and thereafter reiterated in popular analogies between plants and animals. Speculation on this subject had been fueled by the "sensitivity" of the Mimosas, the so-sensitive-as-to-be-carnivorous plants, *Sarracenia*, the North American pitcher plants, their tropic counterparts in the genus *Nepenthes*, and the Venus fly trap, first identified by Charles Darwin; microscopic scrutiny of the "spiral tendency" inside plant vessels that allowed liquid to rise up the stem; and Goethe's analysis of the way the structure of plants metamorphoses leaf to flower.

But in another sense, the new microscope evidence of cellular structures and activities suggested yet more interest in possible affinities between plants and animals. If members of both kingdoms are made up of cells, some asked, what other affinities might exist? A.-L. de Jussieu's surprising defense of what Franz Stafleu called "the old analogy between the plant embryo and the hearts of higher animals,"[89] speculations about the spiral tendency in plants (so suggestively akin to the human circulatory system), and Robert Brown's discovery of Brownian motion (which twentieth-century physicists confirmed in cell nuclei)[90] were all occasions for asking how plants might be like animals.

A number of these speculations concerned the Cryptogams, whose obscurity invited the language of the sublime, rerouted here as elsewhere from the monumental to the minute. An 1805 review of J. E. Smith's *Flora Britannica* noted that the cryptogamic plants . . . "have long been involved in confusion and obscurity and have powerfully solicited the patient scrutiny and minute elucidation of the nicest observers."[91] Brown apparently persuaded the naturalist Dawson Turner, an expert on lichens and mosses, that one species, *Fucus peniculus*, linked plants to animals.[92] An 1834 review of John Lindley's *Ladies Botany* and Sydenham Edwards's *Botanical Register* (to which Lindley was a major contributor) suggested that "the Cryptogamic family descends to the very verge which divides the vegetable and animal kingdoms."[93] Although it is difficult to know what this claim meant, one modern taxonomist speculates that the reviewer may have been referring to "unicellular algae that are motile (via ciliae and flagellae, like Euglena) and thus combine plant (photosynthesis) and animal (motility) properties." In the current "Five Kingdoms" scheme, these plants belong to the Kingdom Protoctista or Protists. The other four kingdoms are Animals, Plants, Fungi, and Monera (all prokaryotes; that is, bacteria and cyanobacteria, or blue-green algae).[94] Modern biological systematics retrospectively suggests that early nineteenth-century speculations about the cryptogamic plants recognized creeping hybridity when they saw it.

Whiffs of plant animation survive in English common names that present plants as birdlike or insectlike, hence "animated" in ways plants are not supposed to be animated. In the entry for "Botany" in the 1842 edition of the *Encyclopaedia Britannica*, its author (possibly Sir J. E. Smith) neatly captures the impeccable logic of the nineteenth-century interest in plant-to-animal resemblances:

> The distinction made by Linnaeus between plants and animals consisted principally in the power of motion in the latter. Many animals have, however, now been discovered, which seem to be unable to remove themselves from the spot on which they first made their appearance; and, on the other hand, there are many plants, as the duckweed (*Lemma*), ball conifer (*Conifer aegagropila*), and others, which, if they have roots, do not send them into the earth, but float about as if in search of food; and our distinguished countryman Mr Brown, whose philosophical observations all must respect, has within these few years demonstrated that the component particle or molecules of all matter whatever, whether organized or not, when suspended in a fluid, and viewed with a suitable microscope, are found to be in motion without any visible agency. Perhaps the true differences are to be looked for in sensation and an intestinal canal in the animal kingdom, into which the food is collected; whilst plants are endowed only with irritability, and receive food through many canals or mouths.[95]

From providing examples that undermine motion as the feature that distinguishes animals from plants to speculating about a more viable basis for making such a distinction, the author repeatedly invokes figurative resemblances. Animals are sensitive (never mind the sensitive plant), but plants are merely irritable; unlike animals, plants receive "food through many canals or mouths." A reader might be forgiven for imagining plants as polymorphously hungry and more anatomically adept than animals (with their single mouths) at satisfying their hunger.

To be sure, other botanical discoveries demonstrated that what had once seemed animal-like activity in plants was not. Brown and others showed that the complex mechanisms of the ovule and plant fertilization, especially in orchids, need not be explained by a hidden plant agency or, even more invitingly, by a conspiracy between insects and the flowers they fertilize. In the 1860s, Wilhelm Hofmeister's confirmation of a two-generation life cycle in lichens and life-cycle relationships between the lower (cryptogams etc.) and higher (i.e., flowering) plants significantly expanded botanical inquiry about these groups.[96]

The career and achievements of Robert Brown convey how the work of botany had changed since the end of the eighteenth century, when a gentleman amateur botanist like Sir Joseph Banks could and did become a "center of calculation" in the discourse and exchange of natural history after a youthful voyage of exploration. Whereas Banks had been and remained an amateur, albeit a socially and politically adept amateur, Brown was a university-trained professional whose botanical career was among the most remarkably accomplished of his era. Roughly a contemporary of William Wordsworth, Brown lived a long life (1773–1858) that encompasses the taxonomic debates to which he contributed a great deal, as did many others, some of them gentlemen, but many others working-class gardeners and artisans, including some women. Like Banks before him and Charles Darwin after him, Brown launched his scientific career by serving as a naturalist on a voyage to the new world. In Brown's case, Matthew Flinders's ship the *Investigator* took him back to the scene of Banks's famous voyage on the *Endeavour* to Australia and adjacent lands in 1770. Brown sailed to Australia and Tasmania from 1798 to 1805 and there began to classify even more of the exotic plants of Australia that Banks had seen during his voyage. After his return, Brown became the librarian of the Linnean Society, inherited Banks's home and collections after his death, and became the curator of those collections after they became housed at the British Museum, a position he kept from 1827 to the end of his life.[97]

A friend and colleague of Sir J. E. Smith, Brown nonetheless redirected fellow English botanists from the Linnaean to the Natural System. Despite Brown's never having presented a theoretical summa of his botanical discoveries and their implications, Alexander von Humboldt declared him the "Prince of Botany" (inevitably, the tribute was in Latin: "*botanicorum facile princeps*").[98] Brown reargued and reordered the one hundred plant families of A.-L. de Jussieu into a sequence and hierarchy that established the model for published floras ever since.[99] By the end of his career, he had confirmed the presence of a nucleus in each plant cell and identified the phenomenon of cytoplastic streaming. Other botanists had detected these phenomena and even Brownian motion, but Brown integrated these findings into systematic and floristic knowledge.[100]

Relying on his extraordinary microscopic skill[101] and cellular drawings by botanical illustrators, principally the brothers Franz (or Francis) and Ferdinand Bauer, Brown distinguished the detailed structure of plant ovules for the first time. He determined that naked ovules (that is, ovules not enclosed in an ovary) are the distinguishing taxonomic feature of Cycads, Conifers, Ephedra, and Gnetum, and as such the trait that specifies their key difference from the angiosperm or flower-bearing plants in which ovules are enclosed in an ovary.[102] In 1744, Collinson had written about current discoveries in "Minute Creation."[103] By 1810, Brown had already examined creations far more minute than Collinson and other eighteenth-century naturalists ever imagined.

From microscopic details finer than before possible to see to giant flowers to exotic plants from far corners of the globe (the Australasian plants and the Protea of South Africa), and perhaps including the larger systematic implications of his analyses, Brown's investigations invoke the scalar extremes that haunt the pictorial aesthetics of this period—the attraction of miniature and gigantic forms as allied, if contrastive, features of nature that deliver the sublime in different registers;[104] and, remembering Mary Shelley's Frankenstein and the sustained nineteenth-century fascination with freak shows, the unsettling status of deformity and monstrosity in discussions of species.

Fascinated with botanical monsters, above all with the *Rafflesia arnoldi* from Sumatra, still billed as the largest plant flower (42 inches in diameter) in the world,[105] Brown extended his work on ovules to the other end of the scale. In successive monographs on this flower, accompanied by engraved plates that he often took on the road when visiting Continental botanists and gardens, he hinted at its implications for discerning differences between angiosperms and gymnosperms.[106] For Brown as for other taxonomists, plant "monsters" represent disruptions in the normal processes of plant development. Like those "monsters" James Wallace managed to carry home from Darien in 1700, such plants may point to morphological and genetic complexities that would otherwise go unnoticed, such as the way an insect gall might stimulate odd or anomalous growth patterns.[107] In the case of the *Rafflesia arnoldi*, its remarkable size became part of its exotic identity: exotic plants are bigger than life, alluring, poisonous (the so-called poison tree of Java was a favorite subject of botanical notice), or smelly when fully in bloom, and erect (the Titan arum). Whatever the trait, exotics commanded notice because they exceed European norms and not infrequently the classificatory schema Europeans developed to encompass plant diversity.

The difference between Brown's attention to plant monsters and a more wary taxonomic view is suggested by an exchange between characters in Maria Edgeworth's novel *Belinda* (1800). Responding to a taxonomic description of bad clerics as "this species of animal," Dr. X replies, "but I consider them as monsters, which belonging to no species, can disgrace none."[108] The logic of this preservation of clerics as a class reiterates a frequent claim in discussions of natural history: monsters are taxonomically aberrant forms, individuals that exist outside a species (or as Victor Frankenstein decides about his monster, outside "the species" of humankind). In 1790, Kant argued that something is monstrous if it exceeds our capacity to represent it, making the monstrous an occasion for the mind's recognition of its own sublimity.[109] The purported referent is colossal forms in nature, but a similar argument could be made for plants or animals that exceed accepted taxonomic categories, like those plants that Wallace had said would utterly confound botanical method. The distance traveled from Aristotle's argument that monsters (*terata*) lack specific form is instructive. Aristotle argued that monsters are "amorphous" because, Atran notes, "they appear to combine two forms that should not ordinarily combine."[110] Omitted here is the taxonomic recognition embodied in Brown's attention to botanical monsters, not because they lack form but because their specific forms may indicate unexpected or unrecognized information about a species.

Monsters, or "curiosities," were a staple of voyages of discovery and those "cabinets of curiosity" designed to house such artifacts and curiosity about the monstrous, aberrant, or merely strange underpins discussions of racial and sexual categories and fears from the early modern period until the end of the nineteenth century.[111] Brown's taxonomic interest in aberrant forms echoed earlier interest in how to display forms or creatures that did not fit known categories. Sir Hans Sloane's 1573 bequest of his natural history collections for the creation of the new British Museum prompted extensive public debate about how a public museum ought to manage and display its possessions. At issue was the nature and purpose of curiosity. Adam Smith argued for a balance point between seeking out the curiosity as something odd, strange, novel, hence wonderful or repellent or both—a practice that flourished in the taste for public spectacles and human deformity or aberration from European norms throughout the nineteenth century—and attempting to understand curiosities in a systematic or anthropological context.[112] Public discussion of how the collections of the new British Museum would be organized and made available worried about what one writer in 1801 characterized as the "curiousity of the multitude."[113] The phrase implies that the masses have not been cured of their addiction to those old curiosity cabinets. But when Peter Collinson told John Bartram, as he often did, that he had sent or received a "curious" plant, he meant that it prompted interest and investigation. For these Quaker botanists and their correspondents, being "curious in our way" means looking for patterns and resemblances that might place a new plant or feature in relation to others. Their sense of this project is suffused with theological wonder at the variety of divine creation; there is nothing suggestive of finding cheap thrills in weird, deformed, or monstrous things. Indeed, Collinson is wary of what he and others call "botanical enigmas," such as the animal-like properties of plants like corals, which housed microorganisms then called "polyps." His reasoning is telling: "I profess myself no Botanist neither am I fond of Novelties. The Science of Botany is too much perplex'd already."[114] What Collinson meant by *Botanist* would include Brown and other taxonomists who supposed that a monstrous or odd plant might advance taxonomic inquiry.

Whether botanists dealt with "monsters" or attempted to organize specimens into species and genera, the difficulty of aligning particulars or individuals within a taxonomy registered a persistent and exfoliating problem with a strong romantic pedigree: how to align the individual with a supervising category; how to align the different pressures of the material and aesthetic concerns that mark this era; how to, in short, read local evidence for indicators about larger, theoretical claims, and how then to determine the reach of exemplary distinctions and categories. This cultural logic impels arresting parallels between Brown's investigations of minute and enormous plants and monsters (topics that also interested his scientific contemporaries), and allied motifs in the visual culture of romanticism. In Thomas Bewick's woodcuts, in J. M. W. Turner's vignette steel engravings and mezzotints, and in John Constable's explorations of different degrees of minuteness for representing the same scene on a "six-footer" oil canvas, romantic artists (and many who caricatured high art for a living) collectively registered the task of aligning what cannot be aligned, those disparate levels of form and idea. For botanists, the work of bringing such extremes to order was intricate, material, and never finished.

NOMENCLATURE

In 1745, Linnaeus published a pamphlet in which he reported his treatment of a scholar from Uppsala, Sweden, who had fallen into a sleeping sickness: the man awoke when the doctor put a Spanish fly to the back of his neck. Alert, but unable to speak Swedish any longer, the scholar spoke, as Linnaeus described it, "a foreign language, having his own name for all words." Linnaeus was particularly taken by the fact that the man had lost memory of all nouns, including the names of his children, his wife, even his own name.[115] What I find notable about this anecdote is Linnaeus's reaction: what else would amnesia entail but the forgetting of nouns and names? Linnaeus's notice of this amnesiac event suggests a shadow of anxiety for himself, another Uppsala scholar, were he similarly indisposed as he faced an onslaught of new discovered plants in the material form of specimens brought back. The antidote and its curious psychic trace obliquely register Linnaeus's contribution to the sciences of nature: a stable, binomial system of names that can be easily remembered, easily folded into a system, whether Linnaeus's or someone else's. For Linnaeus's predecessors, contemporaries, and successors, not having a universal method for naming plants would insure a world in which the amnesiac Uppsala scholar's coping strategy would be multiplied and refracted: foreign words, understood by no one else, for plants, birds, and animals.

As specimens arrived from around the globe, as often with "foreign" names roughly transcribed, attached to them as not,[116] Linnaeus like naturalists before him struggled to find a way to forestall a confusion that was inevitably laced with allusions to the tower of Babel.[117] It is perhaps not surprising, then, that since the mid-eighteenth century the protocols for naming plants have become ever tidier, more rigorously scientific, more absolutely latinate (with occasional rebellions from the rank and file) than they were before. Although earlier naturalists had proposed two- or three-word plant names, sometimes in Latin, Linnaeus established specific ground rules for using and creating Latin binomials instead of a longer set of terms that might well be more descriptive but would also be much harder to remember. Pavord's example registers the difference: before Linnaeus, the plant *Plantago media,* or hoary plantain, was called "Plantago foliis ovato-lanceolatis pubescentibus, spica cylindrica, scapo tereti," or "a plantain with ovate lanceolate leaves becoming softly hairy, a cylindrical head, and a smooth stem."[118] Since Linnaeus, taxonomists have mostly agreed that plant names should be in Latin, or that species of Latin that serves this botanical purpose,[119] and that they should be brief, consisting of just two words, the first being the genus name, the second the species name. This is not to say that Latin binomials won without a battle. In the 1830s and 1840s, English naturalists again debated the rule of Latin, which they regarded as retrograde, given the de-Latinization of pedagogy and the desire to market natural history to a large, interested, and buying English public that did not read Latin.[120]

But even in the midst of successive nomenclatural decisions on which modern principles have been meticulously constructed, the rule of Latin does not rule out a spirit of difference that jostles plant categories and suggests "affinities" that take the Natural System discussion of "affines" into realms they were never intended to include. My discussion of botanical nomenclature from Linnaeus forward is frankly pitched to make the most of these untoward, transgressive names as vehicles for a metaphoric waywardness that uses botanical terms and names for quite other purposes. Linnaeus himself gets this project under way. Some Linnaean binomials follow a practice that had long existed, and still exists, in common plant names: *hoary plantain* means a species of the genus plantain that is hairy. In the Latin binomial, the genus leads; in the English, the species. The difference is subtle, but indicative of a problematic, leading impulse in modern taxonomy: to order plants in ways that emphasize their membership in higher taxa, from the genus on up. In contrast, common plant names foreground their quite literal specificity and thereby suggest that these particulars matter. They also proliferate when a single plant is accorded more than one common name.

When Linnaeus and his students adopted Latin binomial nomenclature, they did so at a great rush, gathering epithets from classical sources, German mythologies, names of plant discoverers or naturalists who had first identified specific plants, descriptive features, and the old common names, all Latinized, although other linguistic origins might still be apparent.[121] Linnaeus's contemporaries were not altogether pleased. Collinson complained to him that Linnaeus would "perplex the delightful science of Botany with changing names that have been well received, and adding new names quite unknown to us." The Earl of Bute, a superb amateur botanist, was more blunt, if also more indirect. Writing to Collinson about Linnaean plant names, Bute snarled, "I cannot forgive him the number of barbarous Swedish names."[122]

Although one philosopher of science has urged that Linnaean names be expunged from the nomenclatural rolls, current nomenclatural practice retains many of them.[123] From 1867 to 1958, successive efforts to define an international system of botanical nomenclature extended and refined protocols for naming plants that A. P. de Candolle had first offered in his *Théorie élémentaire de la botanique*.[124] Although he evidently urges the adoption of his natural system over Linnaeus's artificial one, Candolle repeats two ground rules of Linnaean nomenclature: plants should be given Latin names and the right of priority for naming a plant should be given to the discoverer or first describer.[125]

And yet, after nearly a century in which naturalists had been mostly in agreement about the value of Latin binomials, the specter of multiplying common names continued to trouble systematic efforts to insure that a single species or genus would have only one name and that species names would not double as genus names, despite the usual exceptions derived from already established names. In 1867, a botanical congress meeting in Paris took specific aim at multiple plant names of the sort that had dogged botany (and enriched it) for centuries. The new Paris laws were just four: one plant species could have no more than one name; no two plant species would share the same name; if a plant has two names, the one that is valid would be the earliest one published after 1753, when Linnaeus's *Species Plantarum* introduced a binomial name to identify each species; and the author's name would be cited after the plant name to insure the priority of that name over others.[126]

The slight redundancy in this 1867 set of rules conveys in brief prior efforts to eliminate the creep of repetition as plant names migrated from species to genus and beyond when what had looked like simply a species was later recognized as a member of a genus and so on. Thus it is not enough to say that a single plant species can have only one name; the rules must also stipulate that no two plants can share a single name. Whereas John Ray had urged that the old plant names be preserved, together with the plant groups identified by earlier naturalists,[127] the rise of modern nomenclature marked, more or less, the demise of some of the old names and, above all, ended the practice of tolerating many names in many languages. De Candolle, Linnaeus before him, and numerous botanical congresses after them collectively target both the proliferation of common names that could in principle exceed the proliferation of newly discovered species and genera and, behind those names, disparate, even competing naming protocols and orthographic variants that would, if unchecked, swamp the task of naming names.

Romantic era disagreements about plant names and the protocols for creating those names suggest how hard it is to let not only old, quirky names go but also to let go of the possibilities for linguistic and figurative play that common names offer, especially those that described plant features imaginatively or by comparison with other things, and even by some Latin names that hilariously reproduce common plant descriptions in Latin disguise. Consider these examples, happily supplied by Linnaeus himself: the two-lobed or twin-leaf plant *Bauhinia* is named for two brothers who were also distinguished botanists, John and Casper Bauhin; the plant Linnaeus calls *Dorstenia*, whose flowers the nineteenth-century commentator Randal Alcock described as "obsolete" and "devoid of all beauty," commemorates the "antiquated and uncouth book of Dorstenius." Linnaeus was equally willing to target contemporary adversaries and challengers: the name *Buffonia tenifolia* [slender-leaved Buffonia] refers to what Alcock refers to as the "*slender* botanical pretensions of the great French zoologist," the same Buffon who often and publicly objected to Linneaus's systematics. Warming to the practice, Alcock suggests that the late eighteenth-century Viennese botanist N. J. Jacquin might have named *Hillia parasitica* after "our pompous Sir John Hill."[128]

Even after the various nomenclatural debates ended in 1867 with an agreement to use Latin names, the descriptive bias of common plant names survived as systematicists debated whether plant names should be descriptive or might simply be arbitrary. In any language, descriptive common names can convey what we might call a metonymic array—a series of meanings or ideas or figures that fan out, rippling in a way that claims the plant name for local cultures and dialects. Consider, for example, these common names: for the snapdragon, or *Antirrhinum majus L.*, "calf's snout";[129] for the houseleek, or *Sempervivum*, "welcome home husband however drunk you be"; for *Viola tricolor*, "meet her i'th'entry kiss her i'the'buttery" or, more concisely (and imperatively), "leap up and kiss me"; for the balsam *Impatiens noli tangere*, "touch me not," where the Latin name simply translates the common English name.[130] Linnaeus was particularly incensed by the last example in this list, which he included in a list of religious names among those "absurd names" generated by private individuals (aka not real botanists).[131]

Absurdity is one thing; proliferation another. Two wild flowers that John Clare frequently mentioned illustrate this second liability (from one perspective) of common plant names. D. Gledhill reports that *Caltha palustris* is called "marsh marigold" but also "kingcup" and "May blobs" and another 90 local British names, plus more than 140 German vernacular names and 60 French names. Clare calls the same plant "horse-blobs," among other names.[132] *Arum maculatum* is wilder still, not for the number of common names but for the evocative character of its names: "priest's pintle" (referring to its phallus-like stalk, admittedly less spectacular than those of the exotic Arums); "cuckoo pint" (from "cuckoo pintle"), "wake robin," and the various spellings recorded on a single page of William Turner's *A New Herball* (1551–68): "cockoupynt" and "coccowpynt," evidently close orthographic kin.[133] Then, too, at least ten different plants are called "cuckoo-flower" in England simply because they flower at about the time that the cuckoo is first heard in the spring.[134] In a modern dictionary of plant names, the cuckoo flower is identified as *Cardamine pratensis*, but an 1867 work declares that the cuckoo-bud of Shakespeare is the creeping crowfoot, or *Ranunculus repens*. Clare calls a variety or species of orchid he treasured "pouchd lip cuckoo bud."[135] A series of plant names that feature cranes present a more diffused array: crane flower is *Strelitzia reginae*, the bird of paradise, its Latin name an honorific Joseph Banks invented for Queen Charlotte, erstwhile wife of George IV and like her husband a patron of natural history.[136] But a host of cranesbills follow. All are species of Geranium (aka Pelargonium), and one is doubly a "bird-plant": cranesbill, dove's foot, or *Granium columbium*. Here the Latin binomial intensifies rather than mutes the figurative character of the name.[137]

The nineteenth-century debate about whether names should be descriptive or arbitrary is at times chary of the scandalous, or merely distracting, property of descriptive names. In private, Sir J. E. Smith could be jocular on this point. Writing to his friend and colleague James Sowerby in 1801 about the use of the prefix "Phalli" to identify plants with phallic floral structures, Smith declared, "I'm glad you agree with me about The Phalli—The old trivial names are unnecessarily indecent—As to the things themselves, if Nature would be jocose she must take the blame."[138] In modern nomenclature, species names may be constructed arbitrarily, or with reference to a location or whomever discovered or first identified, or descriptively. In all cases, Latin now rules: if a name is descriptive, the reader either reads Latin or reaches for a glossary of botanical Latin.[139] Even for those who do read Latin, some epithets permit what Gledhill unhappily recognizes as "a degree of latitude" in translation.[140] The problem is hardly new: when the sixteenth-century botanist William Turner translated older Latin names into English so that his *Herball* might be used by a wider public, he often needed to create a common name that is oddly evocative: for the Latin *barba hirci* comes the English "goat's beard"; and from the Latin *hieracium* comes "hawkweed,"[141] also known familiarly as "mouse's ear."

Despite the drive to craft and foreground Latin binomials, because a high degree of conservation of both plants and traditions stipulates that any earlier published name for a plant be preserved and listed among its "synonyms,"[142] the old common names never entirely disappear. One 1833 contributor to the journal *Field Naturalist* complained: "The language of zoology and botany is necessarily changing. And what is the consequence? We are overburdened with synonymes, . . . [which] create as much, if not more, confusion than did the provincial terms."[143] If the specific target of this complaint is Latin synonyms, the charge remains true, perhaps truer, for common English synonyms. The common names for certain species of the *Orchis* or *Ophrys* genus were a persistent irritant for botanists from J. E. Smith to Charles Darwin. Identifying such names as "trivial," meaning popular rather than scientific,[144] Smith could then afford to be magnanimous: "In unimportant matters strict propriety is sometimes obliged to give way to common custom."[145] The rub of impropriety, taxonomic or otherwise, is precisely the interest of botanical nomenclature throughout the long arc of romanticism, for it is by such means that the proliferation of botanical figures occurs in different registers, with different ends in view, be they poetic or political.

CONCLUSION

Romantic and postromantic discussions about botanical species, systematics, and nomenclature suggest a version of what Neal Curtis, after Jean-François Lyotard, has called "heterogeneous finalities,"[146] outcomes that do not or cannot cohere in a unified field theory of, in this instance, biological difference. Romantics did not embrace or acknowledge cumulative evidence of this incoherence, which Curtis ascribes to the postmodern condition of our time. Yet romantic writing on botany registers the possibility that other or many other systematics might converge with or suggestively disturb each other. The late eighteenth-century compendium of such possibilities is Darwin's "Loves of the Plants," where affines are presented by Darwin, as had Linnaeus, as brothers or sisters in complex family hierarchies that resemble extended lineal relationships in other cultures. The heterosexual and homoerotic couples that constitute the ground for these botanical lineages hint at relationships that Linnaean classes only appear to organize and thereby limit. Plant sexuality may be from this perspective a symptom for other, incommensurable plots—in the imperial project, in the economy of gender, in the displacement of one systematic by another.

The romantic pairing of philosophical and material perspectives on botanical matters that I have described here must reckon with plants and species that exhibit affinities across as well as within categories and account for plants within a given species that do not altogether match the species type. At the ends of species, individual plants may look as though they belong to another species or to two at once. The material pressure of these adjustments on the species as a manageable and coherent category undoes the logical project of philosophical differentiation and hierarchy that had been since Aristotle the basis for classifying nature. Romantic botany inhabits this collision between the logic of categories and the work of nature as material difference. This friction at once weighs on and enlivens botany as a scientific and figurative enterprise in romantic culture.

CHAPTER 3

Botany's Publics and Privates

You are interested in botany? So is my wife.

—*Napoleon to Alexander von Humboldt (1805)*

HARD TIMES AT MALMAISON

In January 1814, the month and year in which Wordsworth and others wrote poems of thanksgiving for Napoleon's defeat at Waterloo, someone who signed himself as Botanicus sent to an unidentified correspondent a copy of a letter addressed to Josephine, whom Botanicus rather unnecessarily describes as the "ex Empress" of the French. The Tory bona fides of the poem and its transmitter are clear. Botanicus says he got the poem from Mr. Goldsmith of the *Anti-Gallican* and that its author is the Marchioness of Hertford, at this time the mistress of the Prince Regent, whose Tory inclinations she encouraged.

The *Anti-Gallican* appeared in print in various incarnations from the final decade of the eighteenth century until at least 1804. Its last iteration seems not to have made its way into print, circulating instead among a coterie of Tory and aristocratic friends who took the kind of delight in Napoleon's final defeat that some of us today might take in the fall of a despised chief of state (take your pick).[1] The letter masquerades as an offer of plants to Josephine, whose garden at Malmaison was renowned for its exotic and native species.[2] The plants on offer are transparently not gifts but figures of speech that bite.

Noting that Napoleon had not been able to satisfy Josephine's "urgent demands for Laurel," the marchioness regrets that she, too, has none to offer, for the Marquis of Wellington had "monopolized" all the English laurel. She helpfully suggests that perhaps Josephine could ask him for some. The marchioness's figure is bluntly self-evident: as the ancient crown given to heroes who win, the laurel belongs to the English general. Next she offers to send a "Crown Imperial," although she cautions about planting it "in a French soil at this changeable season, else it will be very apt to tumble from the stalk, and be replac'd by a Fleur de Lis de Bourbon."[3]

English roses the marchioness cannot send because they "will not bear transplanting," with the unappetizing exception of the "Dog or Full Blown Cabbage Rose," this last unavailable because it is much sought after by "Maids of Honor," all this in pointed contrast to Josephine, who is no maid and an ex-wife. Unable (or unwilling) to give Josephine "Hearts Ease," the marchioness offers her "Rue" instead and, with a flourish of noblesse oblige, she included an offer from the prince regent of plants from his pleasure garden: "Honesty, Devil in a Bush, Love in Idleness, French Beans," but not, with a soupçon of imperial regret, "Thyme" for "he has lost all of his."[4] Pursuing the logic of stinging botanical substitution, the marchioness claims that he instead offers Josephine the Balm of Balsam which he found so soothing after suffering from a compound medication that included "Hemlock, Dogstooth, Dragon, Ratsbane, Stinging Nettles," and other toxic plants, a prescription that nicely captures the thin line separating medicine from poison. Moving on, the marchioness lists other English plants and donors, including the lord chancellor's offer of the trunk of an "old Sloe Tree from which branches of "Weeping Willow spontaneously shoot," a "Lusus natura" that will come, it is promised, with a learned paper on that subject; from one Mr. Whitbread, a real British Heart of Oak and so on.

This putative letter parodies a minor genre and cultural practice: the exchange of plants between correspondents, usually accompanied by a letter to explain what had been sent or would be sent. These exchanges were by no means exclusively aristocratic: they occurred among amateur and professional botanists regardless of rank and not infrequently across ranks. The Philadelphia nurseryman John Bartram sent hundreds to correspondents for much of the eighteenth century, among them Peter Collinson in London and Carl von Linnaeus in Uppsala. The poet and itinerant agricultural laborer John Clare sent them to his friends, some aristocratic but most of them not, and received letters back. British botanists in India sent them to each other or to Joseph Banks and others in England and Scotland. The marchioness's offer (or threat) of a botanical pamphlet on the weeping willow sprouting spontaneously from another tree is equally mimetic of a manuscript and print culture in which accounts of plant monstrosity as the play or game of nature were thick on the ground.

The marchioness's letter gestures toward the argument of this chapter: that romantic botany is a site where material and scientific fact and figural possibility intersect not just once but repeatedly, such that a map of these intersections would look less like a schematic grid than a nest of filaments whose materiality and figurality are difficult to separate. This difficulty and its implications for romantic language and culture are the exfoliating plot of the book as a whole. I specify the contours of that plot here by asking how that material and the cultural practices of botany become public and private occasions for figure, beginning with plants themselves.

Taxonomically, a plant is both representative of the species to which it is said to belong (some plants do this better than others) and a thing—that is, a material particular whose identity as such, Bill Brown has argued, exceeds or is at least not limited to its value as a commodity or reified object that is useful or profitable.[5] One might be tempted to believe that romantic era botanists were blind to this doubleness, either because they were so obsessed with classifying plants or they thought of them only as commodities to be traded around the globe. Yet even those who were very much involved in classifying and thinking about plants and profit seem to have been thoroughly aware that plants are also material things. Writing in 1831 to Robert Wallich, a fellow Scottish botanist who was at the time working in London, R. K. Greville described what he, along with Robert Wight—home on furlough from working on Indian flora as an East India Company (EIC) surgeon—and other companions had found during a botanical excursion in the Scottish Highlands:

> If you had been present, instead of vegetating in your dusty workshop—bad man that you are—you would have been present at the discovery of Phaca astragalina [now Astragalus alpinus]—the first time that Phaca has made its bow to a british audience—our friend Wight performed wonders, and to the gazers at the foot of the precipice, seemed Wightia scandens rather than gigantea [Wallich had just described the new genus and species Wightia *gigantea*]; especially as being seen in perspective little more than his anonymous end visible. He & I however really, having less regard to broken bones or the sighs & tears of those we left behind, carried off most of the specimens. I beg that you will not turn up your nose at so much being said about a new british species.—Man of a thousand discoveries deign to recollect that to find a new phaenogamous plant in this exhausted receiver, as it were, of vegetation, is like finding a new species of Elephant or a Palm tree in Nepal.[6]

The rhetorical flourishes of Greville's letter signal both his voluble excitement about Wight's discovery of this plant in this place, a discovery Greville puts on a par with finding new plants in more exotic locations, and his admiration for Wight. Greville's wit about *Wightia scandens* suggests that this climbing plant (*scandens* means "climbing" in Botanical Latin) ought really to be named *gigantea* (Botanical Latin for "tall, or imposing") to signal Wight's stature as the botanist who found the plant. It is worth noting how easily Greville mixes three distinct entities—the plant itself, its species name, and a figurative use of "*gigantean*"—to signal Wight's botanical expertise, knowing well that his correspondent will get this joke and the others that travel with it in these sentences: the comparison of the "vegetating" Wallich to the climbing Wight with his "anonymous end," *Phaca astragalina* as the plant that makes "its bow to a british audience," and the "sighs & tears" of the personified plants left behind (not many) to mourn those taken as specimens.

Greville's infectious wit makes several points. However much Wallich, Wight, and their botanist contemporaries were eager to locate plants within taxonomic systems and participate in the developing global economy of plant production and transportation, they were also keen to find single plants, even those they already knew, in unusual places, both because botanists like plants (surely part of Greville's wit in this letter arises from the pleasure of finding this plant) and because knowing where plants grow matters to taxonomic arguments. As clearly, plant names, especially species names, inspire metonymy: a climbing or gigantic Wight comes as easily to Greville's letter as the "sighs & tears" of plants left in the field.

The fact that one cannot do taxonomy without plants was perhaps even more pressing in the late eighteenth and early nineteenth centuries, before genetics and cladistic trees shifted the experimental basis for determining plant relationships. Reflecting years later on his early efforts to identify Indian plants using only the descriptions provided in an earlier catalog of then-known flora from around the globe, Wight complained that he had to "select the name of each plant from among forty or fifty thousand"; had there been an Indian flora available at the time, he would have had to pick out the right description from only "four to five thousand."[7] Like botanists before and since, Wight needed live and dried specimens and he needed good illustrations of them in the event (not infrequent) that the specimens were lost or eaten by insects. Although dried specimens and illustrations were evidently commodities that circulated around the world back to Britain and Europe or between various colonial plantations under European control, in taxonomic writing they also circulated in an odd middle space between materiality and idea. For precisely because botanists neither could nor wished to get away from the plants they collected, dried, and depicted, the culture of romantic botany never arrived at a moment when plants and to a degree botanical illustrations became wholly objectified, reified, and abstracted as "ghostly abstractions" of commodified culture and labor.[8] Embedded in the work of commodities exchange, plants and botanical illustrations nonetheless insisted on their materiality and its invitation to aesthetic pleasure.

The materiality of romantic botany is as curiously circumstanced. In one sense, it is still oddly present. Looking around the world at the botanic gardens, public and private, created for romantic plants, visitors can find traces of some of those plants and records, along with a rich network of affiliations among those who identified them. For example, the Dragon Arum plant, the subject of an especially lurid plate in Thornton's *Temple of Flora*, still grows in Gilbert White's garden at Selborne. Yet the actual plants that romantics studied and depicted are dried specimens, housed not in their original habitats but in cabinets around the world, arranged to illustrate their current taxonomic location. As fragile, skeletal remains with a limited shelf life no matter how well preserved, their spectrality is a version of the difficulty and evanescence that inhabit all archival and historical inquiry. I argue here that even in these reduced circumstances, this spectral materiality calls us back.

This is so, paradoxically, because romantic plants invite figurality of all kinds, from the ham-handed political allegory of the marchioness's letter and the teasing figures of Greville's letter to Wallich to the figurality that permeates romantic discussions of education, race, slavery, gender, imperialism, colonialism, and politics. Running underneath all of this, plant names offer a special instance of the waywardness of figures as tropes or turnings from literal meaning and use value. Debates about nomenclature, the delight and bane of taxonomists, reveal how botanical names may refuse or put aside the sense of purpose that otherwise compels botanical discourse, whether the purpose at hand is taxonomical rigor, the profitability of colonial plantations and plants, or the ready adaptability of taxonomic inquiry to species and race. With the letter to Josephine as a counter instance, my point might be put this way: the figural portability of botanical names and ideas exposes different levels of difference, from the taxonomic to the figural, with anthropomorphic impulses at work all along this trajectory. Erasmus Darwin's hyperreadings of Linnaeus's sexual system in the *Botanic Garden* expresses this point by describing the Linnaean sexual system in terms that invite speculation about how taxonomic categories do or do not fit the plants they purport to describe.

As Darwin and others write about plants, botanical figures at times leave purposiveness behind, like wayward children whose unseriousness and playfulness remain to a degree untouched by the norms of goal-oriented inquiry. For, whatever else botanical figures do, they fool with categories, thereby undoing the work of hierarchical and taxonomic analysis to entertain the resources of figure and aesthetic play. Commenting on how the rhetoric of purposiveness dominates the nineteenth-century British reform novel, Caroline Levine identifies a similar logic in the proliferation of critical discourses after New Criticism (feminist, Marxist, cultural studies, etc.) that espouse "more socially conscious, efficacious modes of reading."[9] Wondering about the utilitarian inheritance and program of critical use value, Levine asks whether there might be some version of aesthetic freedom, Kantian or otherwise, that might argue for a theory of reading more attuned to pleasure or more skeptical of the emancipatory claims that are at times embedded in criticism with a purpose. Romantic botany offers, from its location in a historical moment when utilitarian purpose lived uneasily beside claims for aesthetic freedom, a discourse and material culture that are pitched both ways. In one direction, botany registers its role as a factual array with clear scientific and cultural ends in view; in the other, its potent mix of materiality and pleasure invites figures that challenge the existing taxonomic order.

Below, I consider evidence of the material and cultural presence of romantic plants in the botanical networks in which letters and plants circulated, in popular print culture and in botanical illustration, both published engravings and drawings. Whether social, religious, or commercial, private or official, botanical networks supported a flourishing scientific and popular print culture about botany and natural history. By the last decades of the eighteenth century, the European market for botanical illustration shifted to England, where artists and booksellers deployed new engraving techniques alongside older artistic practices to corner the growing market for plant illustration. Together with botanical drawings, this print culture was critical to the role botany would play in the culture of the era.

This chapter concludes with Erasmus Darwin, whose formal exploration of speculative thought and botanical figure in *The Botanic Garden* is echoed by his contemporaries and successors, among them grandson Charles. If the Linnaean sexual system is the hook for the elder Darwin's figurative and discursive project, he carries thinking about botany and figure well beyond Linnaeus. More flamboyantly than any other romantic writer, Erasmus Darwin suggests the risk of botanic figure. That he does so by dramatizing kinds of sexual difference and conjunction in whose presence monogamous heterosexuality looks like a disappearing species is critical to my argument. For much as queer theory and reading tend to unhinge the stabilities of modern critical practice, so does his exuberant portrait of different ways of sexing plants disturb the scientific, material discourse that is the avowed focus of botany's visible popularity from the late eighteenth century forward. This is so despite the fact that Darwin's suggested couplings remain determinedly heterosexual. As clearly, it is the materially specific, culturally marked work of botany that so obsessively invited this figurality.

NETWORKING

The marchioness's putative letter to Josephine assumes what anyone reading the letter at the time, whether in private or in print, would have known. The former empress depended on the English (originally Scots) nurseryman James Lee in Hammersmith to send her exotic plants for the garden at Malmaison. Lee belonged to some of the many and sometimes overlapping networks of professional and amateur botanists and nurserymen around the globe who exchanged plants and botanical knowledge and speculation. Probably until Joseph Banks returned from the voyage of the ship *Endeavour* that made his reputation and career, he and Lee were professionally acquainted. It was Lee who recommended his Scottish countryman Sydney Parkinson when Banks was looking for a natural history illustrator for the voyage.[10] After that voyage, with the support of royal patronage, Banks instituted the botanical network of far-flung correspondents and plant exchange (Banks appears mostly to have received plants) that is, after Linnaeus's, today the best-known example of such networks. Like Linnaeus, Banks developed a network of correspondents who sent him specimens of plants and animals. Some of them had done so for Linnaeus.[11]

Among aristocrats and even the royals, botany was a major pastime. The wife and daughters of George III received instruction from Francis Bauer, Banks's principal botanical artist at Kew, on how to draw plants.[12] Yet royal patronage and royal pastimes tell nothing like the whole story of the era's botanical networks. At about the time the women of the royal household were learning from Bauer, James Sowerby taught Mary Wollstonecraft how to draw plants.[13] Anne Secord has examined widespread networks that involved working-class men and women; E. C. Spary has demonstrated the role of other such networks in pre and postrevolutionary France; still others have tracked the imperial exchange of botanical specimens and drawings that depended on voyages of discovery, mostly but not exclusively British. In studies of Scottish botanists working in India, Henry Noltie has identified networks and friendships among officials of the EIC and many botanists who were not EIC employees. The literary and scientific circle of correspondents who sent letters to and from the informal "post office" at Mme. de Stael's Coppet Castle in Switzerland included botanists who exchanged information even in times of war. As Collinson and Bartram and their wider circle of botanical correspondents anguished about the onset of the revolution in America because it made old friends enemies of the state, they worked out other connections (via the French!) that would allow them to continue their exchange of plants and ideas.[14]

For professional botanists, the academic career path typically began in Scotland at the University of Edinburgh, which produced most of the major botanists of the period 1750 to 1850. Some were Scottish, including Robert Brown, Wallich, Wight, and William Roxburgh; others came from England to attend university there, among them Sir James E. Smith. Those without academic training learned botany from books and from the ground up as apprentice gardeners, nurserymen, naturalists on voyages of exploration, or as women who pursued amateur and in a few cases professional careers as experimentalists, taxonomists, and artists. Among the working classes, some men and women did botany on the side, when they could, meeting (in the case of the men) in pubs to identify plants. Some published scholarly essays in regional as well as London periodicals; some became members or associate members of the Linnean Society, the gentleman's club for English botanists. Botany certainly did not make for a level playing field among the higher and lower classes, but it did allow a point of contact, interaction, and in a few cases a recognition that botanical expertise could complicate the usual class hierarchies of masters and servants. Put briefly, because this expertise counted for something at all ranks, it made possible some unlikely correspondents and plant exchanges. In such networks, many of them anchored by dissenting circles, botanists like Banks who were accustomed to being, in Bruno Latour's phrase, the "center of calculation" might well be found outside London. When Erasmus Darwin, the very figure of late eighteenth-century dissent, contacted Banks (and other botanists) for his advice on a projected translation of Linnaeus's *Systema Naturae*, Banks loaned him rare works on botany and answered numerous letters in which Darwin tried out problems and solutions for his first Linnaean translation.[15]

A correspondence network of mostly Quaker botanists in England and America functioned in ways that may have been more typical than the official correspondence network that Banks supervised. In the eighteenth century, the sixty-year correspondence between two Quakers who never met, William Bartram in Philadelphia and Dr. Peter Collinson, a well-connected physician in London, anchored the transatlantic arm of a network of correspondents that included Linnaeus, John Stuart, the third Earl of Bute, and many others.[16] Eighteenth-century curiosity was well tuned to a Quaker and, more broadly dissenting, theological and educational tradition that understood natural history and scientific investigation as well as commercial industry as linked endeavors open to talent in ways the universities and other professions were not to those outside the Anglican communion.

During his fifty-year tenure (1722–1770) as head gardener at the Chelsea Physic Garden, still located in London on the north bank of the Thames, Philip Miller developed a remarkable network of correspondents, some of whom contributed to the expansion of the garden's seed-exchange program (begun in 1682). Miller was a Quaker whose network of friends and contacts included dissenters of different persuasions and the Anglican naturalist Thomas Pennant, whose copy of Miller's enormously influential and frequently reprinted *Gardeners Dictionary* includes notes in Pennant's hand. Pennant corresponded with John Aikin, brother of Anna Letitia Aikin, later Barbauld, others at the dissenting academy at Warwick, and the Hammersmith nurseryman Lee, a Quaker who moved to London from Selkirk, Scotland, in 1740. Lee learned Latin to translate the Linnaean system into English in his *Introduction to the Science of Botany* (1760), the first published English translation of Linnaeus.[17]

Sydney Parkinson, the artist on the voyage of the *Endeavour*, was a Quaker from Selkirk who moved to London in 1765 at the age of twenty. There he met Lee and began to teach Lee's daughter Ann how to draw plants. Ann Lee was no mean botanist herself: she knew enough and was sufficiently confident about what she knew to challenge botanical claims made by men whose public rank and stature far exceeded her own.[18] The two men may have been introduced by John Fothergill, a Quaker doctor who later edited Parkinson's journal for publication and belonged to the Quaker meeting that Parkinson attended in Westminster. Fothergill commissioned Ann Lee to create flower paintings on vellum—works now housed in the library and herbarium of the Royal Botanic Garden, Kew. He also commissioned William Bartram's botanical exploration and drawings.

After Parkinson died, Banks took possession of his drawings and personal effects. Parkinson's older, and apparently unstable brother Stansfield was outraged and attacked Banks in print. Fothergill managed to calm the brother and get Banks to distribute belongings that Parkinson had mentioned in his will, among them works on vellum and artist's tools he had specified should go to Ann Lee. Fothergill's account of this intervention in his preface to Parkinson's published journal never mentions their shared identity as Quakers. His earlier eulogy to Collinson had been similarly discreet. Titled *Some anecdotes of the late Peter Collinson, Fellow of the Royal Society, in a letter to a friend*, Fothergill's pamphlet makes no mention of the Society of Friends. The lowercase notice of an unidentified "friend" to whom the pamphlet was addressed may be a remarkably reticent allusion to the Quaker circle of botanists and naturalists to which he belonged.[19]

Other Quakers were botanists, illustrators, or publishers of botanical magazines and books. William Curtis was the first editor of the *Botanical Magazine* and the author of two renowned works on botany, *Flora Londinensis* (1777–98) and the folio edition *Beauties of Flora* (1820). Sowerby drew the plants that Sir J. E. Smith then described in the joint multivolume production *English Botany* [1790–1814], thereafter known as Sowerby's *English Botany*, to Smith's chagrin, and later projects, including *Exotic Botany*, a work that introduced Indian plant illustrations to a wider audience.[20] By enlisting his sons and daughters to draw and engrave specimens of natural history from plants to fungi, Sowerby inaugurated a middle-class, family publishing enterprise.

Although the Bartrams and occasionally Collinson specify the religious fervor of their botanical curiosity in private letters, most Quakers seem to have been circumspect in public about the affiliation between their botanical interests and religious convictions.[21] Fothergill, a Quaker with close ties to other Quaker botanists and a defender of Quakers against the usual antidissenter arguments, said nothing about botany in a widely reviewed pamphlet on the Quakers. But in a letter written from Dublin to the Wensleydale Friends Meeting he invoked two plant figures whose biblical pedigree is indisputable: the "Burning Bush" and some (unspecified) "Plants of Righteousness" that, he argued, "too many amongst the Lord's People" would mix with "the unbridled, fleshly Will and Affections of Man."[22] The generic figurality of these remarks looks very different from private exchanges between John Bartram and Collinson, for whom the phrase "curious in our way" gestures toward their shared regard for an intricacy, pattern, and subtlety in nature that they supposed was divinely ordered. Neither seems to have believed that God's nature was somehow in their private gift or inheritance. It is rather that both believed their curiosity—about the structure and nature of plants and even apparently monstrous plants and animals—was humanly appropriate, even desirable as a form of worship. For them, doing botany could be at once free-thinking, experiential, and yet hallowed.

The free-thinking side of botany was assuredly not exclusively Quaker, but it surfaces among them in intriguing ways. In the commonplace book that Collinson kept for decades, a schoolmaster friend, William Massey, recorded, dated, and signed this declaration about strangeness in plants:

> There are some Instances of <u>Vegetation</u>, that are really amazing; <u>Nature</u> seems in many cases to act <u>lavishly</u> and yet, I believe, it is owing to our <u>Ignorance</u> of her grand designs, when we think so. . . . I think Accts and observations of this Nature ought not to be made public for <u>Amusement</u> only; or to satisfy an <u>idle curiousity</u>; but with a View to show . . . what great care <u>Divine Providence</u> takes, in preserving and propagating and even sometimes wonderfully <u>every species</u> of Beings, animal and vegetable; so that it seems impossible that any of them should be <u>intirely lost</u>, not withstanding the great Destruction of Some, and <u>Neglect</u> of others.[23]

Massey's pious account (dated 1752) edges near speculative cliffs of fall. As had Linnaeus, most botanists in the late eighteenth century and for some decades after insisted that all species were created at the beginning of creation and none had been lost since (Linnaeus, late in life, thought perhaps some had been lost). A half-century later, the Quaker botanist and illustrator Sowerby took a looser and longer view of speculative questions in a note recorded among his manuscript papers on fungi. The topic is *monstrosity*, a term that often turns up when contemporary writers discuss cryptogamics, including fungi:

> Accuracy is ever to be admired yet in some instances it may be wrong applyed Thus were I to endeavour to arrange the creation begining with any individual for instance Man, I might light of one with all the parts belong to a most perfect

> man. Consequently I might make a good and perfect Description–pursuing my theme I might go on acurately describing every creature I met with, and might light of beings like the man I had described with every part somewhat longer & shorter[?] or[?] broader [something crossed out] or with some extraordinary protuberance or monstrosity a limb less &c consequently an acurate description would make a new being or species and a monstrosity or lusus natura might be made a species when it was only a meer variety.[24]

Imagining a situation with an uncanny resemblance to Victor Frankenstein's botched creation of a creature that was to have been beautiful but turned out to have "every part somewhat longer & shorter," Sowerby considers how a taxonomic definition of a man might well go wrong, mistaking perfection for monstrosity, or a new species for a variety.[25] Much is left open for inquiry here: the possibility of taxonomic error, the question of how one goes about defining the human species—a question widely asked throughout the era—and the possibility that one taxonomist's monstrosity might be either a new species (whether new to Europeans or new to the world, Sowerby does not say) or simply a variety. Writing about natural history fifty years apart, these Quaker botanists show how far taxonomic and botanical curiosity could go, even for Sowerby, who reverently dedicated a separate number in his *Fungi* series to Rev. H. Davies.

Sowerby's taxonomic speculation conveys a curiosity that does not conform to the early modern tendency to put odd bits, called curiosities, into a cabinet or taxonomy to contain them. Nor does it convey the reaction against the drive to taxonomize everything in the early eighteenth century. In her analysis of this reaction, Barbara Benedict shows how the category of the monstrous could be used to rationalize imperial gestures as well as social estrangement.[26] Yet by the late eighteenth century and beyond, Quakers like Sowerby and others view nature and kind as occasions for admiration and taxonomical curiosity.

That mix is evident in William Bartram's *Travels through North & South Carolina, Georgia, East & West Florida* (1791), which attracted romantic poets and a large reading public to its vision of a wild, sublime North America. As the younger Bartram describes it, that vision is densely botanical and figurative, as well as Linnaean:

> At this rural retirement were assembled a charming circle of mountain vegetable beauties; Magnolia auriculata, Rhododendron ferrugineum, Kalmia latifolia, Robinis Montana, Azalea flammula, Rose paniculata, Calycanthus Floridus, Philadelphus indorus, perfumed Convalaria majalis [now *Convallaria Montana*], Anemone thalictroides, Anemone hepatica, Erythronium maculatum, Leontice thalictroides, Trillium sessile, Trillium cesnum, Cypripedium, Arethuza, Ophrys, Sanguinaria, Viola uvularis, Epigea, Mitchella repens, Stewartia, Halesia, Styrac,

> Lonicera, &. Some of these roving beauties stroll over the mossy, shelving, humid rocks, or from off the expansive wavy boughs of trees, bending over the floods, salute their delusive shade, playing on the surface; some plunge their perfumed heads and bathe their flexile limbs in the silver stream; whilst others by the mountain breezes are tossed about, their blooming tufts bespangled with pearly and chrystalline dew-drops collected from the falling mists, glistening in the rain bow arch. Having collected some valuable specimens, I continue my lonesome pilgrimage.[27]

These are travel notes on flora and fauna, not florid letters from an overheated sensibility.[28] Yet Bartram moves easily from Linnaean names to apparently female personifications whose company gives him pleasure before he continues on his "lonesome pilgrimage." This mix registers the slippage between plant and figure that makes botany the intersection between a highly regarded approach to material nature and an array of discursive practices in which figuration is rarely a silent partner.

BOTANY IN THE NEWS

Botanicus's letter offering plants to Josephine at Malmaison is not the whole story, either about botanical figures or names. To begin, the marchioness (probably a figure herself) omitted the Latin botanical names for plants that appeared in botanical handbooks and pedagogical tracts of the period. Moreover, whether she would have liked it or not, the marchioness was only one among many, from many ranks and occupations, for whom botany mattered. Class and rank and education tended to insure that how one did botany depended on one's opportunities, but not entirely. Secord has observed that some working-class botanists knew enough to correspond with renowned botanists, among them Banks and J. E. Smith. Nurserymen like Bartram in America and Lee in England learned enough Latin to navigate Linnaeus's books or, in Lee's case, to translate the *Species Plantarum* for English readers. By the late eighteenth century, botanical habits of mind and reading involved all classes and ranks, ranging from scientific treatises to reviews and botanical queries in popular magazines and reviews. These last offer graphic evidence that English reading audiences of the late eighteenth and early nineteenth centuries were attracted to botany and natural history to a degree that most modern readers would find puzzling or boring. This point might be put another way: few of us find the highly differentiated material particularity of the natural world as compelling as it was for Bartram and many of his contemporaries.

First, some leading indicators. In the first two editions of the *Encyclopedia Bri-*

tannica (1771 and 1778), the entry on botany, the same in both editions, covered twenty-five pages. In the much expanded third edition, a rather different essay covered fifty-seven. When the *Annual Register* began publication in 1758, natural history was not among its list of indexed subjects. But in the next year, a "Natural Register" became a regular feature. It included a sharp critique that echoes Buffon's earlier attack on Linnaean categories.[29] In subsequent years, "Natural History" (of all kinds) became a regular section of the *Register*. For a slightly more extended census, drawn from two magazines intended for a wide reading audience and published from the early to mid-eighteenth century forward into the nineteenth century, we can go to the *Gentleman's Magazine* and the *Monthly Review*. Both published essays and notices on botanical topics, although the second, the premier review in England before the *Edinburgh Review* began publication in 1802, was inevitably more devoted to reviewing.[30] Unlike the *Transactions* published by the Linnean Society and by the Royal Society or even the more egalitarian *Philosophical Magazine*—publications that were intended for readers who had some training in scientific inquiry—the *Monthly Review* and the *Gentleman's Magazine* were, as Keats puts it in another context, "caviare to the general," serving up tidbits of knowledge and discussion pitched toward a general reading public whose appetites editors tracked and to which they responded.[31]

By the 1790s, the *Gentleman's Magazine* had become more programmatically pro-Constitution, in Burke's understanding of this concept, whereas Ralph Griffiths's *Monthly Review*, old Whig in disposition, with churchmen and dissenters among its reviewers, was sympathetic to the French Revolution.[32] These differences inflected how each reported works and events, yet both remained attentive, surprisingly so, to botanical topics in the midst of wartime news. In 1770 the *Gentleman's Magazine* published only short notices on botanical topics—a dispute about whether the Spanish or Sweet Chestnut tree is a native species, in response to Philip Miller, head gardener (also Quaker) at the Chelsea Physic Garden in London, and short reviews of books published that year on "Husbandry, Botany, Gardening."[33] The pace picked up in 1775, when the big news of that year was the American colonies' agitation for independence. A long essay on the "Progress of Botany in England" by one signed as Rusticus (the penchant for pseudonyms is a constant in British magazine writing on all topics in the eighteenth and early nineteenth centuries) surveyed English works on botany beginning in 1525. The author of the piece may have been Richard Pulteney, a physician and botanist. In 1783 still more abbreviated pseudonyms engaged each other in botanical disputes: one, signed PBC, defended Linnaeus against English critics in two separate issues. In 1784, a series of pieces on trees ensues, written by TAW, aka (the editor helpfully explains) THW. In the same year, the magazine reviewed Sydney Parkinson's posthumous *Journal of a Voyage to the South Seas, in his Majesty's Ship the Endeavour*, in line with its general practice of printing obituaries of botanists and nurserymen.

Later in 1784, PBC returned to criticize TAW for errors in reporting Linneaus's classification of holly species. PBC called TAW a "novice in botany" and recommended several books to get him up to speed. PBC returned the next month to offer a "scale of beings," with man at the top, above orang-utans and monkeys and other less biologically fortunate beings, making clear the ease with which those who debated the status of botanical species could shift to animal and human ones. In the supplement for 1784, THW (aka TAW) appears to come around, apparently because he had learned that PBC is a "distinguished professor." This nod to authority and professional rank turns out to be a feint: THW went on to restate his earlier claims. In the midst of this shuttle of published letters between correspondents with unpronounceable because abbreviated pseudonyms, the magazine reported William Jones's delivery of a sermon on "The Religious Use of Botanical Philosophy" Jones's deist argument from a text in Genesis urged attention to botany as evidence of God's bounty, a theme that Jones echoed from India, where he worked hard to become an adept amateur botanist of Indian plants.[34]

Over several years, THW continues to send in letters on species of trees, later offering more detailed botanical descriptions and abetted by one JA. Classical allusions abound, along with notice of Rousseau's botanical writing. Someone who signed himself Viator gave an account of the "celebrated Banion tree or India Fig" in June 1787, and THW continued to send in botanical notes on trees. Late in that year, a review of William Wright's *Account of the Medicinal Plants growing in Jamaica* is reprinted from the *London Medical Journal*. These notices, which convey the explicitly global aspect of English botanical interests, share space with an increasing number of essays on the slave trade, especially letters in support of the humanitarian character of said trade from one Thicknesse, who continued in this vein into the 1790s. Along with essays on other curious bits of natural history like "animalcules" and ants, there were more pieces on exotic species like Carolina grass, poems, including some addressed to flowers, and a good deal about, by, or to Joseph Priestley, who defended dissenters and the liberal side of most questions. In the late summer and early fall of 1789, when the English had other things on their minds, there was still more about Carolina grass in separate issues and a detailed, fairly erudite review of three botanical publications that emphasized questions having to do with the Linnaean classification, a review of Price's *Discourse on Love of Our Country*, and many reports on the French Revolution. By 1790, Thicknesse is, inevitably, back, repeating the same arguments, adjacent to reports of more insurgent slave uprisings in the West Indies. In the midst of all this were two detailed botanical descriptions of "Anastatica," or the Rose of Jericho; one of them digressed to African vetches, whose Linnaean name the contributor wants to know. As the editorial policy of the *Gentleman's Magazine* becomes more emphatically pro-Constitution and anti-Jacobin, letters pro and anti Priestley, anti-Price, and increasingly antidissenter (although the magazine continues to publish letters in defense of dissenters), these topics took up more space. Yet botanical pieces continued to appear: a biographical anecdote of one Mr. J. Wilson, a botanist of working-class background who died in 1751 and is here credited with trying to systematize botany "first"—that is, before Linnaeus's publications were translated into English that year.

Cryptogamic plants were given more attention in the botanical notices the *Gentleman's Magazine* published in the early 1790s: a curious fungus, the rediscovery of *Polypodium*, identified as an English Fern, which a subsequent contributor said was in fact *Anemone nemorosa*, an acrid and somewhat poisonous plant.[35] Its more popular name was Vegetable Lamb. Working both sides of the ideological street, the magazine published a poetic epistle to Dr. John Aikin, Anna Barbauld's brother, which attacked him for not having defended Price (by now dead) or Priestley. By 1792, someone had written that the fern earlier in question was misidentified, and letters to and from and about Priestley continued, along with pamphlets on the Test Act and more on trees, swallow migration, and plants, including a species identified as "knee holly," RG remarked on the Smith/Sowerby collaboration in *English Botany*, with specific objections to some of Sowerby's engravings and design choices. In 1794, when the English again had other pressing topics on their minds, the magazine published a long obituary on Anna Blackburne, an amateur botanist whose work more famous English botanists and Linnaeus himself had praised, and the anonymous "A Naturalist's Ramble in the North," mostly about geology and birds. The species-oriented argument in favor of slavery also turned up that year: Africans were slaves in their own country and, besides, members of the Negro race were "but a set of wild beasts when let loose without controul."[36]

From 1796 on, botanical articles were on the increase, with obituaries for major English botanists (as well as minor colonial functionaries who died abroad), reports on the efforts to get breadfruit from Tahiti to the Caribbean for slaves to eat, the cultivation of sea kale, some criticism of Linnaean nomenclature and Erasmus Darwin's English renditions, debates about whether certain plants were natives or simply nativized, and a suggestive comment appended to a notice on the forthcoming publication of a pocket "Flora of British Plants" arranged according to the Linnaean system: the algae and fungi (typically, if nervously, still assigned to the Cryptogamia) were to be omitted because "We know not to what link of the chain of Nature they absolutely belong." This pocket flora inspired a number of replies, in praise of, or not, and at least one call for a Latin edition, together with

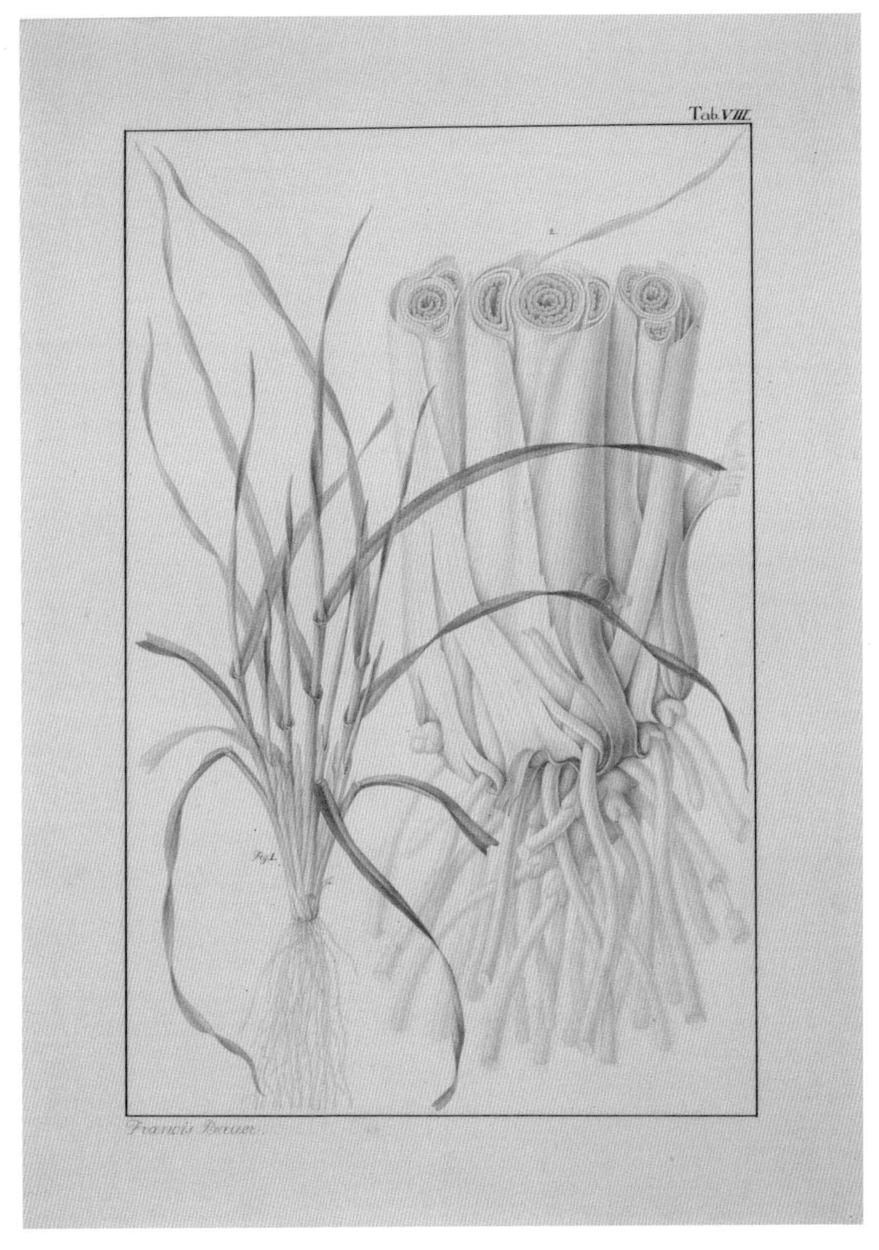

Fig. 1. *Wheat Stalk study,* Franz Bauer, graphite and colored pencil, 40 × 28.5 cm. (15.74 × 11.22 in.). Copyright Natural History Museum, Botany Library, London. Tab. VIII, accession no. 140883.

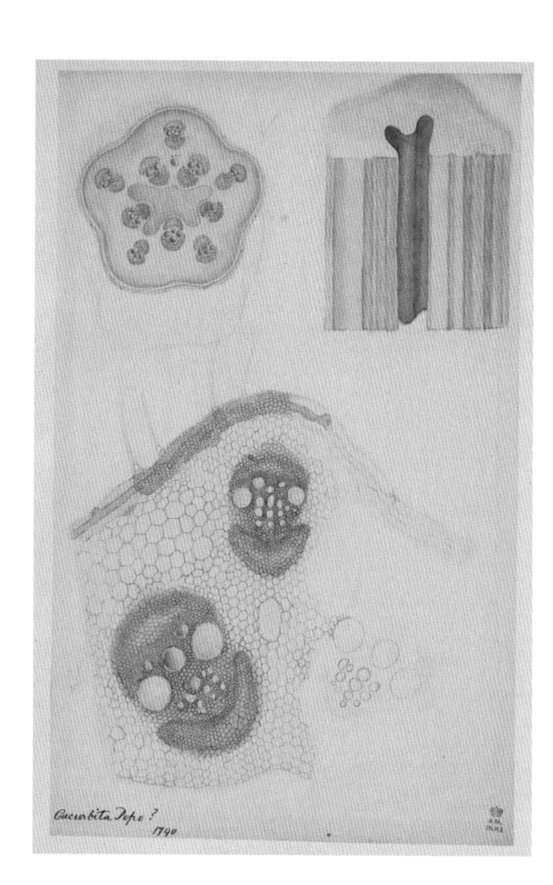

Fig. 2. *Stems and Petioles,* Franz Bauer, graphite and colored pencil, 40.6 × 25. cm. (15.98 × 9.84 in.). Copyright Natural History Museum, Botany Library, London. Fol. 13, accession no. 140879.

Fig. 3. *Rafflesia arnoldi,* Franz Bauer, graphite and colored pencil, 52.7 × 35.7 cm. (20.75 × 14.06 in.). Copyright Natural History Museum, London. Kew plants plate 82, accession no. 368530-1001.

Fig. 4. *The American Cowslip*, Peter Henderson, painter, Warner, engraver, aquatint, stipple and line, 57 × 46 cm. (22.4 × 18.1 in.). May 1, 1801. In R. Thornton, *Temple of Flora*. Reproduced by permission of The Huntington Library, San Marino, California; RB 282000.

Fig. 5. *Dragon Arum*, Peter Henderson, painter, Ward, engraver, mezzotint, aquatint in final state, 57 × 46 cm. (22.4 × 18.1 in.). Dec. 1, 1801. R. Thornton, *Temple of Flora*, plate 22. Linda Hall Library of Science, Engineering & Technology.

Fig. 6. *Curious American Bog Plants*, Philip Reinagle, artist, D. Maddan, engraver, aquatint, 57 × 46 cm. (22.4 × 18.1 in.). July 1, 1806. R. Thornton, *Temple of Flora*. General Collection, Beinecke Rare Book & Manuscript Library, Yale University. Bibliographic Record no. 2007279, image ID 1036523.

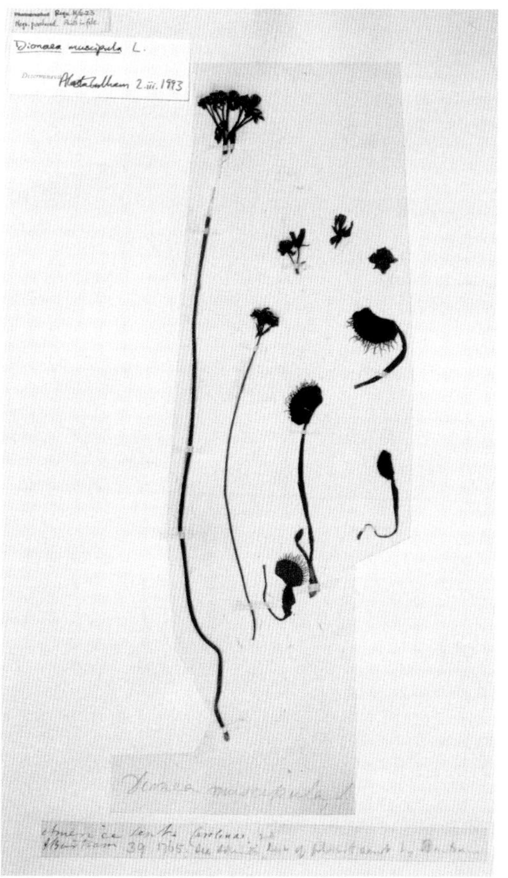

Fig. 7. *Venus Fly-Trap*, William Bartram, Black ink, 39.8 × 30 cm. (15.67 × 11.81 in.), 1769. Copyright Natural History Museum, London. Negative no. 023303, accession no. BM000752964.

Fig. 8. *Bandana of the Everglades or Golden Canna*, William Bartram, black and brown ink, 24.9 × 18.8 cm. (9.80 × 7.40 in.), 1776. Copyright Natural History Museum, London. Negative no. 015948, accession no. 359483-1001, plate 48.

Fig. 9. (*Below*) *American Lotus, etc.*, William Bartram, watercolor and black ink, 24 × 29.9 cm. (9.45 × 11.77 in.), 1769. Copyright Natural History Museum, London. Negative no. 021865, accession number 359483-1001, plate 34.

Fig. 10. *Exhibition Extraordinaire in the Horticultural Room,* G. Cruikshank, engraved, colored etching, 32.1 × 45.3 cm. (12.64 × 17.83 in.). HEW 4.12.6 v.5, plate 55. Harry Elkins Widener Collection, Harvard University.

Fig. 11. *Trapa bispinosa*, Roxburgh Indian artist, watercolor on paper, 49 × 35.5 cm. (19.29 × 13.98 in.). Royal Botanic Gardens, Kew. Roxburgh no. 1345, *Flora Indica* 1(1832): 428.

an increasing number of letters for but also against the Linnaean system. In the run up to 1800, when articles about the rebellious Irish and Jacobinism were thickly distributed, botanical notices more sharply questioned Linnaean nomenclature.[37]

During this period some of the disputes were so narrowly taxonomical that one might have imagined them of interest only to a specialist audience. Not so. Even when arguably far more pressing concerns were at issue such as the American and French Revolutions and the magazine editorship's growing anxiety about dissenters and Jacobin tendencies, the frequency of botanical articles remained fairly constant, even as their scientific range deepened. As clearly, those who reviewed works on botany in the *Gentleman's Magazine* did not pursue the implied affiliation between taxonomic debates about, say, species of oak, and Thicknesse's recycling of the lower-species rationale for enslaving Africans to work on Caribbean plantations. The figurative exfoliation of botanical and taxonomic ideas was evidently not the work of one reviewer, although it is clear that the same correspondent who was so adamant about botanical distinctions also offered arguments about the species of the human with a similar degree of confidence.

The *Monthly Review* was the premier English review of the late eighteenth and early nineteenth centuries. Between 1790 and 1805 it extensively reviewed works on botany in every year except 1793, the year in which England declared war on revolutionary France. In the issues published that year, contributors passionately defended the French and various declarations of human rights and just as passionately criticized Edmund Burke for opposing the revolution. Like the *Gentleman's Magazine*, albeit from the other end of the political spectrum, the *Monthly Review* covered botanical works along with everything else, and in particular works on slavery and revolution. During the same period, it was attentive to key players and their botanical disputes, including reviews of works that its popular competitor never mentions, including the first folio volume of Roxburgh's monumental *Plants of the Coast of Coromandel*, a French study of "natural" plant families, books by botanist Dawson Turner, who specialized in fungi, and tracts on botany, including one on perspiration in plants. Its political and botanical alliances were more coherently aligned: the magazine that favored the French Revolution admired Darwin's *Botanic Garden* (which anti-Jacobins most certainly did not) but disliked William Withering's *A Botanical Arrangement of British Plants*. Reviewers for the magazine took careful note of various disputes about botanical names, Latin and English, charging some authors with the cultivation of problematic names, but praising the etymologies and nomenclature that others offered. Reviewers also noted, more than once, features of the Linnaean classification that were in dispute, the Polyandria Polygynia and the Cryptogamia, which one reviewer suggested someone, perhaps J. E. Smith, ought to revise (recall those problematic fungi and ferns).

Predictably, the magazine's review of J. E. Smith's *Plantarum Icones* (1791) was so respectful that the effect is just this side of cloying. During this period (the magazine ceased in 1810, when English botanists still advocated the Linnaean system), the *Monthly Review* continued to support Linnaeus despite Buffon's searing critique, calling only for Smith or someone else to rectify those parts that need revision or rearrangement.[38]

A 1796 article in the *Monthly Magazine* shows how easily botanical distinctions and names could be exported to or used to supervise other kingdoms of natural history. The article described a new botanic institution in Glasnevin, near Dublin, in which botany and agriculture would be joined, as they typically were in botanic gardens around the world. "Pursuant to act of parliament," its committee of agriculture reported, the garden would be laid out in sections, the first being a "Hortus Linnaeensis" in which plants, shrubs, and trees would each be "arranged according to its class, order, genus and species," from first to last, with labels to match. Unlike Linnaeus, and other botanists who had taken little notice of varieties, the plan of this garden would also include varieties, all properly arranged.

What follows from this beginning suggests a penchant for inventing Linnaean binomials without Linnaeus around to monitor the outcome. The next garden was generically named "the Cattle Garden," for the various agricultural animals (not plants for animal consumption) that it would include:

1. The Sheep Division, or *Hortus Ovinus*
2. The Horned Cattle Division, or *Hortus Bovinus*
3. The Horse Division, or *Hortus Equinus*
4. The Goat Division, or *Hortus Hircinus*
5. The Swine Division, or *Hortus Suinus*[39]

All the other garden divisions (Hay Garden, Succulent Plants, Dyers' Garden, Plant Nursery for seedlings, a projected Medicinal Garden) were presented without Latin tags, except for the Hortus Siccus, the usual term for an herbarium of dried specimens. Presented without commentary, the use of the term *Hortus* and appropriate Latin tags to describe farm animals, arranged as though they, too, were plants in a garden, engineered a slippage from one kingdom to the next. Whatever else the curious nomenclature of this Irish project suggests, it presented the Linnaean binomial in terms that made what was to have been an authoritative, legislated system of naming into something more pliable.

Intended for the nonspecialist reader, these magazines and the *Encyclopaedia Britannica* show that botanical inquiry attracted a wide readership from the last decades of the eighteenth century on. Beginning in the mid to late 1790s, when English readers and magazine editors had a lot on their minds, the number of essays on botanical questions steadily increased and the main encyclopaedia survey on the subject expanded its coverage by one-third of its original length. Collectively, they register the persistence of botany and its unsettled questions in romantic popular culture.

FIGURING PLANTS, FIGURING SPECIES

From the 1780s to the early 1840s, English publishers and their artists used every available means to depict plants for as many readers and buyers as they could find. During this period, careers were made and ruined by big botanical books with expensive plates as new engraving techniques and developments (mezzotint, aquatint, stone lithography, and finally steel engraving) sharply altered romantic print culture. Nowhere was that alteration more visible than in natural history and, above all, botanical, illustration. Whereas engraved copperplates were soft and soon worn out (one estimate claims that the maximum number of pulls from a single plate was one hundred), the new engraving techniques allowed for much bigger print runs. Together with less expensive woodcut engravings, which still dominated book illustration, these innovations entered the market at the moment when the public taste for illustrated botanical magazines and books expanded exponentially, no doubt in fairly precise relation to their availability—both those that middle-class buyers could afford and those that were so costly that few could buy them.[40]

The slight lag in the development of color printing techniques for mass production during these years meant that it was often cheaper to hire others to paint engraved sheets under direction from a botanical artist. Botanical magazines typically used a fixed number of plates per issue, which were engraved and then hand-colored by women and children working off Fleet Street, not precisely coloring by number, but following a template created by a presumably better-paid artist. What is astonishing given this mode of production is the high quality of these individually hand-colored plates, which today show up for sale as single sheets cut out of bound volumes. More expensive botanical books, often large and offered in much smaller print runs, were usually illustrated by well-known and mostly botanical artists, including some women. For the *Temple of Flora*, a double elephant folio work published in fascicles between 1798 and 1807, Dr. Robert Thornton commissioned artists and engravers to create large-scale drawings and then engravings of exotic and native plants.[41] In some of these engravings, the mix of hand and color printing techniques is so subtle that it is difficult to determine precisely which parts are colored by hand and which are color-printed. Some of these big botany books were so expensive to buy that publishers or entrepreneurs issued a few of them at a time, over a period of several years, or they issued expensively illustrated books in limited production with lists of subscribers given at the beginning, either to invite others to join the illustrious purchasers thus mentioned or simply to show that the limited edition now belonged to a limited number of savvy (or wealthy) purchasers.

Behind this print culture and now mostly hidden away in archives and libraries are thousands of botanical drawings as well as botanical images in unusual media. By far the majority of amateur botanical artists were women. Professional artists were mostly male, although the exceptions to this rule suggest that women were surprisingly (given the era's gendered protocols for professional artists) involved in drawing and engraving plants for pay.[42] Prior to the invention of photography, when nature printing was rare and not feasible for certain kinds of botanical information, an artist like Franz or Francis Bauer, the botanical artist who worked for Banks at Kew, painted plants and drew meticulous cell studies in pencil of wheat and other plants. William Bartram was commissioned to draw plants during his travels to the southeastern United States. What supported botanical illustration in public and in private was a long-standing conviction among botanists, especially those who had been Scottish trained, that botanical images were essential. That conviction owed much to one man, John Hope, who was appointed professor of botany at the University of Edinburgh in 1768. Hope created detailed and exquisite drawings to teach his students, among the most prominent botanists of the next generation.[43] They in turn drew, taught, or hired others to draw plants at home and around the globe.

Although illustration was by 1800 a massive commercial as well as scientific enterprise, this practice put added strain on a long-standing debate about whether images were necessary or helpful to the study of plants. Linnaeus used only a few schematic diagrams of different leaf shapes and other plant parts in the appendix to *Philosophia Botanica* (1751), although he later hired artists and draughtsmen to depict plants.[44] In the reaction against botanical illustration that intensified in the early nineteenth century, many rejected the long-standing assumption that botanical images were pedagogically useful. To the contrary, the objection now ran, images distracted readers who preferred to look at figures, especially colored figures, and were unwilling to study taxonomic argument and evidence.[45]

Then as now, botanical illustration exists in a flyway between science and art or, to put this point more philosophically, between species exemplarity—the avowed scientific reason for depicting plants as well as offering verbal descriptions—and the appeal of brilliantly drawn and colored plates. Precisely because botanical illustrations, as published engravings, archived and unpublished drawings, or images constructed with paper or pressed flowers, inhabit this fungible border between scientific verisimilitude (their professed task) and aesthetic production, they offer a compressed visual analogy for the role of botany in romantic culture as at once a material event (or complex of events) and an aesthetic practice with philosophical consequences. Authors of illustrated botanical works have nearly always insisted that their images were faithful to nature or taken from nature, yet the singularity and desirability of botanical illustrations large and small, and even the circumstances of their production, mark them as works of art whose aesthetic appeal obscures their mimetic and exemplary purpose.[46] For this reason they occupy a shadowy middle ground between the exemplarity of a given illustration—the way in which the plant depicted is argued to be typical or representative of an entire species—and the particularity, even individuality of the plant represented, often in living color.

Igor Kopytoff has argued that this dispute turns on the difference between a commodity so individual that its value cannot be exchanged for any other (say, a work of art so exceptional that it could not be exchanged for even another work by the same painter) and a commodity so indistinguishable from its class (say, a steel bar made to a measure consistent with thousands of the same).[47] Botanical illustrations push plants away from the collective, exchangeable status suggested by claiming that this *hieracium*, or mouse's ear, typifies all *hieracium* plants of this species. Presented as individual works of art to cut out of the *Botanical Magazine* and frame as collectors do now, or simply a single plant, like Wordsworth's single tree in the *Intimations Ode*, plants look like individuals.

This difficulty is implicit in the way artists of the late eighteenth and early nineteenth centuries prepared drawings and engravings for publication. Despite claims to the contrary (or silence on this point), botanical artists typically copied parts of different published engravings to create their own. Kärin Nickelsen's comparative analyses of English and German books of botanical illustration published from 1700 to 1830 show that copying, as much as absolute originality or drawing from nature, was a matter of course precisely because the illustrative goal was species typicality: the plant depicted had to convey the taxonomic traits or "character" of its species.[48] Artists copied other artists, sometimes several other artists, to get as close as possible to a full and accurate species image. Fidelity to the systematic responsible for identifying and naming the species in question was part of the bargain: artists who drew or engraved plants for Linnaean botanists were expected to depict its floral parts in detail, typically in separate, schematic drawings on the same sheet or plate. Doing so might require a further degree of schematization so that, for example, the parts of a seed vessel would be represented in separate drawings.

Embedded in the history of the debate about botanical illustration is perhaps the most compelling point of contact between the material practice of botany and its figural, philosophical involvement with questions of aesthetics, typicality, and representation. As long as botanists, as had Aristotle, believed that the number and kind of species were constant, the notion of plant types could invite and has often invited the conviction that they are archetypes. Although modern definitions of botanical types (holotype, syntype, lectotype, and nomenclatural type) steer clear of archetypicality, the history of those definitions began with early modern and modern taxonomic efforts to sort out what plant specimen or specimens would be the reference point for determining and illustrating the traits of a species.[49]

The triangulation between plants, earlier engravings, and artists is visible in the collaboration on the publication of *English Botany* between Sir J. E. Smith, first president of the Linnean Society of London and one of the most respected botanists working in England, and James Sowerby, the publisher of *English Botany* and a premier English artist and engraver of works on natural history. From 1790 to 1814, the two corresponded regularly about drawings and descriptions for the thirty-six volumes of *English Botany* they published during this period. The extant portion of Smith's side of this unpublished correspondence shows that they worked from specimens, although Sowerby almost certainly consulted earlier prints. As one of the principal artists for William Curtis's *Flora Londinensis*, the work whose engravings Nickelsen includes in her analysis of copying practices, Sowerby would have been introduced to this way of augmenting information derived from specimens, already dried or getting there as the artist worked.[50]

For the *English Botany* project, Sowerby mailed plant specimens as well as drawings to Smith, who would write the botanical description with both in front of him and on occasion ask his artist/partner to correct a drawing. To indicate a set of distinctions between two closely related species that an initial drawing failed to offer, Smith explained to Sowerby what the leaves of the species under consideration looked like under magnification so that the drawing might be altered to reflect this information. He also directed Sowerby's application of color, noting in some cases that holding up a plant to the light will show its true color.

On one occasion Smith recommended both sketching a flower (*Glaucium phoenicium*) from a specimen and looking at Curtis's *Flora Londinensis* because it had a "very good" figure for the same plant. Despite the men having exchanged holiday gifts almost every year (Smith sent a Norwich turkey and Sowerby sent Smith a barrel of pickled herring), and despite the middle-class artist Sowerby having paid Smith, the baronet, for his descriptions, Smith could be querulous or imperious when the published engravings fell short in his estimation. Of a wrongly colored engraving, he complained: "It is a pity Hypnum stellatum in the last number is coloured so green—the plant is so very remarkable for its golden colour, surely the sketch was not so green." A poor engraving (Sowerby did the engravings as well as the drawings) prompted an even sharper response: "How came the capsule of the lid of Hookeria t. 1902 to be so erroneously engraved, furrowed, instead of reticulated? The drawing was beautiful & correct. It is grievous that the essential characters should thus be quite perverted. . . . Surely I ought to be able to depend on the drawings being faithfully copied."

Given the perfectionism with which Smith here urges fidelity to nature (and his instructions), his suggestions for tweaking nature are surprising: from consulting other engravings, relying on description of a plant under magnification to adjust one that was not magnified, to (my favorite) suggesting that Sowerby add water to make "fresh flowers" even bigger, on the principle that "a flower must be very highly magnified, & then its beauty will be conspicuous." Magnification, by adding water, by using a microscope: if this is not quite the scalar excess of the huge flowers depicted in *The Temple of Flora*, neither is it a straightforward brief for copying the plant specimen at hand.[51]

The more transparently aesthetic dimension of botanical illustration is color, which strictly speaking had no place in taxonomic protocols, yet artists and botanists were by 1800 obsessed with finding the right color and standardizing colors drawn from nature. Their fervor had little scientific precedent. Echoing Aristotle's distinction between secondary and necessary or essential attributes of objects, Linnaeus had argued that color was secondary, on the grounds that it so varied in flowers of the same species that it could not be counted essential to the species. Nonetheless Dutch, French, and finally English artists in succession made color central to botanical print culture. By the early nineteenth century the Viennese botanist Leopold Trattinnick had clearly had enough:

> I avoid hand-colouring as anxiously as the size of luxurious botanical coffee-table books. It is true that, at first glance, a hand-painted plant gives a far more lively and sensuous impression than a black and white engraving. Yet only on rare occasions does it contribute something essential for recognizing the species and the genus—at no time is it indispensable! All the characteristics of plants that have been stated so far by the great masters were taken from the structures, which are always more clearly discernible in black and white illustrations than in hand-painted copperplates. Wherever colours are in the least bit essential, I will not fail to say so and describe them. The slight effort one still needs to transfer the powers of the imagination into action will be rewarded not only by a reduction in price but also by the purity and clarity of the pictorial representation.[52]

Yet contemporary botanists and illustrators like Smith and Sowerby and many others pressed on, creating sumptuously colored works of art under the rubric of botanical illustration. Variable, difficult to establish, likely to err, of no use for taxonomy—color ought not to have mattered, but it did. In part, its desirability was surely market driven. Color distinguished one book or magazine of botanical engravings from another. Even in drawings, brilliant coloring could convey the splendor of native and exotic plants as uncolored lines could not. Botanical artists like the brothers Franz and Ferdinand Bauer created number-coded color charts to use when sketching a plant in nature so that they could color it accurately in a studio.[53] Most botanical artists worked in color, mostly watercolor, the preferred medium for hand-colored book illustration, usually grinding their own colors by trial and error. European and English artists had long tried to create colors that would be stable over time, with varying success. Some travelers investigated techniques and pigment sources that Indian artisans had used for centuries to create brilliant, color-fast dyes for cotton. None of the surviving European recipes apparently got all the steps or ingredients quite right. As color makers began to sell watercolor cakes to the European and Indian markets after 1770, artists continued to experiment with palettes as new color combinations were invented, recorded, and charted.[54]

Keenly aware of the problem of "fugitive" colors and the difficult of matching material colors to those in nature, several artists, including Sowerby, published treatises on color theory at nearly the same time that Goethe's much better known *Farbenlehre (Theory of Colors)* appeared in 1810.[55] For Sowerby, thinking about Newton's theory of prismatic color offered a putatively more natural system whose ordering might stand in for the vagaries of material color. This artistic turn toward color theory, prompted by material and commercial difficulties with color, turns back to its aesthetic and philosophical character as a feature of nature and plants that had long been declared secondary, nonessential. Yet as the standard bearer for the role of the aesthetic in the science of botany, color directs attention to other aesthetic features of plants or their observation: symmetry, unusual formal eccentricities, and quirky, because seeming unplantlike, features.

The possible examples constitute an embarrassment of riches. Here I consider three sets of examples, chosen for their differences from each other. My argument concerns how each is aesthetically distinctive. I do not offer anything like a taxonomy; to the contrary, their extrataxonomic properties are what invite notice.

Franz Bauer, like his brother Ferdinand, was trained as a miniaturist in Austria. Both came to England to pursue careers as botanical illustrators. Franz, or Francis as he was often called, worked as the botanical artist at Kew for fifty years, under appointment as Her Majesty's painter, but more directly supervised by Banks. Franz Bauer painted whatever came to Kew, included a series of brilliant studies of *Strelitzia reginae*, the bird of paradise plant that Banks named after the queen. I discuss here Bauer's cell studies of wheat epidermis, done in pencil and lightly colored around 1800 and still unpublished, and engravings of drawings he did for

Robert Brown of a Tasmanian flower still said to be the biggest in the world, *Rafflesia arnoldii*, botanically important as the basis for Brown's successful challenge to the Linnaean system in 1810.[56] Evidently scientific in purpose, these works invite notice of an intricate particularity that is also aesthetic.

In Bauer's study of a cross section of wheat stalks that also depicts the entire plant beside it, he offers two scales of vision and invites the eye and mind to view them together: the fleshly, pale, magnified stalks, to the right, cut across their tops to show the inner cell pattern, its roots magnified and bulbous. Beside this view is the undissected, unmagnified plant, its roots now spidery filaments, its stalks depicted with leaves sprouting from the jointed segments, wavy and turned, flopping a bit as it might not in a wheat field. In a single drawing, penciled and colored, Bauer conveys the extremes in scale that botanical writers and tourists seem never to have seen enough of. His cell studies of stems and petioles convey a minute world made large. Bauer's drawings convey the scalar range of botanical illustration. Those drawings that present magnified cross sections with the unmagnified plant offer what a single viewer could not otherwise see—a visual synecdoche in which the minute part is offered beside its larger whole. I wonder, too, whether this kind of visual figure may gesture toward the encompassing vision of the world's flora and its minute particulars that no single viewer could achieve, though many in the era wished to have.

The exotic appeal of *Rafflesia arnoldii*, also presented in cross section in Bauer's drawings, works this point from the other end: a huge flower seen up close, its inner structure visible to Bauer, Brown, and readers of the publication in which the engraving appeared. This remarkable parasitic plant, difficult to find in tropical forests, is extreme by other measures as well. The early twentieth-century botanist Bengt Mjoberg reported that it has "a penetrating smell more repulsive than any buffalo carcass in an advanced stage of decomposition."[57]

If Bauer's botanical art, particularly the more sedately colored works I have discussed, serves a scientific community's need for accurate drawings, Thornton's *Temple of Flora* did not. Thornton intended his publication to be expensive and popular, an ill-advised combination given the financial ruin it brought him. The engravings were printed in color, then finished by hand, in some instances to offset the degradation of the soft copper with each impression. Identified by its subtitle as a set of illustrations to the Linnaean system, *The Temple of Flora* was at once enormously expensive to produce and buy, a capital investment designed to take advantage of the widespread popularity of illustrated books and magazines dealing with botany and other aspects of natural history. Thornton hoped thereby to make an enormous profit from its publication; instead he went very nearly bankrupt and was barely saved from immediate financial ruin by a royal lottery in which the original paintings of *The Temple of Flora* were the first prize; other prizes included bound copies and loose plates.

Many of the illustrations were exotic plants; some were native to England and Europe. All present flowers in the foreground of botanical landscapes that extend the visual tradition inaugurated by Maria Sibylla Merian in the early eighteenth century. Whereas Merian's compositions emphasize ecological relations between plants and insects, those commissioned by Thornton are deliberately dramatic, such that even relatively small flowering plants like the so-called American Cowslip are made to look larger than the landscape features depicted behind them.[58] This visual aggressiveness is especially marked in its illustration of exotic flowers and emphasized in Thornton's commentaries. Of the one called Dragon Arum," he declares:

> This extremely foetid poisonous plant will not admit of sober description. Let us therefore personify it. She comes peeping . . . with mischief fraught; from her green covert projects a horrid spear of darkest jet, which she brandishes aloft: issuing from her nostrils flies a noisome vapour infecting the ambient air: her hundred arms are interspersed with white, as in the garments of the inquisition; and on her swollen trunk are observed the speckles of a mighty dragon; her sex is strongly intermingled with the opposite! confusion dire![59]

Thornton's group portrait of "Curious American Bog Plants" is carefully selected to emphasize their strangeness as plants and names: "Fly Ophrys, Venus's Fly-Trap, Bee Ophrys, Yellow Sidesaddle Flower, Tutsan-leaved Dog's-bane, Stinking Pothos."

Making pointed use of a figurative strategy developed by Linnaeus and extended by Erasmus Darwin in *The Botanic Garden*, Thornton's highly sexualized botanical personifications shamelessly exploit the commercial value of sex in a gothic register. The political message Thornton inserts here (the Dragon Arum's "garments of the inquisition" make it inevitably an adversary to the Protestant English) is a recurrent motif in *The Temple of Flora*. As Kay Dian Kriz and Meghan Doherty have argued, a retro emblematic format (images, verse, and expansively interpretive botanical descriptions) frequently emphasizes the imperial mastery of the British, at times presented as implicitly at war with great big exotic blooms.[60]

Bartram's botanical drawings of American plants, including the first known depiction of the venus flytrap, often present plants in compositional relation to other plants, animals, and insects, looking more like group portraits than still lifes, the nearest generic relative to botanical illustration. The compositional relationships among elements of Bartram's drawings imply that these plants are ready to move or at least interact with adjacent figures in ways that pay little attention to taxonomic differences between living beings and nonliving plants. The division is complicated for thinking about plants like the venus flytrap and the sensitive plant (both appear in Bartram's drawings) that seem to have animal functions. His drawing of "Bandana of the Everglades or Golden Canna" (*Canna flaccida*) places it with several companion figures: a stone pipe bowl given to him by an Indian chief, a gastropod shell, and an ant. The composition is slightly surreal and exciting in its formal rhythm: the fluted, semi-spiral curve of the blossom doubles the smooth geometric spiral of the shell, and the pipe seems to echo the jointed stem of the Canna or the jointed body of the ant. Put simply, even if Bartram was using paper sparingly to sketch, he need not have put these figures together unless he chose. The aesthetic effect is odd and riveting. The canna is elegantly drawn, but without the usual, Linnaean-era cross sections of inner and reproductive parts.

Bartram's most crowded botanical portrait features the "American Lotus" (*Nelumbo lutea*), "Whitelip Snail" (*Triodopsis albolabris*), "Ruby-Throated Hummingbird" (possibly *Archilochus colubris*), "Black Root" (*Pterocaulon pycnostachyum*) and "Arrow Arum" (*Peltandra virginica*), joined by an unidentified snake, dragonfly, and frog.[61] From the snail, depicted from two sides, as though it rolled and flipped from one to the next image, to the snake swallowing a frog from its position, half hidden, behind the American lotus plant on the left, to the other plants and the arrow arum, represented by its blossom in cross section, this drawing defies easy formal analysis. It is busy and unstoried, yet it invites a story about this inert plant and the dead Arum blossom lying nearby. The drawing is utterly unlike the great tulip paintings of the Dutch; it is also unlike Sybilla Merian's botanical landscapes, which seem from a modern perspective ecological portraits of a small segment of the biosphere. Bartram instead offers an array pitched to reveal an exotic profusion of American flora and fauna. These are purportedly botanical drawings, commissioned by Fothergill and intended to advance scientific knowledge of the tropical United States among English botanists and gardeners like himself, who could not get to America. It may have also been Fothergill's effort to support the thirty-five year old Bartram, who had not as yet settled on making a living.

Writing Sir James E. Smith in 1798, Charlotte Smith compared the "delightful and soothing study" of botany to "the present rage for gigantic and impossible horrors, which I cannot but consider as a symptom of morbid and vitiated taste," wondering "whether the simple pleasures afforded by natural objects will not appear vapid to the admirers of spectre novels and cavern adventures."[62] Romantic botanical illustration at times plays havoc with Smith's botanical piety. In books, magazines, and devilish satiric prints of botanists at work and play, the visual impulse is toward strangeness and excess not unlike those gothic thrills that Smith deplores. George Cruikshank's *Exhibition Extraordinaire in the Horticultural Room* is stuffed with jokes. The chairman of the Horticultural Society sits at a desk inscribed "A most respectable Cauliflower always in order," a motif carried elsewhere by presenting the actor Edmund Kean as a little manikin, mostly naked, with a strawberry on his head; and a plate of apples is labeled "Sundry Specimens of forbidden Fruit." Much of what in one way or another passes for depicting plants in the romantic era could be presented as evidence for "the present rage for gigantic and impossible" plants, be they beauties or horrors, and the burgeoning print culture that introduced botany to a wider audience and aesthetic.

ERASMUS DARWIN'S BOTANICAL MONSTERS

> *Poet*: In the gardens of a Sicilian nobleman, described in Mr. Brydone's and in Mr. Swinburn's travels, there are said to be six hundred statues of imaginary monsters, which so disgust the spectators, that the state had once a serious design of destroying them; and yet the very improbable monsters in Ovid's Metamorphoses have entertained the world for many centuries.
>
> *Bookseller*: The monsters in your Botanic Garden, I hope, are of the latter kind?
>
> *Poet*: The candid reader must determine.[63]

Erasmus Darwin was nobody's fool. Knowing that the wild improbability of personified plants conducting sexual adventures might affront the shy reader, in *The Botanic Garden*, quoted here, he made straight for readers whose willingness to entertain improbable monsters on the Ovidian plan makes them more likely converts to the project at hand. There is cunning here of several kinds. This "interview" between the Poet (aka "Mr. Botanist") and the Bookseller, the first of three dialogues interspersed between the cantos of part 2 of *The Botanic Garden*, "The Loves of the Plants," begins with a fairly serviceable distinction between the prose notes and essays of *The Botanic Garden*, which their "stricter analogies than metaphors or similes," and its verse, whose loose analogies between plants and everything else involve allegorical figures as strange and improbable as anything in Ovid, whose verse also offers something like a formal sanction for the rapid-fire narrative succession of personifications and myth offered in Darwin's poem, with an Ovidian disregard for transitions. The relations between Darwin's prose and verse in *The Botanic Garden* are not easily parsed. As notes weave relationships between parts of the poem and other notes and as supplementary essays at the end of each part add to the overlay of analogy, the work of reading this work begins to look unworkable, or workable only if we could map the interstitial arguments embedded in the constant relay between prose and verse whose arguments and intractable contradictions exist somewhere just off or behind the printed page.[64]

Alan Bewell argues persuasively that the free range tendencies of Darwin's poem are also cosmopolitan (with all the complexity that accrues to British imperial cosmopolitanism around 1800) and commercial. The plants he personifies are often exotics recently discovered by Europeans who sought them as much for profit as pleasure. Put another way, the figurative plenitude of Darwin's poem exists inside a discourse about use and profit that is as much a part of Erasmus Darwin's view of industrial change and the world as the improbably figures of the poem are. When Darwin's poet proclaims at the end of "The Economy of Vegetation," part 1 of *The Botanic Garden*, that his "Sylphs" should bring him flowers and balms from Mecca, Arabia, Italy, China, India, Siberia, and the Andes, he specifies the global, commercial world in which the "Loves of the Plants," part 2 of the revised poem, take place.[65] When part 2 closes with a vision of plant marriages and liaisons on the Tahitian plan (or Captain Cook and his men's special version of that plan), Darwin insists one last time on the global perspective of the poem's narrative. My argument concerns instead a counterrhythm and impulse within the poem in which plants challenge the authority of the Linnaean system with repeated and dramatic imperiousness. Paying little attention to the putative Western authority over who they are and what they do, Darwin's plant figures run loose, their personifications slipping from verse to note and back again with an abandon that takes little notice of the poem's purported distinction between scientific and "looser" analogy.

Having set off prose from poetry in his exchange with the "Bookseller," Darwin's "Poet" tries to defend this distinction by invoking the sister arts of painting and poetry, noting that poetry can handle figures that are improbable, unnatural, possibly revolting. Painting cannot because it makes all these traits too distinct, whereas poetry offers "a train of ideas" so pleasurable that we "cease to attend to the irritations of common external objects" (irritated perhaps because they no longer count for much). When the bookseller asks whether it matters that such representations do not correspond with nature, the poet (who knows his Longinus) replies by defending the work of figure, particularly those figures whose sublimity justifies their exaggeration. Still, the question of monstrosity returns just at the end of this first interview, in the lines quoted above, and again in the next "interview" as the Bookseller commends the poet on "the monsters of your Botanic Garden . . . as surprising as the bulls with brazen feet . . . yet they are not disgusting, nor mischievous."[66] Again, the poet says, wait and see. Within a decade of the first publication of "The Loves of the Plants" in 1789, anti-Jacobin mockery of its strange and playful mix of poetic figure and scientific prose managed to insist that it was both a silly poem on an unimportant subject (botany) and so politically suspect that it could be retitled "Jacobin Plants." As Bewell has argued, the anti-Jacobins were not simply or merely being hysterical: they recognized in Darwin's work a spirit of rule breaking that was at once botanical and figurative, grounded in a man whose intellectual commitments were by this time aligned with revolutionary thinking. On all these points, the anti-Jacobins got it right.[67]

The anti-Jacobin attacks recognized Erasmus Darwin as someone who did botany with far less respect for taxonomical hierarchy and authority than Linnaeus might have wished. This is not to say that Darwin was anti-Linnaean, although his differences with that systematic sharpened rather quickly, appearing first in *Zoonomia* (1794), then elaborated in *Phytologia* (1800) and *Temple of Nature* (1803). Darwin's translations of Linnaeus, which were advertised in *The Botanic Garden* (the 1783 *System of Vegetables* and the 1787 *Families of Plants*) were the first to offer fully English versions of Linnaeus's descriptive vocabulary for plant traits, from the best-known—stamens for the Latin stamina and pistils for pistilia—to many other descriptive terms. In the nomenclatural debates of the past two centuries, some have survived, some have not. Darwin's plan was to present the most literal and complete translations of Linnaeus he could manage. He scorned his Lunar Society colleague William Withering for trying to sanitize the sexual character of the Linnaean system.[68]

In life as in thought, Erasmus Darwin had no interest in received or doctrinally safe convictions. His tinkering with the Linnaean system, the outcome of reading that system against his and other examinations of plants, is one facet of a lifelong intellectual habit. Darwin was forever inventing new gadgets, devising unlooked for ways to solve practical problems, trusting speculation, and thinking by analogy as others might trust (whether advisedly or no) the Bank of England or, more probably, Burkean custom. Uglow reminds us of Darwin's entry for Credulitas/Credulity in *Zoonomia:* "Life is short, opportunities of knowing rare, our senses are fallacious, our reasonings uncertain; man therefore struggles with perpetual error from the cradle to the coffin. He is necessitated to correct experiment by analogy, and analogy by experiment."[69] Darwin had used a version of this account of knowing and not knowing in "The Economy of Vegetation" to justify the role of conjecture, or "extravagant theories," in scientific inquiry.[70] It was probably inevitable that he would find himself disinclined to read species or families as immutable categories. The philosophical tug of this disinclination turns up in his translation of Linnaeus's *Genera Plantarum* as *Families of Plants;* it is explicit in *Zoonomia* (1794–96) and the *Temple of Nature* (1803), works that insist, contra Linnaeus, that species are not eternal and fixed.[71] Published between the Linnaean translations and later works, prose and verse, in which the elder Darwin finally declares his evolutionary views, *The Botanic Garden,* so hugely popular in the 1790s and so bizarre to consider today, manages two contradictory projects.[72]

The first of these is the explicit goal of the work: to introduce the Linnaean sexual system to a popular audience. The second is implicit or, more accurately, is hidden in plain sight in the form of the poem, where prose notes and essays and verse arguments work in tandem, but also more subtly at cross purposes, as indeed Darwin's Poet indicates when he tells the Bookseller that the analogies his verse pursues are glancing, open to digressive tendencies (the Ovidian narrative impulse), whereas scientific or, as he describes them, philosophical argument requires "stricter analogies than metaphors or similies" (*BG* 2, 49). In a work that even more expansively animates Linnaeus's figures of males and females in one or more "beds" and a dizzying array of "marriages," this is a distinction without a difference. It is also a distinction that is undone as the poem's argument weaves between prose analysis (typically by analogy) and verse (all analogy). Stephen Jay Gould suggests that the figural work of Darwin's verse can get carried away, mostly by being carried forward into tales from myth and history of women gone wrong or wild. Gould's contrast between the wildness of Erasmus Darwin's figures and his grandson's far more natural (or perhaps more conventional) ones suggests distinct goals.[73] Whereas the elder Darwin seeks by way of figure to break out of the limits of his Linnaean subject, his grandson needs them to make the strangeness of his evolutionary hypothesis seem familiar. Gould's claim that Charles Darwin makes his argument via metaphor might also be said of Erasmus, with this difference: the unruliness of this Darwin's figures gives him the space he needs to write simultaneously about and away from the Linnaean system in the same work.

In *The Botanic Garden*, the botanical monsters hidden in plain sight are those figurative analogies whose spirit of misrule undermines the stasis of Linnaean categories. In part this is so because Darwin's attitude toward analogy is amazingly carefree, or to put it more frankly, careless, about what an analogy might carry along. Early in "The Economy of Vegetation," the poet describes a garden that soothes Philomel's "disasterous love." The adjective is apt, the noun less so if one recalls the myth that Darwin chooses not to specify: Philomel is raped by her brother-in-law, who insures her silence by cutting out her tongue (*BG* 1 49), preferring to craft his own myth that the "scenery" where rocks weep and Philomel (now a nightingale) sings is modeled on "a botanic garden about a mile from Lichfield," where the "modern goddess of Botany," will introduce the "Loves of the Plants." Said goddess, who introduces each of the poem's cantos, is Darwin's homage to Elizabeth Pole, an avid gardener whom he began courting in 1776. The garden and Darwin's genuine scholarly enthusiasm for botany date roughly from the period of this long and ultimately successful courtship.[74]

In an era when speculation about plant to animal similarities and monstrosities was at once rife and uneasy, Darwin pressed the point at every possible turn in the prose as well as verse of *The Botanic Garden*. He was certainly not the only writer on botany to do so. In 1829, John Lindley said this in an inaugural address at the University of London in which he justified the science of botany:

> When you shall have made yourself acquainted with all the principal forms under which Nature presents herself, and shall have studied the various links by which one kind of matter is connected with another, you will probably arrive at the conclusion, that there is no such thing as a definition in Natural History; and hence you will come to doubt: so insensible, and at the same time so complete, will you find the gradations between *men* and *trees*, between yourselves and those gigantic creatures of which the timbers of this building may be called the bones—hence I say you may come to doubt whether any one can truly define the boundaries within which the two kingdoms, that these bodies represent, are mutually confined.[75]

In 1791, when *The Botanic Garden* appeared as a two-part poem with "The Economy of Vegetation" first and the revised, third edition of "The Loves of the Plants," the notes between parts 1 and 2 include a number that present plant functions in terms of human or mammalian systems: the topics "vegetable perspiration," placentation, circulation, respiration, impregnation, and glandulation prompt a cascading series of notes, from lengthy "Additional Notes" at the end of part 1 to briefer, "Omitted Notes," paginated separately from parts 1 and 2.[76] In the reorganized *Botanic Garden*, these notes becomes the conceptual staging ground for "Loves of the Plants," whose figural argument about the animated sex lives of plants continues where the earlier notes left off, acquiring another set of mostly embedded notes that further specify the scientific backstory for what those personified plants do.

Among the plants described in the last canto of the poem is Hedysarum gyrans, or the moving plant, whose Indian name, Chundali Borrum, Darwin provides in a note. "Fair CHUNDA" is kept cool in the "burning waste" by her ten young husbands, who fan her and shade her with an umbrella. There appear to be compensations for females in Darwin's vegetable sexual economy. Noting that the leaves of this plant whirl circularly "by twisting their stems," Darwin suggests that "this spontaneous motion of the leaves, when the air is quite still and very warm, seems to be necessary to the plant, as perpetual respiration is to animal life" (*BG* 2, 168). Here "as" implies more than a slight or partial analogy, insofar as it nudges toward the claim that motion is a common symptom of animate life.

This note in turn recalls another instance of "vegetable spontaneity" from "The Economy of Vegetation," where Collinsonia is a plant whose females engage in "manifest adultery" by bending "themselves into contact with the males of other flowers of the same plant in their vicinity, neglectful of their own" (*BG* 1 197).[77]

Here and elsewhere, Darwin presents the language of sexual crimes and misdemeanors stripped of conventional moral affect. What interests him are precisely those plant "behaviors" that stray from norms, to which he rather gleefully attaches sexual figures. Darwin's faux rhetoric of moral disapproval may be strategic to the degree that less adventuresome (or more prudish) readers might understand it to rein in the poem's adventuresome sexuality. Yet clearly he does not himself disapprove, but instead delights in plants whose motions suggest animation and irregularity, from the *Mimosa* or sensitive plant in canto 1 (*BG* 2, 31), whose popularity as an instance of the genre of plant animation Darwin did so much to advance, to the European water plant "Conserva vagabunda" and its lake-water relative "Conserva aegagpropila," the latter personified in the verse as a husband whose distraught female lover calls "My Lord, my life, my love," as she urges the wind to slumber and the sea swell to cease the motion that carries her lover away from her (*BG*, 2, 170, 171).

Here and elsewhere, Darwin injects a boatload of pathos into his presentations of female personified plants that cherish their lovers and children. Such plants, Darwin's note suggests, "may not improperly be called itinerant vegetables" (*BG* 2, 170–71). There is of course nothing figuratively proper about the phrase "itinerant vegetables." In the way Darwin's Botanist/Poet says such figures do, it brings an image before the reader that carries her far from stationary plants, to (why not?) radishes hitching rides, working fields far from home. This reverie, which owes at least as much, maybe more, to the note as it does to the verse personification, introduces a formal strategy that is at odds with the Botanist/Poet's several suggestions that poetic and scientific analogy function differently.

The figurative two-step performed by Darwin's verse and notes for Turmeric (*Curcuma*) in Canto 1 of *The Botanic Garden* is a case in point. The verse offers a brief, tantalizing vignette:

> Woo'd with long care, CURCUMA cold and shy
> Meets her fond husband with averted eye:
> *Four* beardless youths the obdurate beauty move
> With soft attentions of Platonic love.
>
> (*BG* 2, 65)

The note explains what the verse does not: "One male and one female inhabit this flower; but there are besides four imperfect males, or filaments without anthers upon them, called by Linnaeus eunuchs." But that is not all. Darwin goes on to describe several species of flax that present various numbers of "imperfect males," similar imperfections in other genera and in the class Syngenesia or "confederate males"—as in the florets of the sunflower, which have a style but no stigma and are thus also barren, as well as the "unprolific flowers" of the Opulus. The list of imperfect plant males ends: "In like manner some tribes of insects have males, females, and neuters among them: as bees, wasps, ants." In *Phytologia* (1800), Darwin responded to widespread dissatisfaction with the Linnaean class *Synesthesia* by targeting it in his "PLAN FOR DISPOSING PART OF THE VEGETABLE SYSTEM OF LINNEUS INTO MORE NATURAL CLASSES AND ORDERS."[78] He argued that the numerical bias of the Linnaean system missed more salient information for taxonomic analysis, such as numbers of stamens and pistils that varied with soil and climate more than they do in structure and position.

The second part of Darwin's *The Botanic Garden* note pursues the plant to animal analogy yet more closely: "There is a curious circumstance belonging to the class of insects which have two wings, or diptera, analogous to the rudiments of stamens above described; viz. two little knobs are found placed each on a stalk or peduncle, generally under a little arched scale, which appear to be rudiments of hinter wings," From this analogy, Darwin proceeds to other animals that have "in a long process of time undergone changes in some parts of their bodies" such as "the existence of teats on the breasts of male animals" and then, to a question, or a working assumption, based on the history of the earth's formation: "Perhaps all the productions of nature are in their progress to a greater perfection" (*BG* 2, 7).

The formal differences in these two accounts could not be more clearly marked: verse and prose, one compressed, the other expansive, comprising a more or less alternating presentation that could be said to echo the rhyme of the quatrain, which departs slightly from Darwin's usual (and sometimes mocked) preference for heroic couplets. And yet, as is also the case with Darwin's couplets, these antitheses do not enclose an argument or set of claims. Instead the work of this play of formal resemblance and dissonance seems to be to open the way to speculation. In the case of the verse, it is erotic speculation. How platonic is that relationship between the cold wife and those beardless youths? In the prose, the speculative machinery that begins with antherless filaments ends with a picture of evolutionary progress, the hypothesis that he put under wraps for several decades after being skewered for so thinking by the Rev. Thomas Seward, canon of Lichfield Cathedral, then announced by degrees in *The Botanic Garden* and subsequent works.[79]

Darwin unapologetically reiterates plant to animal analogies throughout *The Botanic Garden*, explaining in a note that the water plant *Trapa bisnosa* (or *Trapa natans*) has "aerial leaves of vegetables [that] do the office of lungs, by exposing a large surface of vessels with their contained fluids the influence of the air, so these aquatic leaves answer a similar purpose like the gills of fish, and perhaps gain from water a similar material." *The Botanic Garden* verse emphasizes the monstrosity of this plant-animal:

> Amphibious Nymph, from Nile's prolific bed
> Emerging TRAPA lifts her pearly head;
> Fair glows her virgin cheek and modest breast,
> A panoply of scales deforms the rest;
> Her quivering fins and panting gills she hides,
> But spreads her silver arms upon the tides.
>
> (*BG* 2, 157–58)

Although it is more likely that Darwin had noticed this plant because it appears as a colored engraving of an Indian artist's drawing in Roxburgh's 1790 *Plants of the Coast of Coromandel*, its emergence "from Nile's prolific bed" gives "Trapa" a head start on deformity, for the Nile mud was said to breed new and bizarre forms of plant and animal life after its annual flooding.

Throughout *The Botanic Garden*, erotic possibilities are standard bearers for a loosening of the Linnaean envelope. Whatever one thinks of the female types that Darwin comes up with (cold wives, harlots, timorous beauties, the virgin lily), the poem evidently orders its view of nature "by budding and full-breasted sexuality."[80] There are personified male plants to be sure, but the poem leans toward female personifications of plants or pistils, even when the plant in question is presented as male, and on occasion traditional gender determinations for flowers are flipped: the young male Rose is among the "Beaux and Beauties" who woo and win their vegetable Loves" in the opening lines (*BG* 2, 2). Although the classes that create the principal architectonic of the Linnaean systematic refer to the number and position of male stamens, as Darwin reminds his readers in the preface to "Loves of the Plants," the loves that ensue involve an array of female characters. And while it is true that Darwin's catalog is both deep and narrow—no bluestockings or intellectuals appear among his "loves"—the sexual prominence of women shifts the ground beneath Linnaeus's feet.

Some of those women are monogamous, as are some of the men; but what gives the poem its erotic charge are its unmonogamous, if still heterosexual, couplings. The sweet blooming *Genista* ("Dyer's broom") in Canto 1 is a fairly typical example. As "ten *fond* brothers woo the haughty maid," and perhaps before we can begin to entertain the possibility that those brothers are fond of each other and not the maid, Darwin launches a note that compares the position of "the males or stamens" in other species of broom. In common broom, the males and females mature in two sets and stages, thanks to the exertions of the stigma or "head of the female" in the higher set, which matures first: "as soon as the pistil grows tall enough to burst open the keep-leaf, or hood of the flower, it bends itself round in an instant, like a french horn, and inserts its head, or stigma, amongst the lower or mature set of males. The pistil, or female, continues to grow in length; and in a few days the stigma arrives again amongst the upper set, by the time they become mature." This "wonderful contrivance," as Darwin calls it, keeps before the reader the image of a highly active female partner in plant flowering (*BG* 1 5). The arum or cuckoo pint offers another. Identified as a member of the class Gynandria, or "masculine ladies," the debate about "the singular and wonderful structure of this flower" turns on the phallic shape and operations of the female receptacle or hood, which prompt Darwin among others to agree that the arum flower "may be said to be inverted"(*BG* 2, 163).

In Canto 3, which produces a succession of "horrid" botanical monsters, most are females. *Impatiens* or touch me not, which flings its seed pod off when it ripens, is given the fury and "fierce distracted eye" of a Medea who "hurls her infants from her frantic arms" (*BG* 2, 102–3). After a full rehearsal of the story of Creusa and Medea's dead children, a train of gloomy and poisonous female plants follow: "Dictamina," "Mancinella," "Urtica" (she that stings), "Lobelia," and "Palmira," who is harmless enough herself but becomes the excuse for a rather wonderful parade of desert beasts, red in tooth and claw, followed by the most beastly of trees for romantic travelers and writers, the *Upas*, which Darwin also calls the "Hydra-Tree of death," more familiarly known to his contemporaries as the poison tree of Java. Indubitably male, *Upas* takes his figurative work seriously:

> Lo! from one root, the envenom'd soil below,
> A thousand vegetative serpents grow;
> In shining rays the scaly monster spreads
> O'er ten square leagues his far-diverging heads;
> Or in one trunk entwists his tangled form,
> Looks o'er the clouds, and hisses in the storm.
> Steep'd in fell poison, as his sharp teeth part,
> A thousand tongues in quick vibration dart;
> Snatch the proud Eagle towering o'er the heath,
> Or pounce the Lion, as he stalks beneath;
> Or strew, as marshall'd hosts contend in vain,
> With human skeletons the whiten'd plain.
>
> (*BG* 2, 111)

Even Darwin appears to have become distracted by stories of this famous tree, forgetting altogether in the note to discuss Linnaeus (who had also heard about the tree) or Linnaean classification, and forgetting, too, to mark its female opposite presented earlier in the canto, the nurturing *Ficus indica*, called Fica in the poem.[81] Because the toxicity of the *Upas* tree makes it act like a rapacious beast or, better still, "a thousand vegetative serpents," it is the perfect figure for Darwinian analogy: rapacious, monstrous, doing apparent damage to basic distinctions between the plant and animal kingdoms. This, of course, is why travelers and artists flocked to Java to see it and probably why, for a different reason, Blake put it into his *Songs of Experience*.

Whether their untoward behavior is deadly or extraordinarily sexual, Darwin's delight in this or that "beautiful monster" extends beyond its reference in this note to plants that produce double flowers and a range of fertile or nonfertile offspring double flowers (*BG* 2, 152). In the organic whole that Darwin had already begun to describe in *The Botanic Garden*, monsters—figures and plants—insist on a curious, astonishing set of connections between plants and animals and across Linnaean classes. Sexual predation (by females and by males) figures a world in which no possibility is absolutely barred. The presiding and legitimating figure for this deployment of sexuality as a figure for the world is almost surely female. In the dialogue between Botanist/Poet and Bookseller that Darwin places between the last two cantos, he offers a Shakespearean simile to show how describing "a single feature or attitude" can suggest an image of the whole to the mind of the reader:

> When we have laugh'd to see the sails conceive,
> And grow big-bellied with the wanton wind;
> Which she with pretty and with swimming gate,
> Following her womb, (then rich with my young squire),
> Would imitate, and sail upon the land.
>
> (*BG* 2, 130)

The simile moves, with acknowledged thanks to Shakespeare's *A Midsummer Night's Dream*, from conception to big-bellied sail to the figure of a pregnant woman moving like a boat in full sail. In the process the simile wraps itself round figure and material/literal ground, making it unclear which is which. This figurative play is the formal project of Darwin's poem, where scientific notes and essays consistently undermine the putative separation between prose and verse and between the analogies or figures said to be proper to each. That the fecundity of women supervises the improbability of figure and this alliance between the verse and the notes of *The Botanic Garden* is another way to mark Darwin's pleasure in all of this. For although he does add to the third edition of 1791 a powerfully directed account of Cassia, the tree that casts its seeds out, in which those "brown seeds" tossed on the water and thereafter mourned by their mother are presented as Africans enslaved and cast across the Atlantic, Darwin is not consistently interested in making strongly inflected ethical arguments (*BG* 2, 124). To the contrary, and despite occasional terms that suggest moral disapprobation, the "loves of the plants" (including some for which the term *love* is at best a euphemism) is the occasion for an unfettered display of sexual possibility as a figure for how to read the affinities between plants, other kingdoms, and figures that assist Darwin's argument about the instruction offered by "unnatural" figures and plants.

The poem ends with a marriage in Tahiti, the island whose sexual possibilities Cook and his men found when they were looking, auspiciously enough, to record the transit of Venus. Darwin finds the precedent for this scene of marriage between one hundred females and males, loosely analogous to the "many males and many females" of the flower he calls Adonis, among the Tahitian society he calls the "Areoi," which "consists of about 100 males and 100 females, who form one promiscuous marriage" (*BG* 2, 178). "Promiscuous" because Cupid's arrows, which tend to work against the claims of monogamous unions, are and because Venus in the end blesses all, knowing all, having watched it begin with side glances and whispers among members of the wedding party, which insure that their marriage vow will be "faithless." Strictly speaking (if that is even possible to say about this poem), these marriages under the sign of Adonis, Cupid, and Venus belong in the Linnaean system to the twenty-third class of polygamous plants (marriages between many males and many females). But the surreptitious intercourse of glances and whispers edges closer to the twenty-fourth class—those plants whose secret unions are legitimized as clandestine marriage. This tendency is more broadly marked across the poem, where plants that are said to belong to this class are presented adjacent to those belonging to other classes. If readers were looking for a linear exposition of the Linnaean system, they would not find it in this botanic garden. Here is its closing marriage scene:

> Licentious Hymen joins their mingled hands,
> And loosely twines the meretricious bands.—
> Thus where pleased Venus, in the southern main,
> Sheds all her smiles on Otaheite's plain,
> Wide o'er the isle her silken net she draws,
> And the Loves laugh at all but Nature's laws.
>
> (*BG* 2, 179)

Promiscuous, licentious, meretricious—none of this matters to "the Loves" or to Darwin, for whom "nature's laws," with all their delightful and monstrous irregularity, license analogies that run astray, glancing beyond the limits of Linnaean classes and kingdoms, to think about the material world as a connected reality and history with the imagined vitality suggested by an array of sexual combinations and figures more explicit and far reaching than those Linnaeus supposed he had described. The improbability and patent unnaturalness of so many of Darwin's personifications of plants, like their Ovidian models, may obliquely mark for popular audiences the point that Linnaeus admitted but few of his popularizers emphasized: that the system of nature based on counting stamens and pistils was itself artificial, a handy counting device against the day when a full map of the "natural orders" of plants might be achieved. Forced women and women forced to inhabit personifications: the persistence of this motif in Darwin's "Loves of the Plants," a poem more evidently about the role of females than males in classificatory argument, suggests that Darwin's goddess of botany supervises a system of figuration that pushes at and away from the epistemological grip of a systematic founded on figurative males as taxonomic as well as social governors.

First published in 1789, the year in which A.-L. de Jussieu published his monumental *Genera Plantarum*, the work whose Natural System helped to supplant the Linnaean system, Darwin's "Loves of the Plants" does not anticipate this systematic shift. When he uses the term *affinity* in his *System of Vegetables*, it refers to the similarities that a plant exhibits to others belonging to the same genus, not those affinities across genera that advocates for the Natural System tried to identify and map.[82] And although he used the phrase *metamorphosis of plants* the year before Goethe published his *Metamorphose der Pflanzen* Darwin used the phrase to refer to the process whereby the "larva" of a plant is changed into a mature plant; for Goethe it referred to successive changes in the leaf structure of plants from seeds to flowers.[83]

Even as Darwin translated the Linnaean system, he encountered questions that carried him away from the sureties of that system, in which the species and kingdoms of nature exist as separate and fixed categories. Like the next generation of systematicists who offered graphs, maps, and some trees of relationships among species and kingdoms, Darwin repeatedly suggested comparisons and differences that work across Linnaean categories.[84] The heterodox habits of thought that inform the organicist and evolutionary view of creation in *Zoonomia* emerged in *The Botanic Garden*, where the sexual congress of botanical figures announces the spirit of analogy on the loose that returns in Darwin's later works with still more heterodox ends in view. In his last work, *Temple of Nature*, also in verse with prose notes, Darwin presents the story of how "Organic forms, with chemic changes strive, / Live but to die, and die but to revive!" as one assisted by Flora's delight in the "triumph of despotic love," seconded by all the flowers who hail Cupid and Psyche as the gods of "Sexual Love."

CHAPTER 4

Botanizing Women

FLOWER-WOMEN

Jacques Le Moyne de Morgues's sixteenth-century drawing titled "A Young Daughter of the Picts" and J. J. Grandville's mid-nineteenth-century caricatures of women as flowers convey in snapshot a long iconographic tradition that personified women as flowering plants and becomes, at the far end of this tradition, the logic behind the Victorian language of flowers.[1] Between Le Moyne in the sixteenth century and Grandville in the mid-nineteenth, botany becomes a cultural and writing practice in which women participate, so much so that, by the late eighteenth century, the phrase *botanizing women* referred more emphatically to women who botanized than it did to women as botanical subjects.

Even so, Le Moyne and Grandville bookend an iconographic tradition that understood women as botanical ornamentation and remained in play long after women became active producers of botanical inquiry and art. It was not for nothing, as Deidre Lynch notes, that Richardson's Pamela remained in dangerous proximity to Mr. B because she was required to "flower" his waistcoat before she was permitted to leave his company.[2] *Flowering* is what Pamela did, even as that verb punningly suggested what she risked in doing so. Men's ornament or women's, flowering textiles was the work of women's hands.

I begin this chapter with Le Moyne's and Grandville's flower women because they convey, more distinctly than Botticelli's *Primavera*, the edge of hazard, authority, and pretty cruelty that attended making women into flowers or women who created flowers in thoroughly domestic settings. Romantic era women who did botany kept at bay the view of women and plants that Le Moyne and Grandville convey.

Le Moyne's small drawing depicts a young girl who is or appears to be clothed in a garment decorated with painted flowers. The rose that covers each breast invokes medieval iconography for a virginal woman; other pairs of flowers decorate her thighs and knees, and what looks like a large mum marks her stomach. The flowers drawn on her body include highly valued European species and varieties and several exotic plants that had been only recently been "discovered" by Europeans: single and double peony, the marvel of Peru (Latinized as *Mirabilis jalapa*; also called four o'clock), delphinium, tulip, hollyhock, mourning iris, hearts-ease, double columbine, orange lily, tazetta narcissus, cornflower, rose campion, and yellow horned poppy.[3] Although a cord at the waist and the neck implies that she wears a fitted garment not unlike the botanically figured Indian muslins that were popular with Europeans from the seventeenth century on, her pubic hair, half-masked with a tiny, flowering grass, makes the point that the rest of the decoration makes more subtly: the girl is clothed only in flowers.[4]

David Richards notes that the painting achieved emblematic status when it appeared as an engraving in Theodore de Bry's *America* (part 1, 1591), where it "purports to represent ancient British peoples" as New World settlers. Aphra Behn darkens the already complicated cultural iconography of this image when she invokes it in the tragedy *Oroonoko* to describe the delicate carving on the body of Imoinda, the woman Oroonoko loves and ultimate kills to protect her against rape and other forms of mutilation that are forecast or shadowed by her tattooed body.[5] Reimagined as ritual tattoos, her floral array suggests either that this young daughter of the Picts belongs to another world or, conversely, that its flowers belong to her. Either way, Le Moyne's drawing looks like a cautionary tale bubbling up in a long history of thinking of women as flowers.

J. J. Grandville's *Les Fleurs animées* (1847) tacks between its purported genre—a manual for the care of flowers—and the figurative argument suggested by its colored engravings of women dressed or caricatured as flowers. The accompanying text, written by others and separately titled, is similarly mobile. As "Botanique et horticulture des dames" it could be a manual addressed to women on the topic of botany and flower gardening, both identified with women's pursuits during the period. The text suggests instead that this is a manual or guide to the botany and horticulture of women, or, more precisely, of those women whose temperament and needs are personified in "animated" flowers whose portraits Grandville scatters throughout the work. The opening paragraphs on the treatment of flowers (*le traité des fleurs*) convey the view of women entailed by thinking of them as well-dressed and coiffed flowers:

> I never cross a flower market without being seized by a bitter sadness. It seems to me that I am in a bazaar of slaves at Constantinople or Cairo. The slaves are the flowers.
>
> Look at the rich who come to buy them; they look at them, touch them, consider whether they are sufficiently youthful, healthy and beautiful. The purchase

> is concluded. Follow your master, poor flower, serve his pleasures, decorate his seraglio, you will have a beautiful porcelain dress, a pretty cloak of moss and you will live in a sumptuous apartment; but farewell to sun, breeze and liberty: you are enslaved![6]

Festooned with petals, headdresses, and colors appropriate to the flower each represents, Grandville's flower women are typically in need of protection: the right kind of sun or shade, nourishment, and so on. If unprotected, they end up in the not so friendly flower market, sold to the highest bidder. The echo of the slave market would have been hard to miss. This tone persists in the presentation of the Sensitive Plant, which by 1847 was a species with no mean claim to botanical fame, though her unsanctioned membership in a group of plants that appear to have animal traits or human feelings is not at issue here.[7] In the text of *Les Fleurs animées*, she tells her own story (as do all the animated flowers), framed by the narrator's introduction:

> [Narrator] The poor Sensitive was not made for the world; this I all too quickly perceived.
>
> [Sensitive] Scarcely had I changed into female costume when my sensitivity caused me terrible torments. I do not speak of love; my modesty must defend me.
>
> I also suffered a great deal on other occasions! At the theater, the music makes me fall in a faint; the emotions of the drama threw me into extended faints; the least change in temperature affected my nerves.[8]

Like Charles Germain de Saint Aubin's eighteenth-century caricatures of women as overblown floral grotesques, Grandville's flower portraits implode the genre of floral compliment to female beauty.[9] Read through the lens provided by these caricaturists, wearing petals looks like a dangerous fashion statement.

It is unlikely that women who did botany from the late eighteenth century on would have read images like those of Le Moyne or Saint Aubin or, looking to the future, those of Grandville, as convincing or compelling self-portraits. Yet from the last decades of the eighteenth century into the early nineteenth, women who painted flowers and then began to think and write about botany did so at a time when it was still common to think of women as flowers, and to say so. When Thomas Moore characterized the harem of the Mughal emperor Jehangir as "a living parterre of the flowers of this planet" in his immensely popular poem *Lalla Rookh* (1817), he appended a note explaining that "in the Malay language, the same word signifies women and flowers,"[10] a point made material by the Taj Mahal, the great Islamic mausoleum inlaid with floral decoration, built by Jehangir's father, Shah Jahan, to honor his beloved wife, Mumtaz Mahal. Moore capitalizes on a

Western as well as Mughal affiliation between women and personified plants that Darwin's *Botanic Garden* and Thornton's *Temple of Flora* reiterate.

Women who wrote about or illustrated plants during the romantic era did so, then, at a time when the relation between women and plants was understood to be ornamental, although not yet as emblematically codified as this relation would become in the Victorian era. Pierre Bourdieu's *habitus,* the rule of established values in shaping culture, renders precisely the degree of habit, habitualness, and inhabitation at issue in the normative view of women and botany.[11] Ann Bermingham notes that even when botanical drawing became one of the desirable accomplishments of women in polite society, ornament was still all or most that mattered. Women were expected to look ornamental when drawing plants. Being ornamental also invited criticism: women could be and were charged with calling too much attention to themselves by displaying their accomplishments.[12]

As Ann Shteir, Bermingham, and Anne Secord have demonstrated, whether welcome or not, women were by the early nineteenth century engaged in all areas of botanical production: dissection, microscopic analysis, books that introduced botany to children (and other women), and above all botanical drawings and engravings. Some kept that art in the family; a few circulated it to a small group of friends until their work became widely known, even if it remained outside the culture of print. Other women exhibited, engraved, and published their botanical art. At least one wrote on color theory, as did several other botanical illustrators.[13] Women writers and artists who contributed to the print and manuscript culture of botany challenged the supposition (or hope) that theirs was merely an amateur, casual, and ornamental pastime.

This chapter seeks to work out the edge of difference that marks how a few women—among them an experimental botanist, several writers, amateur artists, and a poet—sidestepped the iconographic tradition that rendered women as passive flowers in need of care (or abuse). As I read them, these botanizing women deflect normative accounts of how women could be identified with plants, in part by creating botanical figures instead of themselves becoming such figures. The work of botanical difference that Elizabeth Grosz, following Deleuze, identifies as new with Charles Darwin has a prior and eccentric location among romantic era women who botanized just at, and then just beyond, the edge of acceptable womanly behavior.[14]

I begin with Jean-Jacques Rousseau's *Lettres élémentaires sur la botanique à Madame de L*—a widely read little book whose rhetorical structure implies a clear hierarchy in which men teach botany, women learn it and then (at best) teach their children. Published posthumously in 1782, it was soon translated into English, German, and Russian and became a template for pedagogical writing on natural history.[15] *Lettres élémentaires* expands a shorter series of letters Rousseau wrote to Madeleine-Catherine Delessert in which he presented botanical anatomy and classification, illustrated by dried specimens, in a way that he hoped she could use to teach her daughter. Along with *Reveries of the Solitary Walker*, other short works and letters to other serious amateurs, including Margaret Cavendish Bentinck, Duchess of Portland, the *Lettres* convey Rousseau's preoccupation with plants in the last decade of his life.

Thomas Martyn and John Lindley each published translations of the work that extensively amended or expanded Rousseau's original to reflect the botanical principles they preferred to foreground. Martyn's 1785 translation, *Letters on the Elements of Botany, addressed to a Lady* [etc.], was emphatically Linnaean: it expanded Rousseau's original set of eight letters to twenty-four to echo Linnaeus's division of plants into twenty-four classes. Martyn also published a set of plates to illustrate Rousseau's work that was based on a similar set of illustrations in an earlier French publication. John Lindley's *Ladies Botany* (1834–37) was as emphatically un-Linnaean in its discussion of plant structures and tribes or families of the Natural System. Both works were popular and often reprinted.[16]

Although Rousseau praises Linnaean nomenclature for its concision and relative accessibility, he is more interested in teaching plant structure and similarities among plants classified as belonging to tribes, among them liliaceous, papilionaceous, and umbellate plants. He also urges Madame Delessert to teach her daughter how to dissect and examine floral structures as well as the entire process of floral development. Whereas Sophie, the intended spouse for Rousseau's Émile is taught only enough to make her dependent on his greater understanding, *Lettres élémentaires* makes a different case for teaching women about a field of inquiry. The internal contradictions between the Rousseau(s) of *Émile* and *Lettres élémentaires* (and arguably some other Rousseaus as well) help to explain why a writer and educator like Wollstonecraft in *A Vindication of the Rights of Woman* would find him a complex adversary.

If the underlying premise in *Lettres élémentaires* is that women and children are equally capable of learning (just so much) botany, this is not the message that women writers on botany took away. Taking up Rousseau's pedagogical approach and epistolary format, Priscilla Wakefield, Charlotte Smith, and Maria Jackson, among others, published works in which epistolary instruction invited children to draw their own conclusions about what they see as well as what they are told.[17] This inversion of Rousseau's gendered scene of instruction puts women into positions of pedagogical and moral authority, inasmuch as lessons about plants nearly always prompt lessons about a good life. It is no accident that the narrators of these books are nearly always women and, in keeping with the familial/familiar scene of moral education that extends from Rousseau to Wollstonecraft to Maria and Richard Edgeworth and beyond, they are women who instruct their children or those of close family members. The public reaction against women who wrote about plants for a general public did not mistake its target.

By the 1790s, roughly the midpoint of several decades in which women took a public interest in writing about plants or depicting them, conservative attacks on Darwin's "Jacobin" sympathizers and others triggered a polemic against women who did botany, or more precisely women who did botany in public view. This reaction belongs, as Alan Bewell has rightly emphasized, to a broader cultural anxiety about women as public figures and authorities.[18] In his verse tirade against women in the public sphere, *The Unsex'd Females* (1798), Rev. Richard Polwhele rather hysterically charged that women writers of botanical treatises were engaged in teaching the "bliss botanic."[19] In the coils of this reaction against women is the figure of Rousseau, by his own account seduced as a child by an older woman and late in life the author of a set of practical instructions for how women, especially mothers, might teach botany to children. The backlash against women who taught botany to the public may phantasmatically blend these two Rousseaus into one dangerous precedent. Yet even after Polwhele's tirade had given a local habitation and a name to a long-simmering uneasiness about women who took botany seriously, women continued to identify plants and illustrate them, in public and private.

Two vignettes, one biographical, the other fictional, suggest the uneasy public reception of this activity. The first of them is drawn from Shteir's *Cultivating Women, Cultivating Science*. In the early decades of the nineteenth century, Agnes Ibbetson published nearly two dozen reports on her scientific dissections and analyses in contemporary scientific journals. In most cases she was the first woman to do so. She also kept a detailed journal of her investigations and surmises and wrote a long theoretical essay on what she called "phytology," the logic of plant structures. Encouraged by an intermediary to send this manuscript to Sir James E. Smith, she did, enclosing drawings of what she had identified under microscope. Smith probably did not read the entire document—Shteir reports that his marginal notes end about midway in the manuscript—and after two years, and some prompting, finally responded critically. It may well have been that Smith decided Ibbetson was wrong, as she apparently was, to argue against current notions of plant "perspiration," the claim that plants, like animals, have circulatory systems that carry nutrients to all parts of the plant. Ibbetson's poor microscopes probably distorted what her accompanying drawings record.[20] Her lack of interest in taxonomy and nomenclature, matters of essential importance to Smith throughout his career, would not have recommended her more morphological emphasis to

Smith. Ibbetson was neither professionally trained nor welcomed by a community in which botanical dissection and analysis were men's work.[21]

My second vignette is a scene in William Godwin's novel *Fleetwood* that begins as Fleetwood, the first-person narrator, reads to his wife Mary. This is, as Julie Carlson makes clear, no ordinary scene of reading.[22] As Fleetwood understands this moment, it is the measure of their marital cohesion, or lack of it. Even before this scene in the novel, the space of reading has already become a possible bone of contention. Arriving at his paternal Welsh home with his wife, Fleetwood imagines this moment as the real beginning of their marriage. But when she asks that a room that had been his boyhood reading haunt become hers, he refuses, unwilling to relinquish a space he has long associated with solitary reading. All seems well again, or at least better, when the two settle down, at his invitation, so that he can read her Fletcher's play *Wife for a Month*. Reporting the scene, Fleetwood says that as he read the first act and began the second, Mary shared his admiration for the play, that their "pleasure" increased, that she was "roused to an extraordinary degree" as they read the climax of the plot. Yet at this moment, they are interrupted by the arrival of a peasant boy with whom Mary had earlier arranged to be her guide on a botanical excursion for a second look at "three plants of a very rare species," noting too that she did not ask Fleetwood to join her because she knew he did not like such interruptions." His internal assent to the truth of this statement rehearses the role of botany in the history of their courtship:

> I had received with pleasure the early lessons of Mary in botany, aided by the beautiful drawings with which she had illustrated them, and relieved with all the piquancy of courtship; but, by some fatality, I could never shape my mind to the office of herborisation; it appeared too pinched and minute an object for the tastes I had formed.[23]

The "fatality" Fleetwood invokes to explain why he can take no pleasure in what he calls "herborisation" (a botanizing excursion, according to the *OED*) arises from its preoccupation with "pinched" minutiae. His tastes are, he implies, wider, grander, and thus more consequential. When Mary gets up, saying they can finish reading his "beloved Fletcher" later, and goes off to return to the rare species of plants she had earlier spotted on Mount Idris, Fleetwood is disturbed, even insulted: " 'Is this the woman,' said I, 'whom I have taken as the partner of my life, who is more interested in two or three blades of grass, or a wretched specimen of mosses, than in the most pathetic tale or the noblest sentiments. . . . Oh, it is plain she cares only for herself.' "

What he imagines he has at the beginning of this scene, but then loses, Carlson suggests, is a conjugal twinness undone by Mary's departure to look at rare plants.

But then, for a moment, Fleetwood revisits Mary's botanizing, praising it as evidence of her "generous affection" for God's creation and her feminine character: "God bless her, even in her caprices! . . . They are symbols of that truly feminine character, which so delightfully contrasts with my own sex, and constitutes the crown and essence of the life of man. I will turn botanist myself!"[24] As he imagines joining Mary in her "herborizing excursion" (heaven help those rare species), Fleetwood "herborizes" his wife: "How beautiful will the carnation of her cheeks, and the lilies of her soft fingers, the fairest blossoms creation ever saw, appear amidst the parterre of wild flowers that skirts the ridge." From Petrarchan cliché to the pun of the verbal *skirts*, this self-satisfied reverie of woman as parterre supposes a ready figurative shuttle between women and flowers that misses both the specificity of his wife's botanizing and the work of difference that activity supposes.

Although Fleetwood might have reacted in much the same way to any interruption of his scene of reading for marital pleasure and instruction, the minute particularity of this botanical interruption is telling. Going on a botanical walk means above all looking for differences; finding them makes it possible to determine the identity of plants. Fleetwood's vision of married life imagines difference and identity as opposites. Mary's botanical walk assumes that you need to recognize differences to determine identity, and to have one. Fleetwood keeps his childhood reading room, and Mary goes out botanizing. His disparaging account of what she might find—some grasses, some mosses—suggests just how far afield women now botanized, going beyond the flowers of a domestic pleasure garden into plant families (grasses and cryptogamic plants, including mosses) that were of special interest to botanists.

When Mary returns to find her husband apparently unwell, he takes refuge in a philosophical discourse on marriage, directed to his reader, in which he advises accommodation as the price of marriage "between persons who live together in a style of equality" but remain "in many respects distinct" (Godwin, *Fleetwood*, 202). The distinction Fleetwood imagines registers the material weak-mindedness of female domesticity: Mary will be concerned about linen and plate just when her husband wishes her "to be thinking of the caverns of Pandemonium, or the retreats of the blessed." And well might she consider these far regions preferable, given Fleetwood's persistent view of her botanizing as mere feminine caprice. The careful modulation of marital proximity that Godwin and Mary Wollstonecraft retained may be implicit in Fleetwood's advice to the reader and Godwin's novelistic examination of marital difference. If his wife's earlier interest in studying botanical illustration with James Sowerby entered into their marital calculations of separate activities, Godwin may have had some twinge of sympathy for Fleetwood's frustration.

Unlike Mary Wollstonecraft, this Mary engages in no wholesale rebellion against a marital economy in which her husband's identity and understanding would suffice for both. She simply chooses to botanize. Yet this, Godwin implies with unusual narrative economy, is enough. The self-presentation of botanizing women during the era can be as subtly pitched as it is in the scene from *Fleetwood*. It can also be unnervingly direct. Or it may be at once subtle and quite material, as it is when women artists depicted plants in unconventional media. What interests me about the women whose botanical work I discuss here is the way each differently inhabits a culture that is wary—at some times more explicitly than others—about how women do botany. Whether the work in question is a pedagogical tract, poetry, or an image, it offers scenes of reading that convey how these writers and artists did botany in an era when it was by turns or by degrees an appropriate female accomplishment and a disturbing activity.

The inner contradiction that Jenny Uglow identifies in this anxiety turns on sex and difference:

> Botany was often considered the one scientific pursuit suitable for women, yet Linnaeus seemed rather shocking for a female audience. When he organized plants into class, order, genus, species and varieties he chose sexuality as the key . . . female "genitals" [ie stigmas] . . . calyx, became the "nuptial bed." This meant, of course, that some flowers had far more than a single male sharing a bed with the female--and the sexual naming went further, with some structures compared to *labia minora* and *majora*, let alone a whole class of flowers named *Clitoria*.[25]

The problem went beyond Linnaeus. Although some children of members of the Lunar Society, which included Erasmus Darwin, Joseph Priestley, James Watt, Josiah Wedgwood, and several others, received a co-ed education that included natural history, the Lunar Men were just that—men. Darwin's occasional scientific demonstrations included some designed for the ladies, but the few women who turned up at the Lunar Society meetings also served tea. Uglow notes Priestley's expression of relief that after his marriage and years of struggle he could rely on Mary Priestley to handle household affairs and thereby free his time for "other duties of my station."[26] In conveying her view of the temper of the times to William Withering, one of the Lunar men, the Quaker Molly Knowles barely manages to contain her fury:

> *Women* to possess understandings of "masculine strength," is an idea intolerable to most men bred up amongst each other in the proud confines of a College. There indeed they seem to monopolize *learning*, but happily *intellect* cannot be confined there; and as general education increases, Scholars will more & more discover to the confusion of their pride, that genius is shower'd down on heads,

Fig. 12. *Young Daughter of the Picts*, Jacques Le Moyne de Morgues, ca. 1585, watercolor and gouache, touched with gold, on parchment, 16 × 19.9 cm. (10¼ × 7⅚ in.). Yale Center for British Art, Paul Mellon Collection, B1981.25.2646.

Fig. 13. *Sensitive Flower*, J. J. Grandville, *Les Fleurs animées*, colored engraving, 17.15 × 29.96 cm. (6¾ × 10 7/32 in.), 1847. Image provided by Missouri Botanical Garden Library, St. Louis, www.mobot.org. Accession no. GR780.G73, v.2.

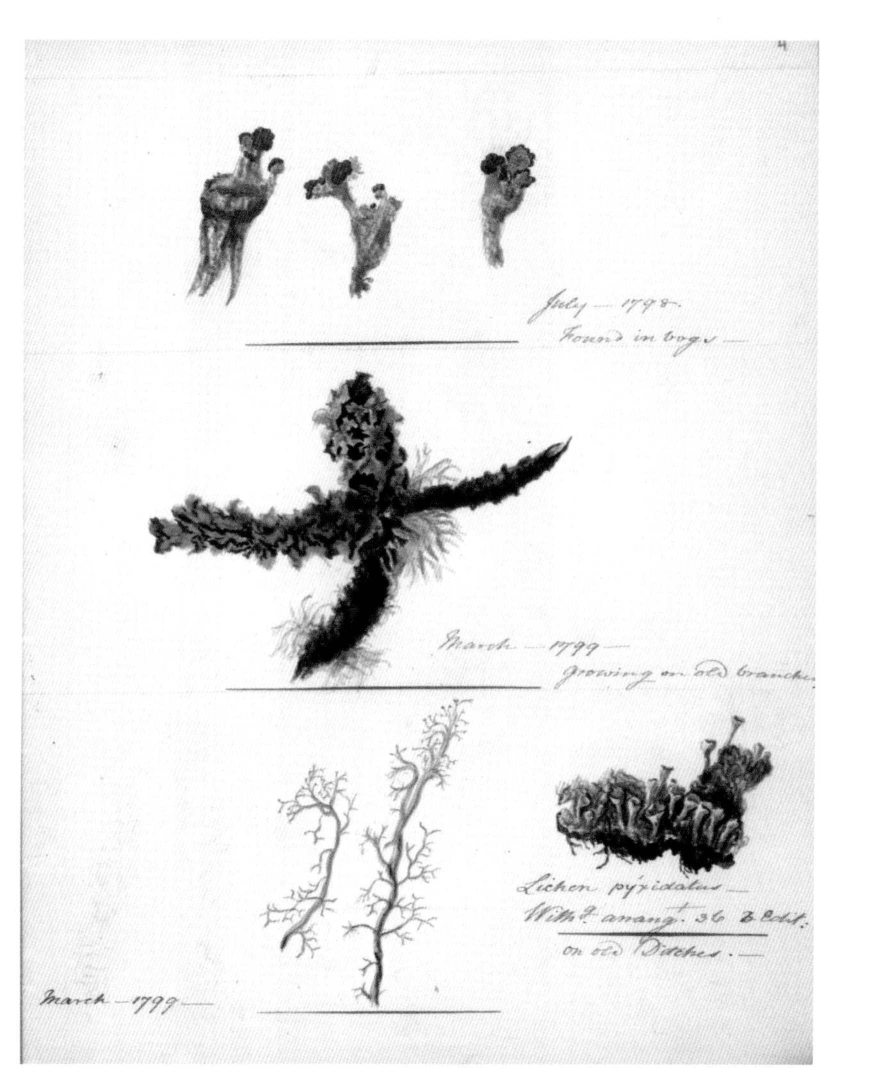

Fig. 14. *Lichen pyxidalus*, Frances Beaufort Edgeworth, watercolor with pen, 23.5 × 18.4 cm. (9.25 × 7.24 in.). Reproduced by permission of The Huntington Library, San Marino, California; ms. FB 59, ff. 4.

Fig. 15. *Ophrys apifera. Bee Orchis or Beeflower,* Frances Beaufort Edgeworth, watercolor with pen, 24 × 17.5 cm. (9.45 × 6.89 in.). Reproduced by permission of The Huntington Library, San Marino, California; ms. FB 59, ff. 35.

Fig. 16. *Dryas octopetala*, Frances Beaufort Edgeworth, watercolor with pen, 22.3 × 18.6 cm. (8.78 × 7.32 in.). Reproduced by permission of The Huntington Library, San Marino, California; ms. FB 59, ff. 78.

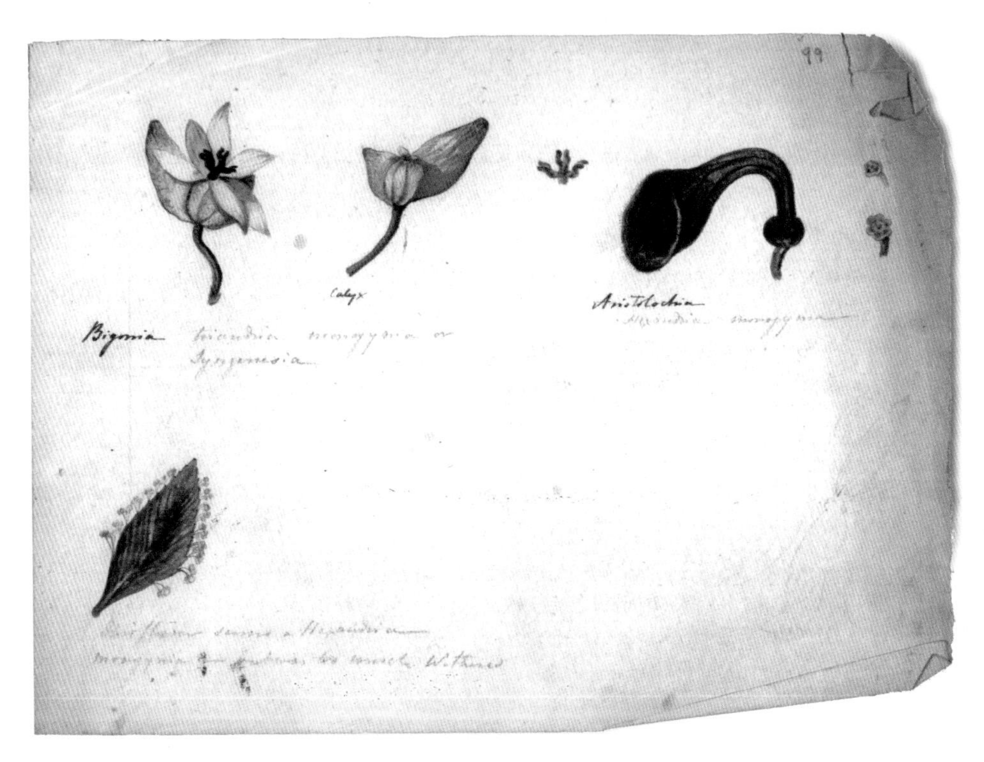

Fig. 17. *Bigonia* and *Aristolochia*, Frances Beaufort Edgeworth, watercolor with pen, 14.5 × 16.7 cm. (5.71 × 6.57 in.). Reproduced by permission of The Huntington Library, San Marino, California; ms. FB 59, ff. 99.

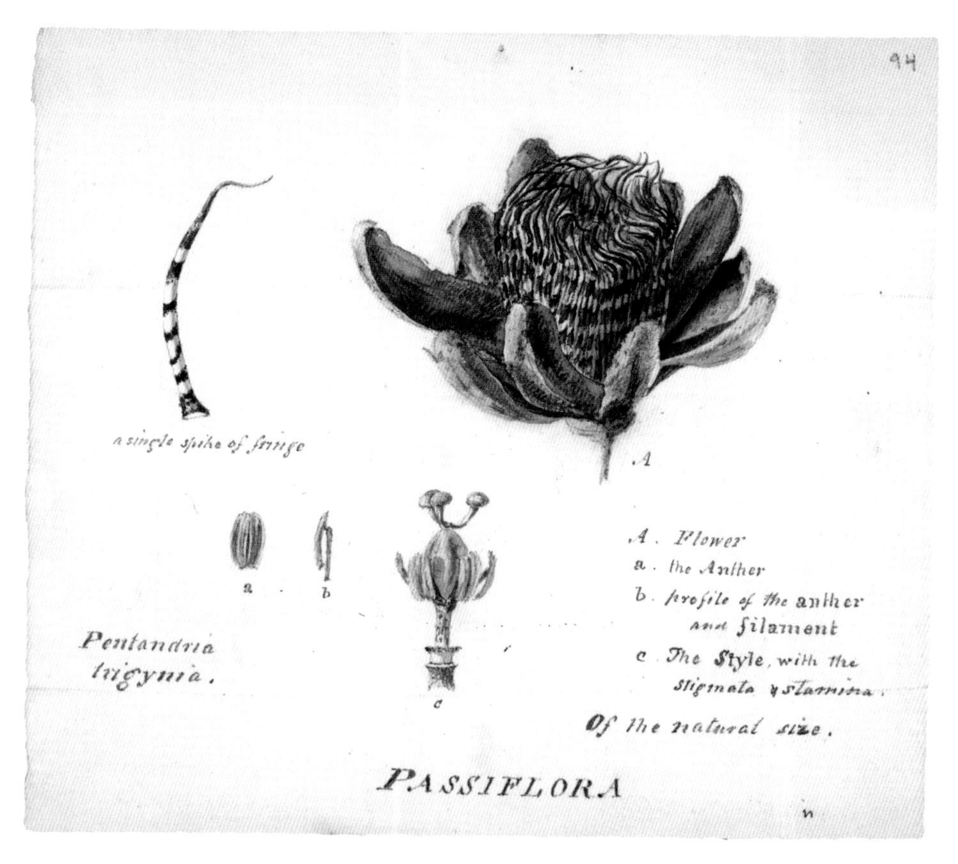

Fig. 18. *Passiflora* (species not indicated), Frances Beaufort Edgeworth, watercolor with pen, 15.1 × 20.4 cm. (5.94 × 8.03 in.). Reproduced by permission of The Huntington Library, San Marino, California; ms. FB 59, ff. 94.

Fig. 19. *Magnolia grandiflora*, Mary Delany, paper mosaic, 33.8 × 23.4 cm. (13.31 × 9.21 in.). © The Trustees of the British Museum. Reg. no. 1897,0505.557, image 00855981001.

Fig. 20. *Passiflora laurifolia*, Mary Delany, paper mosaic, 35.2 × 24.1 cm. (13.86 × 9.49 in.). © The Trustees of the British Museum. Reg. no. 1897,0505.654, image 00037142001.

Fig. 21. *Ixia Crocata (Triandria Monogynia)*, Mary Delany, paper mosaic, 34.1 × 23.9 cm. (13.43 × 9.41 in.). © The Trustees of the British Museum. Reg. no. 1897,0505.484, image 00333100001.

Fig. 22. *Physalis, Winter Cherry*, Mary Delany, paper mosaic, 29.2 × 17.9 cm. (11.5 × 7.05 in.). © The Trustees of the British Museum. Reg. no. 1897,0505.672, image 00032744001.

Fig. 23. *Rara avis*, after Katherine Charteris Grey, engraving, 4.0 × 6.8 cm. (1.57 × 2.68 in.). In James Bateman, *The Orchidaceae of Mexico and Guatemala*. Reproduced by permission of The Huntington Library, San Marino, California; RB 325034.

Fig. 24. *Cycnoches Loddigensii*, Katherine Charteris Grey, pressed flowers, 25 × 19 cm. (9.84 × 7.48 in.). Courtesy of Enville Hall Archives.

Fig. 25. *Pterostylis obtusa*, Katherine Charteris Grey, pressed flowers, 25 × 19 cm. (9.84 × 7.48 in.). Courtesy of Enville Hall Archives.

Fig. 26. *Erica pubescens,* Katherine Charteris Grey, pressed flowers, 25 × 19 cm. (9.84 × 7.48 in.). Courtesy of Enville Hall Archives.

Fig. 27. *Orphrys muscifera, Fly Orchis*, James Sowerby, handcolored engraving, 9 × 16.5 cm. (5.54 × 6.5 in.). Sowerby and Smith, *English Botany* 1:64. Linda Hall Library of Science, Engineering & Technology.

> as seemeth to Heaven good, whether drest in caps of gauze or velvet—in large grey wiggs, or small silk bonnets.[27]

In the climate of opinion that so rankled Knowles, botany and flowers could become rhetorical minefields for women writers. I offer here two examples: the first is the public-private antagonism between Anna Barbauld and Mary Wollstonecraft about Barbauld's praise poem "To a Lady, with some painted flowers." The second is Wollstonecraft's reassignment of manly virtues derived from the word *stamen*, identified since Linnaeus with the male reproductive part of plants, to women as well as men.

The rupture between Wollstonecraft and Barbauld is from a political perspective surprising. Both were writers committed to radical critique, and Polwhele accordingly vilifies both as "unsex'd females." Horace Walpole avowed he could not, would not, read Barbauld because she "curses our clergy and feels for negroes," calling her a "virago" and a fishwife.[28] Like Wollstonecraft, Barbauld strongly objected to Burke's *Reflections on the Revolution in France*. Unlike Wollstonecraft, she never assembled her scattered notes into a formal reply. The dispute began with the "painted flowers" that Barbauld describes in a poem addressed to an unnamed lady. In an earlier poem titled "To Mrs. P, with some Drawings of Birds and Insects," she had offered a similar double gift of poem and drawings to Mary Priestley, a close friend during their Warrington years. But birds and insects evidently do not have the same ideological charge as flowers. Praising this Barbauld poem in the same note that so vehemently dispraises "To a Lady, with some painted flowers," Wollstonecraft argues that the latter reinforces a "false system of female manners" that "classes the brown and fair with the smiling flowers that only adorn the land." This, she adds, "has ever been the language of men" (167–68). In a note, Wollstonecraft quotes the poem in its entirety, italicizing the lines that most emphatically register a more or less conventional view of women and flowers. It is hardly necessary to explain which lines those are or why she also capitalizes other words in the poem to emphasize the view of women that arises from comparing them to flowers, who neither toil nor reap. Here is Barbauld's poem with the Virgilian epigraph that first accompanied it into print in the 1792 edition of her *Poems*:

> . . . . . *tibi lilia plenis*
> *Ecco ferunt nymphae calathis.*
> Virgil

> Flowers to the fair: To you these flowers I bring,
> And strive to greet you with an earlier spring.
> Flowers sweet, and gay, and delicate like you;

Emblems of innocence, and beauty too.
With flowers the Graces bind their yellow hair,
And flowery wreaths consenting lowers wear.
Flowers, the sole luxury which nature knew,
In Eden's pure and guiltless garden grew.
To loftier forms are rougher tasks assign'd;
The sheltering oak resists the stormy wind,
The tougher yew repels invading foes,
And the tall pine for future navies grows;
But this soft family, to cares unknown,
Were born for pleasure and delight alone.
Gay without toil, and lovely without art,
They spring to cheer the sense, and glad the heart.
Nor blush, my fair, to own you copy these;
Your best, your sweetest empire is—to please.[29]

It must have seemed incomprehensible to Wollstonecraft that one of her contemporaries, like herself a woman educator with strong views on female education and British policy, could praise a woman/flower for doing nothing except looking beautiful. At the very least, such sentiments work against the "revolution in female manners" that she calls for in *A Vindication of the Rights of Woman*.[30] Moreover, and more immediately, Barbauld's poem echoes the line of male writing about women that Wollstonecraft works hard to refute in *Vindication*. With this project firmly before her, she takes no note of Barbauld's choice of epigraph and the possibility that the comparison between flowers and sheltering oaks, yews, and pines as figures of military strength carries a negative charge in this poem.

My counterreading begins with Barbauld's epigraph, taken from Virgil's second *Eclogue*, which modern editors translate as "See, for you the Nymphs bring lilies in heaped-up baskets."[31] Lilies have a long pedigree for readers trained in natural history and the Bible, as Barbauld and her unidentified recipient both were. In the Gospel of Matthew, it is of course the lilies of the field that neither reap nor sow (nor, presumably, do they sew). The Virgilian tag suggests that the poem's unidentified recipient is singled out for this gift because it is emblematic of who she is and the pleasure she gives. This is her "empire," in contrast to the one that those emblematically British and American trees seek to secure with their "loftier forms" and "rougher tasks." As Wollstonecraft notes when she italicizes or capitalizes much of what I have just quoted, the diction sounds like the same old opposition of male military strength to female idleness.

But perhaps not. Barbauld's lilies are also the first flower and botanical tribe/family that Rousseau presents in his *Lettres élémentaires*, the extended family of "all true Liliceae" that includes, as he says, "Tulip, Hyacinth, Narcissus, Tuberose, and even onion, leek, garlic, which are also true lilies, although they seem quite different at first glance."[32] It is also the flower that Rousseau uses to explain the structure of a flower. "Take a Lily," he says, and so Barbauld does, offering this flower and its wide botanical family as a gift to someone who is, like them, widely affiliated with a diverse family, even as the lily family includes showy flowers and household vegetables. This praise and allusion operate beneath the poetic diction that Wollstonecraft challenges. If such poems are, like the lilies of this poem, made for pleasure not war, they may nonetheless convey arguments.[33]

The intended recipient of this poem and riddle was probably Mary Priestley, to whom Barbauld explicitly addresses another a poem and gift of drawings. At the time, both were members of the large and close-knit, familial society of students, teachers, and their families at Warrington Academy, probably the most remarkable of the dissenting academies of the late eighteenth century.[34] Barbauld's brother John and Rochemont Barbauld, whom she later married, taught there, as did Mary Priestley's husband Joseph, long before they moved on to Birmingham, the Lunar Society, and much later still to the United States, after the mob riot that destroyed Priestley's laboratory and library. Read against the Warrington community of the early 1770s, "To a Lady, with some painted flowers" argues for a similarly extended familiality of plants and friends who might to outsiders look unaffiliated.

Wollstonecraft's broadside against Barbauld includes none of this. Although she almost certainly knew about the dissenting circle of teachers and families at Warrington, her focus in *Vindication* is to mount a public attack on comparing women to flowers. Barbauld privately countered with "The Rights of Woman," unpublished in her and Wollstonecraft's lifetimes, a poem that satirizes the woman who insists that her "rights are empire," to which the poem responds with a marital union in which "separate rights are lost in mutual love."[35]

Still, when Wollstonecraft considers women who study botany in *Vindication*, she takes a very different tack. Disagreeing with those who insisted that women should not instruct others in the sex life of plants—hardly a secret after Darwin's quite explicit English translations of Linnaeus—she offers a defense of genuine modesty that disentangles this virtue from its implicit iconographic baggage that insists that women are the blushing flowers, the sensitive souls: "Modesty must be equally cultivated by both sexes, or it will ever remain a sickly hot-house plant, whilst the affectation of it, the fig-leaf borrowed by wantonness, may give a zest to voluptuous enjoyments" (*VRW*, 254, 258). The slight edge of risk embedded in this transformation of modesty into its botanical opposite, by way of an allusion to the no-longer-modest, blushing, fig-leaf-covered (but only just) Eve, registers Wollstonecraft's knowledge of hothouse experiments and her willingness to use the figure of the hothouse bloom to suggest female nature unnatured, overheated, nothing like real nature or real modesty.[36]

If thinking about plants invites thinking about sex, Wollstonecraft presses this slippage to her advantage when she takes up the figurative value often then assigned to the stamen as the male reproductive structure of plants and the claim that stamina is a male virtue. In part 2 of *Rights of Man*, published a month after Wollstonecraft's *Vindication*, Thomas Paine argued that under representative government the nation "possesses a perpetual stamina, as well of body as of mind, and presents itself on the open theatre of the world in a fair and manly manner" that is consistent, Paine says a few sentences later, with its "manly character."[37] Arguing that there is no sex in souls and that immortality is accordingly available to women as well as men, Wollstonecraft directly challenges the analogies between botanical and biological identity that stand in the way of granting women equal access to souls and thus to the possibility of immortality: "The stamen of immortality, if I may be allowed the phrase, is the perfectibility of human reason; for, were man created perfect, or did a flood of knowledge break in upon him, when he arrived at maturity, that precluded error, I should doubt whether his existence would be continued after the dissolution of the body" (*VRW*, 166).

Wollstonecraft's uncharacteristic deference ("if I may be allowed the phrase") implicitly grants that writers do not usually speak of the stamen in this way and context. That she does so nonetheless shifts the ground for speaking not only about women and men as biological counterparts to pistils and stamens, but about the very foundation of their capacity for reason: it inheres neither in gender nor in biological bare life but in the degree to which human reason might, in the next world more than in this, be perfected. What needs to be reproduced, the "stamen of immortality" figure implies, is not bare life at all, but an understanding of human reason that is grounded in the hope of a (perfected) immortal life.

That Barbauld and Wollstonecraft sharply disagreed about how to read the figurative language binding women and botany says something about the array of differences in the way that their women contemporaries wrote about plants or illustrated them. The irony of this point is worth noting. The goal of assigning to women and girls "polite" accomplishments that involved plants—gardening, learning about flowers, drawing them—had as an implicit goal a kind of baseline set of botanical accomplishments. What seems to have been assumed or imagined was something like an exchangeability of botanical talent among young ladies such that one would know what they knew and what they could do without bothering with particulars. Failing other amusements and occupations, women could study their own botanical kind: polite flowers. Study flowers they did, but not necessarily with the degree of politeness that would have guaranteed a modest knowledge.

BOTANY LESSONS

Across the romantic era, women wrote about botany more than any other branch of natural history and they fully expected that their authority to do so would be challenged.[38] Although all who wrote about botany cited earlier authorities, women did so with a degree of diligence that may also have served a more covert strategy. By recognizing the authority of others, they distracted attention from the fact that their own botanical knowledge went well beyond an acceptable minimum.

Priscilla Wakefield was the first woman writer to publish a systematic discussion of botany. As a Quaker who published in defense of William Penn when public antagonism toward Quakers and other dissenters was a field sport in the press, she was clearly willing to risk public censure.[39] Yet her *Introduction to Botany, in a series of familiar letters* (1796) presents a surprising reworking of Rousseau's epistolary model. The two interlocutors of Wakefield's book are sisters; neither has much botanical authority. Felicity offers her sister Constance a highly devotional set of general claims about plants. The design of the work, Wakefield explains, is to "cultivate a taste in young persons for the study of nature, which is the most familiar means of introducing suitable ideas of the attributes of the Divine Being," noting that "the structure of a feather or a flower is more likely to impress their minds with a just notion of Infinite Power and Wisdom, than the most profound discourses on such abstract subjects" could be expected to do (Wakefield, *Introduction*, iii).

The ensuing argument is thoroughly Linnaean in its organization, analysis, and conclusions. Indeed Linnaeus is the only botanical authority Felicia urges, on the grounds that although others have made important discoveries, the Linnaean system is "the one universally adopted" (Wakefield, *Introduction*, 39). Although Wakefield insists in the preface that botany is an appropriate because rational accomplishment for young girls, the letters neither invite nor brook argument or experimental inquiry. They also duck the problem of plant sex, gendering Linnaeus's stamen and pistil: "threads or chives" are offered as English synonyms for *stamina*; "style or pointal" for *pistilum* (Wakefield, *Introduction*, 13). This polite terminology echoes William Withering's *An Arrangement of British Plants*, first published in 1776 and in its third edition by 1796, the year the *Introduction to Botany* first appeared.[40] With a degree of indirection that is typical, Wakefield says that because there now exist English translations of Linnaeus's works, students of botany no longer need to learn to read the original Latin. She does not mention Darwin's translations, which mock Withering's asexual English rendering of the Linnaean system.[41] Wakefield is tactfully silent about botanical topics that her contemporary discussed as errors or problems in Linnaeus. Of the cryptogamia, the last and least understood of the Linnaean classes, Felicia says only that it "includes those vegetables which are of the lowest kinds, whose parts of fructification have hitherto escaped the most attentive researches of learned botanists." The two sisters discuss what they describe as the "most obnoxious" cryptogamic plants, as if the topic were too distasteful to warrant inquiry (Wakefield, *Introduction*, 162). Yet precisely because cryptogamic plants were still little understood, botanists and the popular botanical press found them fascinating.

Maria Jackson's *Botanical Dialogues between Hortensia and Her Four Children* (1797), published a year later, could not be more different. It is, for starters, far less deferential toward botanical authority, in particular that of Linnaeus. In two later works intended for adult readers, *Botanical Lectures* (1804) and *Sketches of the Physiology of Vegetable Life* (1811), Jackson devotes still more discussion to Linnaean errors, the achievements of recent botanists, and her own experimental conclusions.[42] Her most extended praise is for Erasmus Darwin, whose arguments about conjecture and plant/animal analogy are closest to her own. Jackson (naming names, as Wakefield had not) identifies Darwin as the English translator who provided English equivalents for Latin plant names and categories.[43]

The children of the *Botanical Dialogues* are, not surprisingly, able readers of the versified botanical vignettes of Darwin's *Botanic Garden*. Harriet explains to her siblings that Darwin's "fair Tremella" is "this star-jelly, which, becoming transparent after it has been frozen in autumnal mornings, is distinguished by this property from other vegetable muscilage." Whereas these will no longer adhere once frozen, the fact that "poor Tremella" can indicates that she is not in fact a vegetable at all but "a substance, that herons parted with, after they had eaten frogs" (*Dialogues*, 140).

Like Darwin, Jackson's young botanists relish those moments when surprising details overturn easy taxonomic conclusions. Harriet's spirit of observation embodies the dictum that Jackson's good and rational mother insists on repeatedly: "see for yourself." This advice turns out to mean as well see for yourself what sexual difference means for your futures. Echoing Darwin's sentiments on female education, Hortensia more than once explains that whereas the boys Charles and Henry may aspire to careers as botanical experimentalists, the girls Harriet and Juliette cannot. Yet each time this topic reappears, it is more oddly circumstanced, uncomfortably positioned between the near-brutal forthrightness with which Hortensia insists that her daughters should not be known for what they know and bitter reflections on the world's limited expectations for women.

At first, the topic is quickly managed, as part of her motherly calibration of the kind of work each of her boys can do. Jackson's genteel authorial persona as the "lady" or "authoress" is carefully reiterated in the portrait of Hortensia's eldest son and child, Charles, who won't have to work for a living but should find rational, improving occupation nonetheless. The second son, Henry, will need to earn a living and so must prepare himself for a profession, like botany. Juliette chimes in that she is learning two things, botany and thinking, as her mother has just recommended (*Dialogues*, 53–54). Hortensia replies: "One is the consequence of the other; your works you learn by rote, like a parrot; the acquirement of them may be called the education of the fingers, that of science, or language of the mind: they are both becoming the female character; but if I was obliged to omit one in my education of you, which do you think I should lay aside?" (*Dialogues*, 55). Harriet quickly sorts out the reply that she is expected to give:

> I know that it would be science and language; because, ma'am, you have always told us, that the first point was to make ourselves useful in the small duties of life, which daily occur, and that we may have many opportunities of putting the acquirements of our fingers to use, both for ourselves and others, before we can those of science and language. I should however be very sorry if I could only work. (*Dialogues*, 55)

"Work" is evidently shorthand for "needlework," a compression that painfully insists that the daughters' work is by definition a matter of rote, parrot-talk, whereas science and language properly practiced belong to rational inquiry.

Yet throughout the *Dialogues*, the daughters are as actively engaged as the sons in hunting for plant specimens and making hypotheses about their identity and floral character. Hortensia is quick though to chastise what she calls female pride when Juliette reports in some frustration that "we cannot class this parsley" (*Dialogues*, 92). Although the reporting convention thus far has been that one of the children will speak for the collective judgment of all four, the mother/teacher here singles Juliette out for blame:

> I have rarely had occasion to reprove you, Juliette; but for the chagrin you give way to, when you do not excel in the degree you expect to do; I fear this disposition proceeds from pride rather than modesty, and much wish you to get the better of it. I should be sorry to be obliged to lose you from our party; but if this discontent is indulged every time . . . you cannot refer a plant to its proper class or genus, it will render you a very troublesome companion. (*Dialogues*, 92)

To be discontented with not having been able to classify what the mother goes on to say is a difficult plant group is in a female a mark of disappointed ambition and excessive pride.

Hortensia more fully declares the cultural logic at work in this amazing and brutal reprimand, which silences the fictional Juliette for a long time, in the second part of the work. But now the tone more clearly registers the mother's distance from the cultural norms she had earlier asserted. Noting that the cryptogamic plants have not been thoroughly investigated, she suggests that Charles and Henry might "distinguish" themselves in this field. Charles immediately asks "And why not my sisters, ma'am; I am sure they generally go before us in whatever we learn together." Hortensia half-agrees, saying she does not "doubt their abilities," that she wishes them to be "as thoroughly informed upon the subjects that they study" as her sons, but that as females they should "avoid obtruding their knowledge upon the public" (*Dialogues*, 238). Her tone shifts, temporarily and abruptly, when she explains what the world expects women to do:

> The world have agreed to condemn women to the exercise of their fingers, in preference to that of their heads; and a woman rarely does herself credit by coming forward as a literary character. The world improves, and consequently female education. . . . and the time will come, when it must be granted, that by improving our understandings, we enlarge our view of things in general; and thence are better qualified for the exercise of those domestic occupations, which we ought never to lose sight of, as our brightest ornament, when properly fulfilled. At this time information in a woman, beyond a certain degree, distinguishes her above her companions, and like all other distinctions is liable to lead her into a vain display, of what she hopes will gain her admiration. Hence she becomes ridiculous, and brings, what in itself might be a credit, into a disgrace. (*Dialogues*, 238–39)

Having come close to Wollstonecraft's carefully limited conclusion in *Vindication* that women should be rationally educated so that they might educate their children, Jackson then backs off right back to needlework, as the best line of defense against the public disgrace of becoming publicly known for what you know. The syntax urges that this judgment is not hers but the world's, as a plural body ("the world have agreed") that has condemned women to the needle instead of the pen, and Jackson to the same old round about women and the needle.

Jackson's woman who knows too much and says so in public may allude to Wollstonecraft, who died in April 1797.[44] Or Jackson herself, who "obtrude[s] her knowledge on the public" in book after book: dancing between praise and blame for Linnaeus and encouraging children to study cryptogamics, plant/animal analogies, and a few plants whose reproductive mechanisms depart from the basic sexual code of the Linnaeus systematic and are for this reason taxonomically distinct. Jackson's botanical writing walks a narrow line between conveying existing botanical knowledge and gesturing toward what future inquiry might offer aspiring botanists.

The sexual politics of Jackson's resistance to the Linnaean system is apparent in her insistence that the pistil is no less important than the stamen, a claim that aims directly at Linnaeus's use of the number and position of stamens to set up his classification. She accordingly lingers in the *Dialogues* with the class *Gynandria* (literally, of female-male, i.e., doubtful, sex), where Linnaeus had placed the orchis tribe, a genus of orchids, the passion-flower, and the arum. When she first presents this class to her children, she notes its defining trait, according to Linnaeus—that the stamens of its plants all grow out of the pistil. She then explains that there are nine Linnaean orders within the class, each identified by "the number of stamens in each flower" (*Dialogues*, 111).

Although Hortensia does not say this, Linnaeus's account of this class and its orders contradicts his most fundamental systematic protocol. Instead of naming the class according to the number of stamens and its orders according to the number of pistils, he does the reverse. Faced with plants whose sexuality is "doubtful," he backtracks by relying on the female part to name the class. When Hortensia offers a more extended analysis of the same class of plants, she emphasizes the uniqueness of their stamen/pistil arrangement. In the case of the arum, that plant group Linnaeus called a "natural prodigy," the sexual arrangement is still more surprising: here the stamens are not raised on filaments but reduced to their anthers, those pollen carrying members that are located right next to the nectary or seed cup, farther inside than the pistils. The taxonomic oddity that had attracted other writers to the arum prompts Jackson's Hortensia to present it in terms of Linnaean authority. Noting that the younger Linnaeus (Carl Jr.) had argued for removing the arum to another class entirely, she comments, "I incline to this opinion myself, but do not venture to remove it . . . till farther observations of respectable botanists have determined more decidedly its proper situation" (*Dialogues*, 221). Having noted this point of disagreement between Linnaeus father and son, Hortensia adds her own judgment, maintaining the barest edge of deference to contemporary authority.

The limit to her deference becomes apparent when her daughter Harriet says she would prefer to study the way a plant is pictured on an engraved plate before trying to work out its traits by examining an actual specimen. In reply, Hortensia launches a series of examples and principles to show why one must question received botanical wisdom: "In examining plates, you take the authority of others; whereas in botany, as in all other things, we can make little progress if we do not see for ourselves." From this point on in the *Dialogues*, it is Harriet, not Hortensia, who urges deference to the botanical authority of "great and wise men" who have studied botany all their lives (*Dialogues* 228–29). With someone, just a child, speaking up for authority, Hortensia can more readily assume the stature of a wise councilor who acknowledges the rightness in principle of established authority while challenging it repeatedly with contrary evidence. When the daughter insists that when botanists like Hedwig or Curtis offer different views she would be inclined to say, "I am wrong, and they are right" (*Dialogues*, 228–29), her mother replies:

> With due limitations, that is a proper way of thinking; but in such cases accustom yourself to state in writing the particulars, in which you differ in your observations, from what you have heard or read upon the subject. You will by this means secure the benefit of being better informed, if you are mistaken, and it may happen, that you may be right; and then you will have the pleasure and honour of improving by your investigations the science of botany. (*Dialogues*, 229)

When Harriet, who has become an unusually tenacious interlocutor, asks whether it is likely that "great and wise" botanists might be mistaken, her mother retorts that it is more probable that her daughter would be mistaken. Yet immediately afterward she again insists: "But as we do not unfrequently see great and wise men err in their judgment and accounts of things, we must not rely upon them as infallible: in whatever you undertake, make it a rule to *see for yourself*. It is the observance of this rule, that has rendered the works of Mr. Curtis so valuable" (*Dialogues*, 229). With one of her preferred botanical authorities invoked to support the principle of questioning authority, she names other esteemed botanists who make mistakes and whose example "is one of the many instances, which ought to deter us from relying upon authority, be it ever so respectable" (*Dialogues*, 237).

Unlike the Rousseau of *Emile*, Jackson's Hortensia teaches her girls what she teaches her boys, although the girls must pay for this instruction in the coin of female reticence. Yet as the children gain botanical expertise, she teaches them how to disagree with botanists, beginning with Linnaeus.

Jackson's next book, *Botanical Lectures by a Lady*, combines even more praise for Linnaeus with still sharper critique. As lectures, not dialogues, in which the author instructs adults, asking them to find certain plants to consider as they read, the form of the work insists more explicitly on a public, rather than familial, voice and authority. She adds more instances in which Linnaeus's system violates its own protocols and lingers with plants and classes that undermine Linnaean assumptions. She argues, for example, that some species of the family Lychnis, which Linnaeus assigned to *Decandria Pentagynia*, a class and order so named because the plants it identifies have ten stamens and five pistils on the same flower, do not in truth belong to this class because their stamens and pistils are on different flowers.

Moreover, she explains, Linnaeus mistakenly supposed that this fact is the essential character of these species within this class. Logically, he should have done quite the opposite. In other words, you don't insist that a genus of plants has a set of traits and then place within it several species whose defining character sets them apart from the rest of the genus (*Lectures*, 73).

Jackson is certainly not the only botanical writer to identify logical slips in the Linnaean system. What makes her arguments so distinctive is their appearance in a pedagogical tract that might have been expected to offer its Linnaeus straight up, without critique, and further that its author is a woman whose public authority would not have been recognized as equal to that of a Curtis or a Darwin, let alone a Robert Brown.

In later works, Jackson keeps careful score of whose botanical authority she will grant and of when she finds Linnaeus insufficiently authoritative. Returning to the arum in *Botanical Lectures*, she repeats what she had written in the *Dialogues*, with one telling addition. Having said of the *Orchis* genus, another in the same Linnaean class, that "it is the business of an experiment maker to be always looking for circumstances which make against his theory, and not for it," she comments that the arum is "a wonderful flower, and seems intended by nature to show us, that she is not confined to any one method of renewing her productions" (*Lectures*, 151 and 157). The arum is "wonderful" because its reproductive structure veers off from the strict sexual economy of the Linnaean model. Linnaeus finesses the point by calling it a "natural prodigy," but Jackson persists, noting that the arum may well be one of those "circumstances against his theory, and not for it."

Her most pungent critique concerns Linnaeus's classification of the cryptogamic plants, which he labeled and group together for what they lacked: evidence of reproductive mechanisms. As the Linnaean class that includes many plants that seemed to many, and certainly to Jackson and Darwin among many others, to be somewhere between the plant and animal kingdoms, the cryptogamics operate like one of those Derridean supplements that may not be in the end an add-on or superfluous category but one essential to thinking through fundamental systematic questions about nature itself. Jackson says:

> The little knowledge, that has hitherto been obtained of these numerous tribes of plants, has been considered a great reproach to the science of botany. Perhaps the system of Linneus [*sic*] may have retarded a more distinct arrangement of them, that being founded upon the parts of fructification, which in most genera belonging to the class Cryptogamia are so difficult to ascertain. (*Lectures*, 100)

At issue here is the preemptive character of a systematic that explains plants primarily in terms of their fructification parts, thereby forestalling discussion of structures that cannot be so characterized and classed. Jackson's argument shows that she has been keeping up with scholarly inquiry, especially that of the German botanist Johann Hedwig, whose scholarly essays and posthumously published *Species Muscorum Frondosorum* (1801) reported his discoveries concerning the reproductive processes in some mosses (*Lectures*, 164–85).[45]

In her last serious work on botany, *Sketches of the Physiology of Vegetable Life* (1811), Jackson examines plants whose apparent motions suggest animal-like properties, echoing many of Darwin's examples: Sensitive Plant, the Venus flytrap, the *Hedysarum gyrans* of the Ganges, *Drosera* (sundew), and *Sarracenia adunca* (a species of pitcher plant). At the same time she insists on her own experiments and hypotheses (*Sketches*, 70). She takes issue with Smith, whose *Introduction to Physiological and Systematical Botany* had appeared in 1807, for failing to recognize the degree of automotion in the habits of a water plant of the Mediterranean (also found in India, although Jackson does not say so), *Vallisneria spiralis*, despite the "ingenious President of the Linnean Society" having made an extensive tour of Italy, where he could have encountered this "curious vegetable."[46]

In an era when other women on botany approached Smith with deference, among them Ibbetson and Charlotte Smith, if indeed they approached him at all, Jackson's boldness is striking. No doubt Smith would have called it effrontery. Her brief, as it had been since the fictional Hortensia taught her children to "see for yourself" in the *Botanical Dialogues*, remained to do just that, and then report her findings to the public in terms that declare her the experimental successor to Linnaeus, an equal to Darwin, Curtis, and a rival to J. E. Smith. This would be a bold stroke from anyone writing on botany during this era. In a woman writer unknown to the public except in print, and evidently not part of the world of professional and amateur botanists who gathered at meetings of the Linnean Society with its insufficiently curious president, Jackson's self-presentation is a riveting public performance. Her repeated claim that children and adults ought to study botany by questioning received opinion and looking at plants under a microscope conveys, in brief, a sharp sense of opposition to conventional views about how much women might learn or teach about plants.[47]

FIGURE AND GROUND

The opening pages of Charlotte Brontë's *Jane Eyre* present a scene of reading that was by the time the novel was published in 1848 familiar to readers of fiction, natural history, and conduct books. Jane takes up Thomas Bewick's *History of British Birds* and finds a secluded spot to "read" its opening pages and vignettes. I say "read" advisedly, for what Jane does with this book, as she remarks, ignores its "letter-press" to concentrate, rather perversely, on scenes and descriptions outside Britain having little to do with birds, among them the bleak shores of Lapland, Siberia, and the Arctic zone.[48] This reading, like the drawings Jane later produces and Rochester examines, is more imaginative than literal, more pitched to record her state of mind and imaginings than it is attentive to the book and topic at hand.

Jane's quirky reading and painterly practices openly declare a rebelliousness that is more covert but equally charged in the work of three women artists I consider here—work privately produced in genteel households, apparently in full conformity with the modeling of female accomplishment that the gentry was supposed to offer for middle-class emulation and that Jane, as the orphaned niece of straitened class and (it is assumed) narrow financial circumstances, is expected to learn from her middle-class aunt, who is at once deeply conventional and mean. The artists are Mary Granville Delany (1700–1788), whose paper cutouts of flowers attracted professional and aristocratic attention in the 1770s and early 1780s; Frances Anne Beaufort Edgeworth (1769–1865, Richard Lovell Edgeworth's fourth wife), who produced an album of unpublished botanical watercolors; and Katherine Charteris Grey (1773–1843), who married into the Greys of Groby, the aristocratic Midlands family of the sixteenth-century Lady Jane Grey of Groby, one of Henry VIII's decapitated wives.

Each of these artists depicted plants in ways that veer off from the usual protocols for this amateur pastime: to draw plants on paper with pencil and watercolors, perhaps ink, not to look much if at all at them under a microscope, where reproductive details would be more evident, but rather focus on outline and a light watercolor wash.[49] Only one of them, Beaufort, used watercolor, but she did so by using a deeper palette and far more microscopic detail that was the norm for amateur women artists. The other two, Delany and Grey, tweaked the genre and its conventions more overtly. After decades of fine needlework, some of it with minute botanical motifs, Delany began to use bits of colored paper to create a form of botanical art playfully adrift between a heightened materiality (not just paper and paint, but paper and paper) and depiction. Grey's pressed-flower depictions of orchids turning into animals, insects, or persons made plant matter itself the means of artistic production. In different ways, each made plant matter and paint stubborn substrates that work against the grain of botanical watercolor as it was practiced and taught to amateur women (male amateurs did not dabble in botanical drawing; only male professional artists did so). The care with which these and other artists experimented with the relation between botanical illustration and its materials turns, surprisingly but perhaps inevitably, on the recognition that botanical matter and paint are remarkably good occasions for creating figures that emphasize the role of representation in the work of romantic nature.

As Michel de Certeau describes it, the transgressive wit of reusing things in unexpected or unsanctioned ways—making while appearing to "make do"—as did Beaufort, Delany, and Grey, subverts norms without kicking over fences. That wit, de Certeau suggested, may be mostly invisible from a global Foucauldian perspective, which has "the disadvantage of *not seeing* practices which are heterogeneous to it and which it represses or thinks it represses." Among those practices not seen, at least during the romantic era, by more than a very few were women's botanical illustrations, mostly though not absolutely understood to be the amateur work of lives in which women's domestic occupation and management were the work at hand. Against this backdrop, the drawings and collages of Edgeworth, Delany, and Grey convey, more or less literally, the possibilities offered by those "indeterminate trajectories" that, de Certeau says, "sketch out the guileful ruses of *different* interests and desires."[50] Each developed a competence assisted by a slightly wayward difference that survived inside a manner of living and working that was unquestionably polite, genteel, and private.

For all three, conventional advice on flower drawing as a polite female accomplishment seems to have had surprisingly little traction, yet they were certainly among those genteel women whose polite accomplishments were the class model for advice to women of middle- and low-class status. All were amateurs whose work was neither paid nor in any other way touched by the exigencies of botanical print and exhibition culture in which other women artists of the period were involved. For each of them, the choice of materials and technique—what I mean by *ground* in the title of this section—plays an arresting role. Had each chosen to work in the light, watercolor style that dominated much of British botanical illustration, their work would have been more easily assimilable, more recognizably mainstream. By working quite explicitly, even wittily, at the edges of this mainstream, each of these artists makes clear the terms of her distinctiveness. Artistic means do not constitute intentional ends, but they do signal decisiveness about what not to imitate and what to do instead. At the very least, this independence is not what the ideology of the conduct manual imagined as the best of all possible worlds—a generalized scene of domestic female occupation.

Frances Beaufort Edgeworth's family history registers several degrees of separation between the polite norm of female education and the education of girls in gentry families where learning mattered. She was the eldest of four children of Daniel August Beaufort, an Irish cartographer of Huguenot ancestry, and Mary Waller. Before she married, Frances supervised the household of her uncle Robert Waller, where she provided his great-niece Frances Anne Stewart with an education that was "more academic" than was then the norm for girls.[51] Harriet Beaufort, the elder of Frances's two younger sisters, published *Dialogues on Botany* in 1819.

The youngest of Frances's six children was Michael Pakenham Edgeworth, who became a botanist and colonial administrator in India. He attributed his facility as a botanical artist to his mother's instruction. Shortly after her marriage, she described the Edgeworth household as a community whose scientific and literary passions and affections she found compatible. They are all, she wrote, "chymists and mechanics, & lovers of literature & a more happy more accommodating more affectionate family never yet came under my observation."[52] Briefly put, Frances Beaufort Edgeworth grew up in, then married into, families whose penchant for learning and teaching matched her own.

In 1798, the year of her marriage, she began a series of natural history drawings, mostly of plants, a practice she continued until at least 1807. Each of the 101 drawings preserved in this collection is dated and the location of the specimen identified. Most are further identified by name, often several names, by their Linnaean class and order, with Latin binomial, and in some cases by a common English plant name. For some drawings she indicates a relevant entry in the first or second edition of Withering's *An Arrangement of British Plants* as her source; for others she notes questions about names and classification. The most compelling evidence of her botanical precision is provided by the drawings themselves, which often include separate, detailed representations of reproductive parts and convey her sustained interest in cryptogamic plants.

Edgeworth's drawings are meticulously detailed, showing ciliae, or root hairs, the front and back of leaves and petals, and above all color, from the blush of a begonia blossom seen whole and in section to the intense black-and-white anther and filament structure inside the fleshy, deep-red passiflora. Early in the sequence, she presents four drawings on a single sheet of two specimens of *Lichen pyxidalus*, one of several cryptogamic plants depicted in the album. Two of them present details of this plant in close-ups that render their material density in thick, dark, textured lines. Near the end of the collection, the density of watercolor and shading becomes yet more subtle and varied. Her interest in lichens (at the time grouped with other cryptogams such as ferns and algae) goes well beyond the expected range of botanical subjects. Her subtle management of light, dense coloring, and fine detail, especially apparent in her cryptogamic drawings, invites attention to the materiality of the plants she depicts. Near the end of the collection, the density of watercolor and shading becomes yet more subtle and varied.

Edgeworth's palette, technique, and penchant for discriminating detail, neither casual nor uninformed, work at the very edge of the conventions used by British botanical artists in the early nineteenth century. Her medium (watercolor) and ground or support (paper) are conventional. But her style is less so, closer to that used by Indian artists who began to draw and hand-color plants for the British in the late eighteenth century, although this resemblance is likely accidental. Unlike most British watercolorists of the period, she uses a thickened, opaque technique. Her choice of botanical subject includes plants on the taxonomic edge, orchid species, and cryptogams, plant groups that were typically said either to be too complex for amateur study (cryptogams) or too salacious (orchids) for women to study in the anti-Jacobin and antibotanical climate of the 1790s.

Mary Delany and Katherine Grey step more decisively outside those protocols, yet they do so via materials and practices that seem at first glance to remain within the expected terms for genteel women. Working with paper, as she had to make needlework patterns (what could be more ladylike, more genteel?), Delany began to create intricate paper mosaics of plant specimens, so intricate and exact that botanists she knew or met via scientific and aristocratic networks made repeated visits to what she wryly characterized as her "hortus siccus," a paper herbarium deftly evocative of real plants. Creating nearly a thousand drawings over about a decade, she developed an array of techniques to achieve the effects she sought: punched outlines, graphite modeling, papers selected from many sources, some likely from India, colored with watercolors she mixed, bits of plants folded into the layers of paper and gum Arabic to make the layering of papers cohere. Modern curatorial accounts emphasize the three-dimensional character of these plant mosaics.[53] Lisa Moore has argued that Delany's sensuousness of line and floral signal her queer affection for women and their bodies.[54]

What interests me is the materiality of Delany's plant curves: those bits of paper—sometimes as many as 150 pieces to depict a single plant blossom—which she cut or tore, with scissors or a knife and other tools, and sometimes painted with colors she mixed to get the precise color, a paper herbarium so precise, so lifelike that botanists consulted it as they would a real herbarium. Marked by the familiar/unfamiliar eccentricity of Delany's paper mosaics, startling in their odd assemblage of matter, color, and botanical detail, difference is here marked by a degree of accomplishment that is at once gendered and arrestingly unexpected because it is also resolutely original in its practice and artistic effects. If this is what female accomplishment includes, the category itself is put into question precisely because Delany's work displays a degree of skill and imagination not typically identified with women's work of the period. If nobody but a genteel woman of her era could have done such fancywork, it is also true that only a woman so skilled could have had the expertise necessary to create an art form so brilliantly mimetic of the plants it depicted. After decades of creating and executing needlework patterns that included a good deal of botanical and floral detail, Delany applied this skill to her new botanical art form with remarkable dexterity. After doing about a hundred mosaics, she worked very quickly, creating as many as thirty mosaics in a single month.

Delany's achievement was as much scientific as technical. By 1769, she had a systematic understanding of Linnaean principles and their application to British flora. Her transcription in that year of an English translation of William Hudson's 1762 *Flora Anglica* numbers 481 folio sheets, including an index, an appended list of Latin genera, and Delany's own notes on species she found at Bulstrode, where she produced this work.[55] The woman who began cutting paper flowers in her ripe old age knew what she was doing. By 1782, when she gave up this work because her eyesight was failing, she had produced about one thousand paper mosaics, most of them created with hundreds of pieces, build up layer by paper layer on a black background. To insure scientific precision, she created her paper mosaics with a botanical specimen before her, using her eye to construct a life-size depiction of the plant and its reproductive parts or using tiny pin holes to guide her application of thin layers of paper. She appears to have dissected some specimens to determine precisely what she was looking at. On the back of each image, she recorded the day, month, year and the source of the specimen and, where there were many pistils and stamens depicted, she would also record the precise number of each on the back. Delany's eye was trained to attend to Linnaean distinctions, and most of the botanical names she used were taken from Linnaeus, but because she depicted specimens of new species, some of their names were as yet unpublished. John Edmundson notes that among them were names for plants Daniel Carlsson Solander had brought back from Captain James Cook's first circumnavigation of the globe in 1768–71.[56]

Although some of Delany's early flower mosaics look very much like paper cutouts, even in those the nuance and range of colored papers and layers are arresting. Later works become so subtle and detailed that in photographs they look very much like paint: *Passiflora laurifolia*, was created with 230 pieces of paper plus a petal and the central flower section, attached to the collage after Delany had dissected the plant specimen; and in *Physalis*, or winter cherry, its netted outer husk is skeined over the fruit. Often built up with tissue like layers of paper that made the plant image perceptibly thicker on the paper, a few mosaics include bits of dried plant matter. Kohleen Reeder has found a leaf pressed to the underside of a cut-paper leaf in a 1777 depiction.[57] Several parts mimesis and more parts paper and plant, the craft of Delany's paper mosaics wittily troubles the boundary between matter and paint, paint and plant, such that they give a new meaning to the term *hortus siccus*, long used to refer to the dried plants arranged in herbaria. Dry, yes, but hardly colorless or withered.

Katherine Charteris Grey created a differently material botanical art in private albums of pressed flowers she assembled in the early 1840s, including one devoted to orchids that depict species from other kingdoms of nature.[58] Identified as "Lady

Grey of Groby" in James Bateman's *The Orchidaceae of Mexico and Guatemala* (1838–43), Katherine Charteris Grey pressed orchids not to preserve them for love notes but to create whole landscapes, some entirely without pen or ink. In the orchid album, she transformed pressed orchids into birds, a springing tiger, the witches of *Macbeth* . . . even as she identified each orchid with its Linnaean name. The swanlike orchid engraving in Bateman's *Orchidaceae* titled RARA AVIS (rare bird) is a visual pun on the image that precedes it, *Cycnoches ventricosum*, or swan orchid. In Grey's pressed-orchid album version of this image, it is titled "Cynoches [sic] Loddigensii," after the plant nurseryman Conrad Loddiges who discovered it. Grey's pressed flowers favor puns: *Pterostylis obtusa* is an Australian orchid, so named because it looks winged (*Ptero*) and has a long style. Grey has flipped the blossom, turning the long style into its legs, and placed this stork-like beast in a moist, grassy habitat similar to those that modern botanical descriptions list for this species.[59] The last Grey image, *Erica pubescens*, is not an orchid but a kind of heather, known for its soft, hairy texture. Although the name is printed under the batlike creatures flying about, the Erica may be the central "tree," which looks very much like Linnaeus's dried specimen of this species.[60]

By working in materials that were quite evidently domestic, making a kind of art that was acceptable work for women because it remained private, known only within a genteel circle or, if it reached print, did so more or less anonymously, Frances Edgeworth, Mary Delany and Katherine Grey worked just within a set of overlapping norms—for botanical illustration, for women working at home. At the same time, they also pushed subtly and wittily away from these norms. If Edgeworth used a more conventional medium for botanical illustration, she, too, was eccentric in the choices she made—plants at issue in the botanical inquiry of her day rather than ones long known and easily identified—and for the way she chose to reflect the density of color or intricacy of parts appropriate to them. Delany and Grey chose media, paper cutouts, and pressed flowers that might be discounted as women's work, had these tweaked versions of that work not been put to such meticulous use. For each, medium worked both materially and figuratively. That Grey made plant matter itself do this work was so far outside the boundaries of polite pastime that it commands notice.

Such transformations or crossovers constitute the poetic work of romantic botany, work that is highly visible and sustained in the poetry of Erasmus Darwin, Charlotte Smith, and John Clare, but present across romantic poetry. What makes the figurality of Mary Delany and Katherine Grey so arresting is its patently literal materiality. Strictly understood, figures are said to come from words, material culture from things, but this too literal account is not what goes on here and commands interest. To move from material things directly to figures is a species of wit that rightly and exuberantly calls attention to how figures work on material culture. The swerve is doubly bound here: a material medium urges the figure in a gesture that simultaneously grounds it and lets it loose—very loose.

CHARLOTTE SMITH'S BOTANICAL INDIRECTION

Charlotte Smith wrote a great deal about botany as well as birds, sometimes as a way to mark character, either pedagogically or fictionally. Indeed, the omniscient narrator and the male protagonist of the novel *The Young Philosopher* are, like Rousseau and the protagonists of his *Julie, or La nouvelle Héloise*, forever thinking about plants, a point made emphatic by Smith's insertion of her poems on botany into many chapters.[61] In the pedagogical works, *Minor Morals* (1798, 2nd ed. 1799), *Sketches of Natural History* (1798), *Conversations introducing Poetry* (1804), and the posthumous *A History of Birds* and *Beachy Head* (1807), talking about plants assists a persistent moral economy. Smith's form of moral suasion, including her use of dialogue to advance the teaching project, follows common practice in pedagogical writing of the period. What is intriguing, even curious (that favorite word of botanical notice), about Smith's use of botanical and other natural history figures is the way they seem both to step away from yet slyly authorize a line of questioning, or narrative shape or argument, where none is expected or apparently sustained.

Smith assiduously points up her botanical expertise in these works and the poems that are frequently embedded in them. In one poem she lists several species of *Erica* (i.e., heaths)—a genus whose traits and species were frequently discussed and illustrated in the botanical literature of the period.[62] In a note to another she reviews what was then a long-standing debate about the nomenclature and classification of plants as Geranium or Pelargonium.[63] Her poetic nomenclature is tactfully split between common names—which usually appear in the text—and Linnaean names, which she appended in notes. The only exceptions to this practice appear to be a few genera that were by 1800 so frequently illustrated and identified with their Latin names that Smith chose instead to put that name in the text and the common name in a note, such as clematis and arum, the latter a Linnaean genus whose classification and species were discussed and frequently illustrated in botanical magazines and books. In the event that the reader misses these cues, Smith's notes cite numerous works on botany.[64]

So much Englishness and seeming deference to Linnaean nomenclature may be strategic. Like other women writing about botany during the 1790s and shortly after, Smith presents her knowledge of botany with as much decorousness and moral fervor as seemed to be requisite after Polwhele's 1798 attack in *The Unsex'd Females* on women writers in general and, in particular, on those who taught the prurient science of botany to boys and girls. Darwin's *Botanic Garden* had already supplied the lubricious foundation for Polwhele's indictment of the "bliss botanic" among the sexual horrors women writers on botany for children could be expected to encourage.[65] As a woman writer and a teacher of natural history to children, Smith fit both profiles. Given the historical moment, Smith's several citations of Erasmus Darwin merit respect (Smith, *Poems*, 287n.).

In "A Walk in the Shrubbery" (a title that sounds like a Monty Python send-up of the English penchant for natural history), Smith presents herself demurely as a "Moralizing Botanist" who uses on personifications or pointed notes, strategies that Darwin had made so familiar in the view of his critics as to be salacious (Smith, *Poems*, 303). And if she is careful not to return to the highly sexualized scenes of his *Loves of the Plants*, her moral dicta are not always, not even in her books for children, politically orthodox. In the *Elegiac Sonnets*, she risks using "the rights of man," an incendiary, prorevolutionary phrase by 1792, to comment satirically that the death of a beggar "insulted" those rights (Smith, *Poems*, 97). In part 1 of *The Emigrants* (1793), published in the year England declared war on France, she invokes taxonomy categories to prompt readers' sympathy for "affliction's countless tribes" (*Poems*, 64, p. 137). Since the end of the seventeenth century, *tribe* had been variously used to refer to classes of plants and animals as well as human groups. Its usage as a classificatory term is instructively labile from 1640 until the mid- and late nineteenth century, when *tribe* settled down to mean a level of taxonomy inferior to order but superior to genus. Until then, it could mean a very general class, as in "the two grand tribes of vegetables are acid and alkaline" or a "tribe of birds whose habit it is to fly as a flock"; a species, or an order that contained a number of genera, roughly equivalent to A. L. de Jussieu's taxonomic use of the term *family* in 1789 to supplant the Linnaean genera as the largest category and division in plant and animal classification.[66] At the end of an extended description of the flowers and birds of her native Sussex in part 2 of *The Emigrants*, Smith issues a sharp critique of recent violence in revolutionary France. Turning back to those who inhabit "this sea-fenc'd isle," the speaker pointedly asks: "What is the promise of the infant year" to those who "Survey, in neighbouring countries, scenes that make / The sick heart shudder; and the Man, who thinks, / Blush for his species?" (*The Emigrants, Book 2, II*, 62–68, in *Poems*, p. 151). By shifting her taxonomic language from *tribes* to same *species*, the speaker nudges the English reader to recognize the French not as some other tribe, but as the same species. For the still unsettled taxonomic location of *tribe* during Smith's lifetime implies that like families tribes may, as Jussieu and others suggested after 1789, have affinities between as well as within these groups.

The taxonomic variability Smith implies here and elsewhere in her poems persistently works a relay between the minutiae of natural history and the human moral lessons those minutiae illustrate. This strategy is especially evident in *Minor Morals,* a series of dialogues between an aunt and her nieces and nephews in which the microscopic observation of botanical subjects prompts the putatively "minor morals" of her title. In the first dialogue, the aunt explains how the underside of a moth looks under a microscope—"an infinite number of feathered quills" more exquisitely detailed than "the finest miniature."[67] She then observes, with an air of digressiveness that I read as part of her rhetorical cover, that the study of botany wards off indolence. The dialogue closes with a story about other children who, having returned from a British colony in the care of a black slave, are at once demanding and dependent little tyrants whose characters have been warped by indolence. The moral scrutiny Smith invites with this tale would translate the minute inspection of a moth into an equally fine moral microscopic capable of recognizing how—to give the implicit moral of this tale its largest geopolitical argument—slavery and servitude weaken the children of masters.

The indirection that here marks Smith's drawing of her moral also guides the narrative logic of "To the fire-fly of Jamaica, seen in a collection," a poem she first published in *Conversations introducing Poetry* (*Poems,* 204–7). In mid poem, the speaker's meditation on an impaled firefly shifts, first to a slave captive who cannot escape and so cannot welcome this firefly to light his way, then to a Naturalist (so capitalized), and finally to a meditation on the difference between Ostentation and Friendship (also capitalized). The poem is a curious poetic performance, encased in intricate rhyme schemes and stanzaic patterns that are, far more than unrhymed iambic pentameter, Smith's preferred poetic rhythm. Although these formal devices call attention to the surface of the argument, the analogies the poem draws between the firefly, the slave, and the Naturalist work, as it were, under poetic cover, but with an assist from Smith's notes, where Latinate names for flora and fauna alternate with notes that describe the suffering of slaves in the colonies.

The poem's concluding stanza about the vanity of ostentatious display and loyal friendship, surely a charge to be laid at the feet of the Naturalist and not the firefly so pinned, scats away from the more politically charged moral embedded in the speaker's odd narrative relay. Whereas some antiabolitionists claimed that Africans belonged to another, lesser species, Smith's taxonomic distinctions are offered without a difference for individuals: whether a dead firefly, a slave, or a Naturalist, all are linked in a chain of being that leads, obliquely enough, to a closing moral about the vanity of ostentatious display. This lesson looks like a cover to deflect attention from the poem's less overt notice of commonalities that work against using taxonomic classification to pin others to a grid.

"To a Geranium," also included in Smith's *Conversations,* addresses a species of geranium that is native to Africa but, once "naturalized in foreign earth" (*Poems,*

l. 21, p. 209), blooms during the English winter. An extended note about geraniums and pelargoniums sketches an argument that goes beyond botanical enthusiasm for naturalized exotic flowers. The numerous varieties of geranium (or pelargonium), the note explains, include native and exotic species. The latter are "crush'd" by "Caffres" in South Africa, but survive first in the brilliant luxury of an English conservatory, and then in the home of the speaker, to cheer her "like friends in adverse fortune true" (*Poems*, 28, p. 209). "To a Geranium" is not an antislavery poem; indeed, its brief notice of South African "Caffres" who crush exotic blooms may be offered in unflattering contrast to those English botanists who cherish them. Yet Smith's phrase "naturalized in foreign earth" runs against the grain of this implied contrast. The "foreign earth" here is English soil and hearth, which have until this line constituted the perspective from which the speaker offers her admiration for exotic geraniums in English hothouses. Although slightly marked, this shift in perspective works in tandem with the global vantage point suggested by the note, where the genus *Geranium* is presented as inclusive of native and exotic species. Like *Minor Morals*, these conversations between one Mrs. Talbot and her children, at once lessons about natural history and about poetic forms and usage, imagine these enterprises as both taxonomic and figurative markers for a moral order in the forms of knowledge they offer.

Smith's "Flora," first published in *Conversations Introducing Poetry* and reissued in the posthumous *Beachy Head, Fables, and Other Poems*, presents an array of personified insects and flowers who guard Flora, among them tiny sylphs, called "pigmy warriors," who use lichen fronds as shields and weapons (*Poems*, 50–78, pp. 281–82). Among the trees that recognize Flora's "soft sway" over them is the Oak, whose "giant produce may command the World" (*Poems*, 132–34, p. 284). Stuart Curran's note specifies the allusion here: oak was the mainstay of English naval shipbuilding, a growth industry throughout the 1790s and the Napoleonic Wars (*Poems*, 284n.). I wonder about this military strain in a poem whose luxurious botanical catalogue is seducing, perhaps artfully distracting. Like *The Emigrants*, whose date of composition is given as November 1792, two months after the September Massacres, "Flora" was published when English naval vigilance against the French navy was a constant topic in the press and in private, with anxious debates about possible French landings on English coasts, among them the Sussex coast. Placed beside hints in *The Emigrants* and *Beachy Head*, poems that specify natural and geological histories as exempla that urge some future reconciliation between England and the Continent to which it was once joined, the miniature, then giant battle array of Smith's "Flora" may pick up the thread of an argument that runs aslant this poetic brief for the realm of Flora. In his retrospective notice of Smith's use of botanical figures in the poem, John Clare quotes an enjambed description of the flag iris, "Amid its *waving swords*, in flaming gold / The iris towers," that belongs to an overtone series in which danger and the possibility of violence is figured in plant and bird names like "Adder's-tongue" (Smith gives the synonym "Hart's-tongue" in a note) and "Arrowhead" (*Poems*, 156–57, 285).[68]

The more telling taxonomic risk of these names and others like Reed bird (a bird, not a plant) and Crowfoot (a plant not a bird)—that these names cross from one phylogenetic kingdom to another—is hardly unique to "Flora," but telling because Smith lingers over them. In a note appended to an earlier notice of the plant Hypericum, she explains its Latin name by quoting lines from Cowper's *The Task*: "Hypericum all bloom, so thick a swarm / Of flowers like flies clothing her slender rods / That scarce a leaf appears."[69] Prompted by the flower-to-flies analogy, Smith expands it to create a personified figure of a female flower whose slender stalk is so covered that its leaves are nearly all invisible. And Smith is only getting started. "The Horloge of the Fields," a botanical survey of flowers that close and open at different times of day, mentions "Goatsbeard" (*Poems*, 38, p. 297) and "Wheatear." The title of a poem in *Conversations Introducing Poetry* is a bird's name with a double cross to plants and human anatomy (*Poems*, 194–96). Clare was most interested in what Smith did with orchises, a native English genus of orchids. In the "The Kalendar of Flora," first published in *Minor Morals* and reissued in *Conversations introducing Poetry* as "Wildflowers," the botanical catalog includes "the Orchis race with varied beauty charm, [to] mock the exploring bee, or fly's aerial form" (*Poems*, 23–24, p. 191). Treated as a plural subject, "Orchis race" includes both the species she names in a note, the "Fly and Bee Orchis," and other species, including one whose Linnaean classification even Sir James Smith had disputed in *English Botany*. In the same poem, she lists *Arum maculatum*, the plant whose Linnaean classification and morphology Maria Jackson, Sir J. E. Smith, and many others continued to dispute into the nineteenth century.[70]

Read against the era's sustained, at times almost fevered, efforts to classify and name plants, Charlotte Smith's figurative attention to plant names that appear to wander off into another kingdom of nature is at once playful and instructive about how such figures can and do wander, despite being, as it were, tethered to a material nature. In *Beachy Head*, the posthumously published narrative poem that now seems among the most puzzling and compelling of Smith's poems, the category slippage of orchises participates in a series of narrative deflections that in many ways constitutes the evasive argument of this poem, which begins with an extended reflection on the geological event that separated England from France, then entertains a series of unpursued narrative directions and arguments, from the image of the ship of commerce plying the ocean in imperial trade and exploitation to the suffering of many peasants struggling to survive near Beachy Head, at times engaged in smuggling contraband, an activity the speaker laments, to speculation about why shells appear on its summit—speculation that is coyly and quickly dropped as the speaker meditates rather repressively on the toils and snare of Ambition, then turns from "thoughts . . . / By human crimes suggested" to local and natural history, settling her attention on plants that mimic, and mock species that belong to other kingdoms of nature (*Poems*, 439–40, p. 235). The gesture looks, and probably is, something of a retreat from the grander geological and human histories with which the poem has thus far been unsteadily concerned.

Yet the path of this retreat is informative about the poem's experimental argument by analogy: its insufficiencies, its surprising reach, and its mimetic ambition. For although Smith twice in this poem pushes back at Ambition, whether in the figure of Napoleon as tyrant or, yet more abstractly, as the work of War, *Beachy Head* is an ambitious poem, with epic designs on nature and human history. It is also a poem that self-consciously resists those designs. What interests me here is the role that botanical and other natural history figures play in assisting and limiting the reach of poetic ambition in this last poem, where the mimicry suggested by plant names may obliquely register a larger claim that the poetic imitation of nature and human history is what poets, including this woman poet, can do. Although Smith's poetic experiment invokes a common practice of constructing analogies from nature, *Beachy Head* is altogether singular in the way its conducts itself toward, then partly away from, this poetic task. It is as if Smith recognizes in the work of such figures a diversionary line of march that gives so much over to conjecture that it becomes difficult to rein it in. That difficulty is formal as well as conceptual, in ways that Darwin's *Botanic Garden* had already made visible in its spinning of affiliations between verse and notes. In *Beachy Head*, Smith is more cautious, yet its more muted dance of verse and notes also includes moments when history—human or natural—spins off. At such moments, conjecture becomes the textual mark of too much figurative free play, and Smith's narrator draws back. Yet precisely because the poem has already made clear its formal and analogical ambitions, this pulling back marks their limit, not their disappearance.

The botanical vignette that specifies the wider work of mimicry and analogy in the poem begins with a gift to the otherwise unidentified "wanderer of the hills":

> shepherd girls
> Will from among the fescue bring him flowers,
> Of wondrous mockery; some resembling bees
> In velvet vest, intent on their sweet toil,
> While others mimic flies.
>
> (*Poems*, 443–46, pp. 235–36)

Smith's notes provide the relevant details: Orchis apifera, or Bee ophrys/orchis, and Ophrys muscifera, or Fly orchis, like several other species in this genus look like the insects or persons for which they are named. Smith's note quotes Sir James Smith in *English Botany*: "Linnaeus, misled by the variations to which some of this tribe are really subject, has perhaps too rashly esteemed all those which resemble insects, as forming only one species, which he terms Ophrys insectifera."[71] Faced with proliferating and hybridizing *Orchis* species that look like many different insects, Linnaeus had tried to reduce them to a single species, attended by variants with which his systematic is not in the end much concerned. In Smith's poem, the flower-to-insect cross that these two species introduce leads more or less directly to two others, bird's foot trefoil and hawkweed, the latter identified in a note as, inevitably, "many sorts." In between these, the "mountain thyme" that "purples the hassock of the heaving mole" prompts a much longer note in which Smith argues that sheep that eat wild thyme do not taste better for doing so, as is commonly supposed, but because the soil and growing conditions that favor thyme also favor sweet pasturage. Mimicry, animal-to-plant affinities claimed or disclaimed—all this marks the ease with which one set of analogies invites another.

Yet this is not sanctioned ease, a point Smith obliquely makes in a long note triggered by the image of listening to "evening sounds," among them those of the "night-jar, chasing fern-flies" (*Poems*, 239n.). Citing John Aikin, brother to Anna Barbauld and the author of *An Essay on the Application of Natural History to Poetry* (1777), on the hum of the "Dor Beetle (*Scaraboeus stercorarius*)," an analogy justified only by the fact that it too makes evening sounds, Smith launches a reflection that is astonishing for its meandering among common names of birds and insects that she involves in a train of associations. Smith's wild mix of scientifically precise description and a careening list of alternative common names and local prejudices blends what Aikin had argued ought to be kept apart. Here is Smith, continuing on:

> I remember only one instance in which the most remarkable, though by no means uncommon noise, of the Fern Owl, or Goatsucker, is mentioned. It is called the Night Hawk, the Jar Bird, the Churn Owl, and the Fern Owl, from its feeding on the *Scaraboeus solstitialis*, or Fern Chafer, which it catches while on the wing with its claws, the middle toe of which is long and curiously serrated, on purpose to hold them. It was this bird that was intended to be described in the Forty-second Sonnet [in Smith's *Sonnets*]. I was mistaken in supposing it as visible in November; it is a migrant, and leaves this country in August. I had often seen and heard it, but I did not then know its name or history. It is called Goatsucker (*Caprimulgus*), from a strange prejudice taken against it by the Italians, who assert that it sucks their goats; and the peasants of England still believe that

> a disease in the backs of their cattle, occasioned by a fly, which deposits its egg under the skin, and raises a boil, sometimes fatal to calves, is the work of this bird, which they call a Puckeridge. Nothing can convince them that their beasts are not injured by this bird, which they therefore hold in abhorrence. (*Poems*, 239n.)

Fern Owls, or Fern Chafers, kingdom-crossed names, are the least of the imaginative excursions this text makes available as it swings to superstitious Italians and back to English peasants, then loops back to a sonnet Smith had written earlier to correct an error about the season in which this bird of many names appears. What bird did appear in November in Smith's 42nd sonnet if it was not the night hawk, now known to be goatsucker bird? The stream of names and seasons and associations Smith unleashes is foreign, well beyond the conventionalities of Aikin's argument that in poems about natural history scientific precision is preferable to extravagant figures drawn from nature (and specifically not from "vegetable nature"). It is difficult to construe Smith's allusive practice as at least as calculating and satiric as it is free-wheeling.[72]

Smith's note, referring as it does to lines that describe what the "tir'd hind" sees or hears as he goes home at night, including the sounds of smugglers "Who sought bye paths with their clandestine load" (*Poems*, 517, p. 289) may also speak to the metanarrative logic in *Beachy Head* whereby different, yet putatively contiguous, narrative paths get taken up, put aside as hazardous, then smuggled back in, as though glimpsed rather that overtly pursued. In the end, Aikin's extended meditation on the strange, quasi-mythical large beasts described in ancient poetry, which occurs in the section of his *Essay* that Smith cites, may be the very hook on which she hangs her account of those "wondering hinds" who imagined that "enormous bones" were those of giants that once wandered the hills (*Poems*, 417–18, p. 234).

The narrative eccentricity of Smith's *Beachy Head*, at once ratified and diverted by botanical figures and notes, marks the most surprising intervention of botanical figure in Smith's poetry, surprising in part because it rises to levels of indirection beyond anything Smith offers in other poems that feature botanical or natural history figures. To urge that this is so by default, that the narrative oddity of the poem arises from the fact that Smith died without completing it, may miss the mark. Making such a case would also logically involve making a separate case from that for Byron's *Don Juan* or Keats's *Hyperion* poems or any of those long romantic poems that look unfinished. Smith's figure of "wondrous mockery" reaches to the quasi-epic, then anti-epic and quasi-pastoral performance of this poem, from its "ship of commerce" echo of Milton's famous epic simile for Satan looking like a ship afar off, hanging on the horizon, to the strange, perhaps truncated ending lines, which offer an inscription for a hermit found dead that may as likely be an epitaph for the reach of epic and other genres (pastoral, ballad, and so on) as an epitaph for Smith herself.

I have argued elsewhere that the poem's tendency to scat-sing its way through and outside epic conveys a genuinely unsettled relation to the epic task of telling a history. Smith's strategic use of botanical figures and notes that go elsewhere may speak to wider generic concerns, marked by an unease with the work of argument and closure that genres help to encode. From this perspective, the use of such figures marks an errant figurality that moves on, slivering through names and identities that will not allow the speaker or reader a resting place or conventional sense of arrival. If Smith here "looks away" from formal resources she might have earlier used with more ease, she does so by making another eccentric path in which botanical figures help her question narrative and pedagogical sobriety.[73] Reading bones, reading plants, may require precisely the aggrandizement of imagination that those untaught hinds (and most geological thinking of her time) were already doing about shells and giants.

The work of differentiation I have emphasized in the way that Smith, as a poet, and other women writers and artists do botany turns repeatedly on their versions of the relationship that Deleuze recognizes between difference and repetition, wherein what seems to repeat does not. This outcome, which I have argued is also the underlying engine of botanical differentiation, works like a saw, widening the sense of differences even as similarity or sameness—in this instance the fact that women were expected to talk about, garden, and draw plants as much as look like flowers—is the work at hand, whether openly declared or seeming to be the work of all female hands.

CHAPTER 5

Clare's Commonable Plants

BEATING THE BOUNDS

> I hunted curious flowers in raptures and mutterd thoughts in their praise I lovd the pasture with its rushes and thistles and sheep tracks I adord the wild marshy fen with its solitary hernshaw sweeing along in its mellan[c]holy sky I wandered the heath in raptures among the rabbit burrows and golden blossomd furze I dropt down on a thymy mole hill or mossy eminence to survey the summer landscape as full of raptures as now.[1]

> When I think about my unweaned experience of the Chilterns, and what I made of it in my writing, they both have the air of ritual acts of possession. There were fixities in it, a kind of vocabulary of place: a valley with woe-water, supposed only to flow in time of trouble; bluebell woods and dragon trees; the ludicrously romantic lynchgate to my old school. But it was the rites and ceremonies through which I folded this landscape into my young self that counted then. The marking of the first swifts, over a particular meadow on May Day; the libations poured on bee orchids; an obsessive walk I took maybe three times a week, following the same route, touching trees, beating my own bounds.[2]

For John Clare in the 1820s and Richard Mabey writing 150 years later, thinking about plants means, as Mabey puts it in the second of these passages, "beating the bounds" to mark how plants are distinctive in themselves and in habitat. Mabey's recognition that he made, rather than found, his sense of place in the Chilterns where he grew up makes explicit a strategy in Clare's writing about poetics and natural history. The lines from Clare quoted above describe plants in brief vignettes that are set off by extra spaces rather than end punctuation and are then looped together by "rapture," repeated in different syntactic environments. In Clare's writing this pattern recurs in multiple, echoing variations in poem after poem, a calling forth or calling out that often names plants as the occasion and work of verse, as in the April section of his *Shepherd's Calendar*, where the order of plants is appropriate to the season as well as the place: "the first fair cows lip," violets, daisies, crowflowers, primrose, and the "first broad arum leaves," stitched together by a repeated, echoing "first."[3]

Although plants are not the only things Clare chronicles in this way, they tend to anchor an interactive, shifting poetic map that also includes places, events, persons, and creatures. Vocality shapes the singularity of this poetic address: its orality, repetition, dialect words and syntax, and punctuation impede or hasten the speaking voice. These features in turn mark the poetic locations he gives to places, but also to persons, plants, animals, and practices.[4] They are Clare's way of "beating the bounds."

Clare's botanical poetics does not seek to disable the effects of land enclosure that so troubled him and continues to exercise his modern readers. Instead this poetics creates a counterrhythm to enclosure that habitually imagines itself in opposition to the rigid boundaries and prohibitions that enclosure produced. As poetic speech might be expected to do, and as the traditions of oral poetry had long assumed, this speaking from and about plants, birds, and the other places and things implies a speaking to that goes on and loops back across many poems in its modes of address as the work of the poetic voice. This, no more but no less, is what impels his poetics of natural history and place.

Clare's diction is *common* in every sense of the term available in the early nineteenth century, beginning with those that Raymond Williams offers: popular, vulgar, and lower-class.[5] These meanings engage two others in Clare's writing, each with its own lexicon: commonability, or common right, the system of land use that enclosure effectively erased, and the use of common plant names that Linnaeus's invention of the Latin binomial system was intended to replace. Mabey's "bluebell woods," "woe water," and "dragon trees," like Clare's "hernshaw," "furze," and "thymy mole hills," belong to the wordhoard of English place- and plant names that had for centuries varied from region to region. In the late seventeenth century, the English taxonomist John Ray recorded many of them, which he solicited from correspondents across England. Many since, including Clare and Mabey, have kept these names in circulation and in print, as if loathe to give up any of the linguistic variability that Linnaeus had been eager to disperse, along with variations below or troublesome to the species category.[6]

Mabey's phrase "beating the bounds" invokes an annual ritual in which parishioners walked around the boundaries of parish common lands to memorize them, using willow sticks to beat the boundaries, before enclosure, whereafter ordinance maps indicated boundaries between common or parish and privately owned land. In "The Opening of the Pasture," Clare describes this event from the perspective of someone who stood to gain from access to cut pasture land and reopened fen land.[7] It was not only the bounds that were beaten with willow sticks: children, usually boys, were beaten at key markers so that they would remember them for the future of the community and in case a sly neighbor shifted a marker—for any number of reasons: to increase his own land, to put poor inhabitants "into" the care of another parish, and so on.[8] When Mabey beats the bounds of his walks, he assumes the right to create his own boundaries for his own purposes, remembering them with his prose, not because of welts on his back. As he suggests in *Nature Cure*, those boundaries are loosened, together with a sense that boundaries matter, when he moves to East Anglia from the Chilterns of his childhood. In the watery fen lands of his new home, a possessive view of the landscape in terms of fixed boundaries is futile because the tides dissolve contours and boundaries on a daily basis. Nonetheless, he does buy land there, and builds a house, but now with a more fluid understanding of where home might be.[9]

Viewed against Mabey's ability to move on and settle in once again, even in a landscape where boundaries can shift in a day, Clare's reiteration of his childhood attachment to Helpston parish looks nostalgic. It seems that he either would not or could not let go, although it might have been easier for him had he been able to go on to Northborough, just three miles away, much earlier, as he finally did in 1832. The paradoxical logic of Clare's poetic adhesiveness to place warrants notice. He could not let the landscape of his childhood in Helpston go for at least two, contradictory reasons: it was, as an article of possession, beyond his grasp, already enclosed and owned by others; precisely for that reason, to "own" it would require a different strategy for beating its bounds. To the loss of commonable, economic, and social particularities destroyed by enclosure, Clare responds with a poetic speech that is resolutely, stubbornly common in its presentations of local natural history, human habitation, sociability, and exclusion as singularities netted together by Clare's writing.

To mark singularity, Clare pushes the instability and variability of common names of plants to the limit and beyond, insisting more than occasionally on differences in spelling that preserve, or simulate, voice and dialect pronunciation. Consider just one set of examples, the names he gives to the cowslip, as it is now usually called: Clare uses, for no apparent lexical reason: "cow slip," "cows lip," "cow slap," "cow slop," "cows lap," and assorted plural forms.[10] A cow in the act of slipping is no more the lip of a cow that it is the slap of one and none is anything like the lap of a cow. As a language and a grammar, Clare's commonable plants do not obey the rules, preferring to preserve variability on the hoof. The poetic as well as taxonomic logic of this variability objects loudly, impolitely, to rigid systems for naming plants and writing correctly. Eric Miller has rightly suggested that Clare's talk of grammar and restriction moves easily from Linnaean taxonomy to enclosure because both are systems for mapping names for things that brook no departures, no localism, no unsettling particularities, and, I would add, no singular voices.[11]

Whereas Linnaean names assume fixed positions in a systematic grid, Clare delights in plant names that seem ready to fly or creep off toward some other kingdom of nature, among them "crowflower" and the orchis variety that Clare preferred to call "pouch lipd cuckoo bud." The systematic relationships imagined by such names convey another way to structure commonable practices and use land without enclosing it, such that the right to exist and live there might be extended even to those species that seemed to exist at the limits of what is held in common. The singularities of voice and name call to Clare both for what they are and as markers for his poetic calling.

Joseph Henderson, who worked for the Fitzwilliams as the head gardener at Milton Hall, just west of the fens, in the 1820s and early 1830s, insisted that Clare could spot plants that no one else saw. In June 1827, Henderson wrote to Clare, "I recd your Note and the Basket this morning . . . in the first place where the Devil did you find the fly orchis for the muscifera it certainly is I have never seen it before in this neighbourhood"; and in July 1830, "My Dear Clare, I am delighted with the orchises you sent this morning. The two small ones with whitish flowers, I believe to be O. albida a species I have never found here. I have not yet examined it, but I have no doubt it will turn out to be that Sp. The plants of Apifera are splendid. Where did you get them?"[12] Clare's was a seeing from among, a seeing in which the seer was equally unusual and particular. It was also, I think, a relentless and obsessive looking, as if to ward off a sense that he is at the very fragile edge of the human community that lives on, off, or near the land, much as Clare's gypsies do as they move across common and private lands, speak in their own dialect. Like them, Clare is an outlier who cannot be assimilated within a system, be it taxonomic or commonable.

As Clare canvasses and recanvasses the Helpston landscape, cutting across its enclosed lands, emphasizing the portions of fen and waste that were open and the rhythm of planting and grazing on fields that were agricultural commons, he repeats the names of flowers he finds in the fields, season by season, poem by poem, sometimes pun by pun. This poetic mapping of place and speech implicitly challenges the published maps that made the local effects of the enclosure acts visible on paper as well as on the land. A modern census of enclosure maps notes that those for Northamptonshire, among them an 1819 map for Helpstone (an alternative spelling then in use) and adjacent parishes, "are generally more informative than the average . . . in their portrayal of foot and bridle paths, turnpike roads, quarries and pits, windmills, inhabited and uninhabited buildings, woodland, owners and acreages of old enclosures, field names, and owners and acreages of new enclosures."[13]

Like the annual practice of beating the bounds of common or parish land, Clare's poetry is of necessity and like commonable practice recursive, returning again and again to a pun on *poesy* as floral bouquets and poetry in these lines from different poems:

> & now I meet a stoven full
> Of clinging wood bines all in flower
> They look so rich & beautiful
> Though loath to spoil so sweet a bower
> My fingers hitch to pull them down
> To take a handful to the town
>
> So then I mix their showy blooms
> With many pleasant looking things
> & fern leaves in my poesy comes
> & then so beautifully clings
> The heart leaved briony round the tree
> It too must in a poesy be.
> ("Walks in the Woods," *MP* 3: 572, ll. 127–38)

> Nor could I pass the thistle bye
> But paused & thought it could not be
> A weed in natures poesy.
> ("The Progress of Rhyme," *MP* 3: 495, ll. 90–92)

Although the pun is hardly Clare's invention, and he seems never to have tired of using it, he at times makes a more sustained poetic use of it than he does in the second passage, which insists that wild flowers and weeds are among the vulgar names and plants that belong in his poetics. In the first passage, the pun anchors a widening figural net in which "clinging wood bines" launch an ambiguous syntactic looping in which the verb *clings* has *poesy* as its subject and either takes *the heart leaved briony round the tree* as its object or turns the figure once more by aligning the way the briony wraps round (another clinging vine) the tree to its necessary winding in a poem ("It too must in a poesy be"). In his willingness to recycle the same pun, Clare is as willing to risk truculence, as he does here:

> My wild field catalogue of flowers
> Grows in my ryhmes as thick as showers
> Tedious & long as they may be
> To some they never weary me
> ("May," *Shepherd's Calendar*, *MP* 1: 65, ll. 193–96)

Although Clare's early editors and some of his friends objected to such repetition, it marks the intensity of his poetic focus on things in ways that recall Wordsworth's better-known brief for repetition in his 1842 note to "The Thorn."[14] Clare simulates the topographic inscription of enclosure as a repeated process in language, instead of a completed legal statute imposed on the land. With this substitution, Clare counterfeits the logic of enclosure and overrides it. This strategy is less nostalgic than it is insistent on recursive difference as a principle of poetic figure and thought. In a very odd sense, if enclosure had not existed, Clare might have had to invent it so that he could unravel its claim on thought, thereby effecting what Deleuze calls (repeating after Proust) the work of virtual repetition: "'real without being actual, ideal without being abstract'; and symbolic without being fictional."[15] This outcome is figurative in the way that Clare prefers: its figurality is anchored in common names, so attached that it hazards only the degree of abstraction needed for figure, wary of pulling far from its figurative ground.

Peopling the landscape with shepherds, schoolboys, a village "doctress" who specializes in curative plants, and others, Clare replays and refigures the sociable, inhabited landscape of his youth. This inclination goes deep: sociability provides a needed and expansive ground for his identity as a poet who writes about commonable land and plants. Many years before the asylum years when he imagined that he was Byron or Burns, Clare published poems that he "fathered" on other authors, creating, as Alan Vardy suggests, a circle of writers so sociable that he could speak for them and they for him.[16] Critics have disagreed about whether a parallel ventriloquism in the later asylum poetry registers delusion or simply boredom with the sameness and restriction of asylum life. Early and late, Clare throws his voice to create a diverse company, much as he does in the "Journey out of Essex," which begins with gypsies who help and an imagined "army" quickly transformed into "troops."[17] In the 1830s, his capacity to trust in his sense of a companionable nature and society faltered or ran aground in a cycle of depression that probably had to do—to a degree that we cannot know—with the friction between his poetic and writerly ambitions and decades of economic want that went beyond rural poverty to the edge of scarcity. Lord Fitzwilliam's 1831 offer of a cottage to rent in Northborough came too late or was too far away or both. Between 1820 and 1840, when Clare wrote much of his major poetry and a great deal as well about ambition, fame, want, and natural history, his friendships gave him reason to think that a different commonability remained possible after enclosure. That he wrote to so many friends for so long suggests that he understood the task of commonability as his own.

SITUATEDNESS

Timothy Morton's phrase "ambient poetics" captures the double-jointed amble of Clare's situatedness as he walks and wanders to secure his ambience, as both his surroundings and his movable location.[18] The humming songlines that Aborigines perform as they travel across lands to identify and recognize who or what is there suggests what Clare seeks to do as he writes about plants, birds, and animals where he finds them. That the final act of enclosure to affect Helpston occurred in 1819, the same year Clare began negotiations to publish his first volume of poems, is not, I think, accidental. For Clare to become the romantic poet of place, he had to reckon with the massive displacement of those who had no private land from lands they understood to be held in common. His botanical poetics respecifies, as a recursive event, a different version of commonable practices that had identified who could do what when and where on land around Helpston, both that held in common and that made available for common use on a more or less regulated basis.

For David Simpson, who has written often about Clare and more recently about situatedness as the peril of localism and the liability of speaking from a place, Clare's nostalgic adherence to a place and its particulars is a problem, so much so that Simpson avoids it.[19] Yet it may be premature to write off Clare's nostalgia. Kevis Goodman has argued that in the romantic era, nostalgia was understood to be a disease of body and mind, attributed to being far from home (soldiers and sailors were among those often so diagnosed) or being displaced by economic hardship and war.[20] Clare had, then, good reason to be nostalgic in the romantic sense: the enclosure of Helpston was one of many internal events that forced economic and social displacements and realignments as work in towns supplied what rural agriculture without benefit of common lands could not, unless one became a wage earner working for a local farmer, which Clare and his contemporaries did without enthusiasm because the pay was low and lacked the opportunities for pleasure (and minor theft) that common lands had offered.

For the adult Clare who recalls his youthful excitement about "going out of my knowledge," things had changed.[21] Now there was more risk than anticipation of becoming—as he writes of his return to Helpston after escaping from the asylum for the mentally ill in Epping Forest—"homeless at home," so deracinated that home seemed somewhere else or nonexistent. This sharply internalized version of dislocation invokes the inward turn Goodman traces in romantic nostalgia. As a poet, Clare pursued a more flexible version of the inhabited, speaking, and commonable practices that he knew as a youth that might also keep its exclusions in mind. The question of what can be said for localism and particularity as the watchwords for situatedness needs to address how Clare crafted a poetics that was indebted to commonable practices and even to enclosure, as it moved along on its own transgressive, figurative track.

Sustaining the constructedness in language of a poetic and writing self is always a balancing art for readers of romantic poets, for whom it might be narrowly said that modern subjectivity was invented. The task is especially difficult for readers of Clare, whose life circumstances invite a pathos that tends to efface or override questions of craft and poetic deliberation. Although wary of a phantasmatic originary society behind Heidegger's situatedness, Fredric Jameson also urges that subjects be understood in terms of how they operate in but also on their worlds, in ways that Michel de Certeau and Bruno Latour have described.[22] Observations and practices, including literary practices, may on this view be said to constitute our access to who subjects are as we watch them at work, at play. Simpson's argument about situatedness is, to be sure, differently tuned: he is impatient with the tendency to substitute speaking from one's subject position for a striving for knowledge that is aimed toward something beyond oneself or one's cultural community. Yet the risk of this polemic is precisely its desire to jettison the singularity and particularity that Clare spoke for and that he used to speak against a tendency toward a systematicity that is pitched against difference.

Clare's friendships, including those with Helpston neighbors with whom he collected plants, worked, or drank as well as the wider network of friends and patrons that developed as he began to publish poems, suggest he had ample opportunity for recognizing a commonability that could and did accommodate individual differences, although the limit of that accommodation is also marked by the figure of the gypsy in his poems. As he corresponded with those that lived beyond Helpston and began to write about local friends and habits from childhood, Clare managed the differences in rank and conviction among his correspondents with acuity and tact. He knew with whom he could be direct and with whom he needed to be circumspect. He registered occasional impatience with some of his patrons and friends in franker letters to others. The evangelical Lord Radstock, for example, urged that he was Clare's friend and surely meant it, but his reforming religious temperament put Clare on edge. The effusive Mrs. Emmerson praised Clare's poetry, but then urged him to adopt her poetic advice. She also tried to direct his conduct vis-à-vis Radstock, whose rank she kept firmly in mind. Clare was astute about the gains and liabilities of these friendships. He found in Marianne Marsh, a correspondent with whom he could be thoughtful and politically acute, someone whose view of the problems that enclosure had produced were not those of her husband, Herbert Marsh, bishop of Peterborough, whose office required orthodoxy.[23]

I understand Clare's poetic sociality as a version of what Martin Heidegger calls situatedness (*Befindlichkeit*). Its moods or "attunements" (*Stimmungen*), Sianne Ngai has argued, show how a speaker's voice inhabits a register, or what Raymond Williams called "a structure of feeling" that extends beyond the personal and subjective to a social or public situatedness. As a poet, Clare often traversed the space between being solitary, walking alone, and sociability, as he invented poetic speakers and characters who recall what it meant to live in a parish bound by common right. In suggestive ways, this transit is what commonability had also required: a sense of what individuals could and could not do vis-à-vis common lands and rights. Heidegger's specific sense of the tenuousness of situatedness, the way it emerges at moments when speakers feel that they are just barely hanging on to a place that is social, cultural, and physical, is especially apt.[24] The risk that Heidegger alludes to when he suggests that one's being is delivered over to the world registers Clare's position as common lands are enclosed and he is thrown against his will into the task of crafting his sense of place by identifying and marking coordinates, be they people, birds, plants, or trees.

COMMONABILITY

Common land and right vanished, juridically speaking, with enclosure—a legal and parliamentary mandated process that accelerated from 1750 on and was virtually complete across England by the early nineteenth century.[25] Common right governed both the access commoners had to lands that were farmed in the "open field" system and to lands that were open waste, fen, or forest, including private forests. Because gypsies and other wayfaring strangers were not and could not be commoners, anything they took from common land constituted theft. Even for commoners, there were restrictions that Gary Snyder misses in this reflection: "The commons is a curious and elegant institution within which human beings once lived free political lives while weaving through natural systems. The commons is a level of organisation of human society that includes the non-human."[26] Although *commonability* describes a set of practices that involved the nonhuman—one cannot otherwise imagine Swordy Well as the speaker of Clare's poem—the freedom commoners had was regulated by a complex system of unrestricted and restricted land use.

As an unlanded laborer whose common rights came with the cottage he lived in and thus had to be shared among several families, Clare lived near the edge of common rights and was apparently willing, like the gypsies whose rogue activities he found congenial, to pilfer where he could, cantankerous about village agents like the pinder (sometimes spelled *pindar*) whose job was to nab those who didn't follow the rules.[27] As someone who lived, whose family had always lived, on the edges of commonability, Clare was readier than the more privileged farmer or magistrate or local enforcer to take profit from the half-lawful edge of commonable practices, like someone (there were apparently many) who exercised the right to pick up fallen wood in a lord manor's forest and then illegally bagged a few rabbits or unfallen branches along the way. Yet he did not agree with radical protesters to enclosure who argued that all lands should be held in common, not just those that had been enclosed.[28] His account of what was left of that right in "The Bounty of Providence" (and not the bounty of Parliament), dated 1832, seems by comparison meager: "Though money nor land fortune ever left me / Yet Ive that which belongs as a free common right / to us all. . . . To see the sun . . . [to] go out in the meadows & see / The healthy green grass."[29] Clare's edgy, poetic version of commonable practices is linguistic and figurative: capturing word and names that launch possibilities not precisely anchored to the letter of a public syntax or grammar.

Deliberations about the rights and limits of commonability were local rather than national or international. Who had to right to graze which animals, to farm which furlough of an open field of common land—these were determinations made for those who had cottage rights to common land and even for landless commoners who did not occupy structures with attached common rights. The right to farm a section of common agricultural land extended to landless commoners, other laborers, artisans, small tradesmen, the poor of the village, the old, and widows with families to support, and in some villages even immigrants and squatters. Commoners also had rights to other sources of income: berries or nuts, food, or clumps of wool that grazing sheep left behind, and furze. They could also graze their stock in open fields after the harvest, and they could, in the view of many commoners, roam the fens, heaths, and even the forests owned by, in the case of Helpston, the Fitzwilliam heir Lord Radstock, looking for food or fuel they could forage, much as they did on common land. Commoners also bartered with farmers, who might shelter and feed commoners's stock in exchange for manure and sometimes furze, or gorse (for making hot wood fires).[30] Such quid pro quo arrangements greased the degrees of class difference. Even lords of the manor needed furze. The freedom afforded by common right was in this sense of two kinds: the freedom to look across a vast landscape and the freedom to make one's living by a highly developed and socialized pattern of planting, hunting, gathering, and herding.

The degree of freedom was a matter for dispute and some friction, particularly during Clare's day, when those who gained by enclosure more often contested the more murky points of common right, such as gathering wild food from private land. Clare's defensive account of gathering rotten wood (it seems to have been part of the code to say that wood gathered on private lands was rotten, even if some of it was still on the tree) suggests that even before enclosure this practice, often conducted after nightfall, was, in the view of farmers and other landowners eager to defend their property, on the edge of the law. Wood that "down with crack & rustle branches come" may not be rotten and is certainly not lying on the ground, the usual requirement for license to gather fallen wood (*MP* 4: 334, l.15). In principle, gathering berries and nuts should have been more innocently understood, yet Clare's repeated poetic notice of this activity suggests that it was not. In these lines from the "October" section of *The Shepherd's Calendar*, Clare aligns all these activities:

> The inly pleased tho solitary boy
> Journeying & muttering oer his dreams of joy
> Haunting the hedges for the wilding fruit
> Of sloe or black berry just as fancys suit
> The sticking groups in many a ragged set
> Brushing the woods their harmless load to get
> & gypseys camps in some snug shelterd nook
> Were old land hedges like the pasture brook
> Run crooking as they will by wood & dell
> In such lone spots these wild wood roamers dwell
> On commons were no farmers claims appear
> Nor tyrant justice rides to interfere.
>
> (*MP* 1: 139, ll. 35–46)

These activities were, Eric Robinson remarks, necessary to the subsistence diets of commoners; they were also debated in the local press as commoners and landowners disagreed about the extent of commoner right to forage.[31] The lines from "October" register this class tension. The phrase "harmless load" nicely deflects the question of whether gathering it harms an owner. The notice of gypsy camps in the wood joins their presence, without further comment, to the activities of "wild wood roamers." The last lines of the passage suggest that all this happens on common land, safe from the claims of farmers and tyrant justice, yet those "old land hedges" and other natural boundaries helped someone lay claim to separate fields at some earlier time.

"Landless" commoners were often not so much landless as land poor, in the sense that they had too little land to be subject to the land tax. Even so, they could have a "pightle," a small enclosed garden, field, or croft. The language of commonability gradually slipped away. Chambers's *Cyclopaedia* (1786) includes "pightle," but modernizes it or mistakes it for the homonymic "pigtail," calling it a "provincial" term for a "strip of ground generally in the state of grass or a "small assarts." The OED preserves a distinction between the two that appears to have emerged in late medieval times: "assarts" refers to land converted from forest to arable land by grubbing up the trees and brushwood (more fuel for fires) or, more generally, a clearing in the forest; "pightle" refers to a small enclosure. The lexical shift from pightle (aka pigtail) to assarts echoes the decline of commonable right, which had specified the kind of clearing and who had the right to have one.[32]

So that sheep and cattle did not overgraze the land, they were "stinted": commoners were given a per acreage allotment that limited the number and kind of livestock that could graze per acre. Pigs were not stinted, but "custom limited their numbers," and some pigs had a nose ring to prevent them from damaging the ground by snuffling. Even the edges of streets and lane commons might be stinted, as in one horse per person. Stinting regulations were monitored by wardens who counted the branded stock, a practice called "drifts." Those who enforced the field orders were called "pinders" and "reeves." The practice of farmers taking in cattle or sheep and feeding them for owners who had no land, too little land, or no winter fodder was called "agistment." "Lammas" meant August 1, the date when the corn field would be opened for cattle and sheep (at different times) for six months, until closed again for spring plowing. In addition to grazing fields planted with hay after the harvest and before winter plowing, cattle were permitted to graze on the "joynts," carriage ways that workers and carts used to cross the fields. "Balk grazing" (Clare sometimes spells the word "balk," sometimes "baulk") regulations governed how many and what kind of stock could graze the land between plowed and planted rows where wild plants might grow, as they surely did in the green and pleasant land of the English Midlands.

Not surprisingly, the commoners' word for enclosing wastes was "encroachment." Even those who supervised the actual enclosure of land and the creation of new roads kept their eye on precedent-setting, on occasion by invoking particular plants. In planting new hedgerows beside the new roads, they advised the new owners not to plant medlars (*Mesipilus germanica*) because idle commoners would collect it for fuel and would then be "less inclined to work," or, more accurately, be less inclined to work for a farmer for a daily wage.[33]

The allocation of common right was local and particular in the extreme. In 1725, the steward for the Lord Manor of Deane was given the task of dividing a five-acre cottage pasture within an old enclosure of the manor among five cottagers. Instead of making an equal division, the steward proposed this adjustment: "Widow Sutton I think deserves more than Richard Wilkins, for though he has 3 children he is better able to work for them than this woman, who besides her own 2 small children maintains her husbands mother who might otherwise be an immediate charge to the parish." Flickers of local resistance to advocates of enclosure bills who sought their signatures in support could also be craftily specific. Some West Haddon commoners resisted enclosure by advertising, then staging a "West Haddon Football Match" on the site of a large field (nearly eight hundred acres) recently enclosed by parliamentary decree. The "match" involved tearing up the new fences and the newly planted field.

The Haddon protest recalled a highly volatile political precedent. During the English Civil War, a dissenting group that called themselves Diggers responded to worsening conditions for the agricultural poor by insisting that the land was "a common treasury for all," digging up commons in Surrey and elsewhere as a material and symbolic act of protest and writing about this activity in biblical and apocalyptic terms.[34] Another commoner in West Haddon, Ann Tabernar, offered only a simulacrum of support for local enclosure. Her terms were reported to the authorities: "she had some trees growing on her land and if they would defer the inclosure till they were fully grown she would Consent but wou'd not till then." The trees were young oaks that would mature many years after her death.[35] As did those who ruled in her favor, Tabernar understood her right to trees planted on common land as one that would bind all parties during her lifetime and that of the young oaks. Few anecdotes of wily efforts to protest, halt, or delay enclosure register as fully as Tabernar's protest the character of commonable right as a Burkean contract between the living, the dead, and the yet unborn (or not yet mature).

Clare registers what losing all this to enclosure meant in terms that are as specific, a resemblance that follows from his resistance to abstract or abstruse systems. In "On Visiting a Favourite Place," from the *Midsummer Cushion*, completed in 1831, the speaker insists that the "sweetest flowers it owns" are "mine / Free gifts" from the place where "Trees their friendly arms extend" and only pismires mark him an "intruder" (*MP* 3: 562). But in "The Flitting," the poem he wrote about moving to a cottage in Northborough, he characterizes weeds and blossoms he knew back in Helpston as "all tennants of an ancient place / & heirs of noble heritage / Coeval they with adams race" (*MP* 3: 485, ll. 129–31). Their noble lineage marks their difference from Clare, no longer a tenant of an ancient place though more apparently of Adam's race than are plants.

John Barrell has argued that enclosure changed rural life, and Clare's life, by reshaping what English poets from Denham to Pope to Thomson, as well as writers on the picturesque, had seen and described—an England with open prospects that conveyed the freedom to look and wander. Before enclosure, agricultural commons in parishes and adjacent forests and fens, wastes, or heaths were spread out before what Clare calls "the following eye."[36] After enclosure, common agricultural lands were sold and parceled out, the view interrupted by new roads as well as fences. On open agricultural lands, each commoner who had the right to cultivate a section worked beside neighbors in open fields that were more or less laid out from a starting point near the village. As in Euserene, the depression in southwest France where farmers and vine growers have grown and irrigated crops since the Middle Ages, Helpston's common agricultural land and the paths that were used to get to it from the village maintained an intricate, unfenced pattern of individually plowed and maintained sections.

Barrell does not oversentimentalize the consequences of enclosure for Clare and other inhabitants of Helpston. Whatever was lost, tax rolls and estate records indicate that the economic effects of enclosure did not adversely affect the Helpston community.[37] To the contrary, the population of the village (atypically in the general history of enclosure) increased, probably because there was work to be had; as a result, the Helpston rolls listing people on poor relief were not as long as they were elsewhere. The Fitzwilliam holdings across six parishes were regularized with enclosure, although, Barrell surmises, their heath holdings to the south of Helpston were probably not enclosed with a fence. Yet because larger landowning farmers in the neighborhood got larger holdings with enclosure, agricultural laborers, most of whom who could not survive without common land access, went to work for farmers.

Clare emphasized both the loss of a wide prospect that Barrell notes and the loss of commonable practices that had long governed how common lands were used, practices that Clare and others before him had both honored and on occasion transgressed. Various agreements, negotiations, and disputes that village councils resolved specified how unenclosed common land could be used for different purposes at different times of the year. Among those agreements were systems of crop rotation across open agricultural lands (Barrell finds evidence that a four-way crop rotation existed in Helpston at the time of enclosure); the use of agricultural land for grazing sheep and the occasional cow; and the exchange of stubble for the sheep (after gleaning) and manure for the farmer.[38] These and other arrangements for the use of forests, fens, and wastes stipulated how commoners, both landed and not, could live off the land according to a system that also permitted a degree of market flexibility. Kinds of crops or livestock might be expanded or diminished, as harvest conditions and local demand made these shifts practicable and profitable. These practices and the local regulations devised to enforce them and settle the inevitable disputes insured both that commoners could make a living without necessarily becoming wage earners for local farmers and, further, a degree of economic and social cooperation among different classes.

Farmers who complained before enclosure that they could not find enough people to work for pay insisted, as did many who took up the cause of enclosure, that commoners were not sufficiently "biddable."[39] Once they became wage laborers, they had to become more biddable or lose their livelihood. The link between enclosure and slavery that Clare makes in "The Mores" targets the ascendancy of wage labor after enclosure. It was above all labor bought and paid for by farmers, for whom the enclosure of more private land created a needed labor force from the ranks of local commoners and others who arrived from elsewhere and depended on wage labor ("The Mores," *MP* 2: 347–50). Although Clare could in most years find employment on the land, he could no longer make legal use of common rights once enjoyed by cottage tenants and others who were both landless and poor.

Clare's repeated notice of particular changes wrought by enclosure is the negative mirror of the network of allusions that bind his botanical poetics to place. In the poem "Helpstone," which he began in 1809 or 1813 in the midst of local felling, fencing, and the laying and mapping of new roads to enclose common lands in Helpston, he complains that this once local version of "those Edens by the poets sung" is "Now all laid waste by desolations hand."[40] The poem "Emmonsales Heath" (probably written in 1823) imagines there remnants of its preenclosure and pre-lapsarian landscape:

Furze oer each lazy summit climbs
At natures easy will

Grasses that never knew a scythe
Waves all summer long
& wild weed blossoms waken blythe
That ploughshares never wrong
(*MP* 3: 363, ll. 3–8 and 611n.)

Writing in 1828 about the "Pleasures of Spring," Clare recalls "The vagrant sward/That claims no fence or ownerships regard" (*MP* 3: 62, ll. 346–7).

The second change enclosure brought to Clare and his neighbors was less immediately visible but nearly as corrosive:

> The fields were places where people talked while they worked, and they worked together. . . . You could tell the time of day by the regular comings and goings of common flocks and herds along the village roads, and the time of the year by their disposition in the fields and meadows. Fieldsmen, pinders, and haywards were often about. Twice a year they made field orders to manage the fields and pastures, and a jury sat to ratify them and to hear complaints. Jurors and fieldsmen met at an inn, in public, with an audience of commoners. They drank together with the rest of the company, or in earshot of them. Then they had the orders cried round the village, before they nailed them to the church door. Once a year the whole parish met together and walked the bounds naming the field marks, remembering the line between what was theirs and what belonged to the parishes around them. Every year after the harvest the field officers opened the wheat field to the gleaners and cried the hours of gleaning round the village. Gleaners came in procession, the women and children led by their Queen. After that the herd came into the stubble, followed later by the sheep. And all through the harvest and afterwards the pigs and geese picked up fallen grain in the lanes and streets.

> So much of the land was in some way shared. You could walk across the parish from one end to the other along common tracks and balks without fear of trespass. Your children could seek out bits of lane grass and river bank for the geese or the pigs; they could get furze or turf, go berrying or nutting in the woods or on the common.[41]

With persistent irony, Clare registers the loss of these shared activities in piecemeal fashion. One poem describes going to pick berries; another looks at a seasonal ritual connected to how common lands were worked and its disappearance; yet another notes that stock that used to graze stubble or waste no longer do so. Clare thus chronicles the habits and practices that ended with enclosure much as they were constructed—gradually, over time, as a loose affiliation of traditions associated with common land.[42]

Gone, too, with enclosure were those occasions when sociability cut across rank and class. In *The Shepherd's Calendar*, finally published in 1827 but written earlier, Clare notes the decline of old rituals like the sharing of "frumity" (hulled wheat or corn boiled in milk, spiced, and slightly fermented) among laborers and farmers and the class distinction between them, newly marked by the way farmers now dress ("May," *SC*, *MP* 1: 83, ll. 155–66). In "October" (*MP* 1: 137) he refers to the grazing of free horses in the stubble field and to sheep, ducks, and geese grazing on cleared fields "at will"—all practices that more or less ceased with enclosure.

"The Opening of the Pasture" (1825) records another departed practice: on "Break day" the "Fen commons used to be broke as it was calld by turning in the stock it used to be a day of busy note with the villagers but inclosure has spoiled all" (*NHPW*, 236). In "The Mores," written in the early to mid-1820s, the tone is sharper still. The moors that never before "felt the rage of blundering plough" . . . "Nor fence of ownership … To hide the prospect of the following eye" now feel all of this because "Inclosure came & trampled on the grave / Of labours rights & left the poor a slave" (*MP* 2: 347–49, ll. 3, 9, 19–20, 36). In the "Lament of Swordy Well," Clare intensifies the personification implied in the phrase "felt the plough" by giving to the place so attacked a name and a voice. Even when seen in the rearview mirror of Clare's poem (and history), the now-dispersed network of plants, person, and open paths insists on its negated presence.

In step with the effort to make enclosure the coming reality, the reputation of commoners dropped sharply between 1750 and 1840, the peak years for parliamentary acts of enclosure. Now being a commoner invited racial epithets like "a sordid race," as alien and uncultivated as the land they had claimed. Fenland commoners (like Clare) were the worst: "so wild a country nurses up a race of people as wild as the fen." Forest commoners (such as Clare, since Helpston was adjacent to forests) were said to be lazy, annoyingly independent, and morally lax.[43] The gradual redefinition of *gleaner* from the seventeenth to the late nineteenth centuries assisted the moral rhetoric in favor of enclosure and regular wage earning. Gleaning was among the most significant rights given to commoners. On wastes, heaths, fens, and even in manorial forests, they could gather what they needed; on common agricultural lands, they were allowed to move in after the harvest to glean what remained and, after them, stock animals grazed the stubble. By the end of the nineteenth century, the term *gleaner* referred to someone who was lazy because he or she didn't work for wages, choosing instead to live off the edges and in the byways of a market economy, a way of life that common right had once protected.[44]

As a set of relationships among individuals, particulars, and a collective idea of land use, commonability addresses questions about how to align individuals with states, abstract principles, or whatever larger system is said to supervise or speak for individuals. These questions became more pressing during the romantic era, when revolutionary discourse and slave revolt made it nearly impossible to ignore (though some did) problems that Rousseau had attempted to resolve by claiming that the will of the individual and the general will would inevitably be aligned. That they were not was among the painful recognitions of the Reign of Terror. The long history of modernity keeps bringing the same recognition home again, where it lives in the heart of political and cultural work. As Jean-Luc Nancy observes, sorting out how being singular also means being plural is the unyielding dilemma at work in thinking about individuals and particulars as part of or outside a community.[45] What Clare takes from commonability is a readiness to attend to particular conditions and a disposition to argue for common rights that may vary with the seasons and the productivity of the land. He finds both in the natural history of Helpston, where seasonal and local variability in plants and their common names invite relationships that provide opportunities for figural dispersal.

GYPSY TALK

He took those opportunities as a gypsy might, from the edge and in stealth. Even if he had not from his childhood been attracted to their camps, the juridical expansion of the category of the gypsy to include anyone who wandered about near or on private lands gave Clare good reason to think that, as far as the law was concerned, he was as suspect as they. In the mid-sixteenth century, gypsies were banned from migrating to England, which they of course did anyway, risking expulsion or death. In 1743, the "Justices Commitment Act" declared that "All Persons pretending to be Gypsies, or wandering in the habit and form of Egyptians, or pretending to have skill in palmistry, or pretending to tell fortunes" would be prosecuted. In 1783, the new "Rogues and Vagabonds Act" transformed the already imaginary category of Egyptian Gypsies again, making their earlier status as rogues and vagabonds the catachretical ground for outlawing all forms of roguish vagrancy.[46] Then as now, the figure of the gypsy remained a lightning rod for public (if no longer juridical) fears about foreigners and those social outliers who also wandered the land instead of working it, or who wandered and worked it illegally.

This sharpened hostility toward anyone who wandered about, clearly refracted in the linguistic demotion of gleaners over the same period, points toward the figurative role gypsies play in Clare's poetry. Precisely because they are perceived as distinctively alien in speech, action, and dress, they become standard bearers and extravagant figures for the widening category of abjected persons, those recalcitrant wanderers and illegal gleaners who refuse to embrace the new hegemony of wage-earning poverty and alienated labor. And precisely because they treat common land as their own (though never legally so), their "song," as Clare sings it, celebrates a liberty to wander that is already foreclosed, unless one becomes or imagines oneself a gypsy, as Clare does when he sings in their voice:

> & while the ass that bears our stamp
> Can find a common free
> Around old Englands heaths we'll tramp
> In gipsey liberty[47]

In "Going to the Fair," gypsies teach "gleaning," now illegal, to the fictional horse Dobbin, a safe stand in for Clare himself and every other commoner on the loose without a legal right to use lands that had been common but were by now enclosed:

> When loose from geers he roved as freedoms mate
> Hed find all gaps & open every gate
> & if aught sweet beyond his pasture grew
> No fence so thick but he would blunder thro'
> His youth from gipseys did these tricks receive
> With them he toiled & worked his wits to live
> Bare roads he traced all day with nought to bite
> Then stole with them to stacks to feed at night
> Tho now a better life was Dobbins lot
> Well fed & fat youths tricks he neer forgot
> Still gaps were broke & dobbin bore the blame
> Still stacks were pulled & Dobbin felt no shame
>
> (*MP* 3: 98, ll. 141–52)

John Goodridge and Kelsey Thornton wryly note that these lines constitute "a condensed guide" to the kinds of trespass Clare practiced, together with a cagily deflected notice of its illegality (and Clare's) in the verb *stole*, which refers to slipping off surreptitiously to haystacks at night, and is as such a preemptive figure for theft everywhere else in these lines, which are embedded in a linguistic surround that regularly reminds the reader that "wits" and "tricks" are the work of mouths and hands.[48]

The 1830 sonnet "The Gipsey Camp" grants just how little protection they gain from their shelter and from their poetic site in the sonnet itself. Endstopped throughout, and unrhymed until the last couplet, the sonnet's untempered verse registers its tonal neutrality. Clare here presents, far less exuberantly than in earlier poems about gypsy encampment, the safety of a "squalid camp, half hid in snow," protected by an oak and bushes. The mealtime preparations inside are as bluntly offered, "stinking mutton" roasting and a "half-roasted dog" that gets too close to the fire, hoping to grab a morsel. The last four lines report gypsy want flatly, beginning with the dog who

> Watches well, but none a bit can spare,
> And vainly waits the morsel thrown away:
> 'Tis thus they live—a picture to the place;
> A quiet, pilfering, unprotected race.
>
> (*LP* 1: 29, ll. 11–14)[49]

The poem's concluding adjectives echo talk about commoners on both sides of the enclosure debate. Those who argued against enclosure praised commoners as independent, off the dole, resourceful citizens whose subsistence common right had long protected. Those who argued for enclosure charged that commoners were independent out of laziness (i.e., they refused to work for a wage) and that they stole from (now) private property. The "place" Clare accords to the gypsies in this sonnet offers them a poetic mooring that is fragile and contradictory, but also "attuned," to adapt Heidegger's and Ngai's analysis of mood, to a larger category and community of persons like themselves, if less outlandishly like. That Clare understood their exclusion to be in some sense his own, writing this sonnet from the asylum near Epping Forest, is clear. That he as a poet gained something for poetic speech by placing himself at the margins where gypsies lived, half-surreptitiously gathering local words and sticks as they did, is the argument I want now to consider.

Clare writes just once about gypsy language, in a tone that appears to urge some distance between theirs and his:

> I had often heard of the mistic language and black arts which the gipseys possesd but on familiar acquantance with them I found that their mystic language was

> nothing more then things calld by slang names like village provincialisms and that no two tribes spoke the same dilacet exactly   their black arts was nothing more of witchcraft then the knowledge of village gossips & and the petty deceptions playd off on believing ignorance (*JC*, 83)

In several poems, Clare's speakers insist they are on to the gypsy fortune-teller's game, and so not figures of "believing ignorance." But there and here Clare deflects attention to how much his "mystic" or "mistic" language is also made up of slang, village provincialisms, and dialect words for things that are unlike those found elsewhere. The strange, even mystic power of this language is akin to that of the gypsy as a figure for something at the edge, unfamiliar, a figure whose particularities of dress, living habits, and language are allegorical signs for the legal category called "Rogues and Vagabonds." This is precisely the figurative move that the poet John Clare could love, although he, too, used more conventional allegorical figures and personifications, as he does for the months of his *Shepherd's Calendar.* It is also the move that gypsy talk, a slang responsible for a good many contributions to the English and European wordhoard, is well positioned to underwrite. What Alysia Garrison has called Clare's "nomadic poetics" also includes the wandering, unstable diction his plant names put on show. Recent work on borrowings from gypsy language in modern English and French insist on a point that Clare muffles, but that we do well to recognize.[50] Gypsy language is paradoxically alien to the degree that it is not. Or to put this slightly differently, its apparently alien character constitutes its shadowy Englishness. Gypsy talk, like gypsy clothes and pilfering, at once masks its proximity to English common speech and displays it.

PLANT GRAMMARS

Clare's polemic about "naturalists" and "botanists" is at least half subterfuge:

> to look at nature with a poetic feeling magnifyes the pleasure yet Naturalists and Botanists seem to have little or no taste for this sort of feeling   they merely make a collection of dryd specimens classing them after Lienneus into tribes and familys as a sort of curiosity and fame
>
> I have nothing of this curiosity about me thu I feel as happy as they can in finding a new spiecies of field flower or butterfly which I have not before seen (*JC*, 62)

Beginning with the claim to "have nothing of this curiosity about me," this strategic self-portrait presents Clare as a poet in opposition, unsullied by taxonomic grubwork. Yet in his youth and again during his long friendship with Edmund Tyrell Artis and Joseph Henderson, Clare collected plants—some wild, a few cultivated—with enthusiasm, using the word *curious* repeatedly to indicate special discoveries.

As he learned more about taxonomy, he looked for new species and offered some claims about plants that seemed to him new or before unnoticed. Although he insisted "Im not much of a classifier," he talked about tribes and families (*NHPW*, 18). When he and Artis saw a bird neither could identify—Clare says it was "about the size and shape of a green linnet with wings of a brown grey color and the crown of the head of a deep black that extended downwards no further then the eyes"—Artis took the next step, which Clare reports: "Artis looked thro Pennant, he coud not find any thing resembling it and believes it to be an unnoticed species of the linnet tribe" (*JC*, 236).[51] Artis takes the taxonomic lead in part because he has immediate access to Thomas Pennant's *Genera of Birds*, but what drives the inquiry is Clare's acutely detailed description, not Latin binomials. In a list of plants he sent to a Mrs. Wright of Clapham, he put three Cranesbills together, writing "'Lilac flowerd Cranes bill,' 'Black flowered Cranes bill' and 'Pencil flowerd 'D' [ditto]"—that is, also a cranesbill (*JC*, 233). He writes of finding ferns, including a species of maidenhair fern that may be new or "unnoticed" and proposes to call it "the Dwarf Maiden hair" (*JC*, 201).

Clare allies his curiosity with taxonomic notice: the particular "curious" wildflowers he and Porter seek out are orchises; yellow and blue "headache" perennials (corn poppies, so named because their scent gave people headaches), and "a curious sort of Iris or flag" (*JC*, 38, 51, 74 and 215). Even his notice of which flowers bloom when, throughout the year, might be understood as a seasonally or temporally adjusted taxonomic arrangement in which, to the dismay of naturalists ever since, every plant that blooms about the time the cuckoo returns is called a cuckoo of one sort or another.[52]

Henderson, who after being the head gardener at Milton Hall went on in later years to take up a similar position at the Fitzwilliams's even larger estate in Yorkshire, was a Scot who had received a superb education in natural history. Artis and Henderson introduced Clare to that estate's extensive natural history library and exchanged magazines, literary and botanical, with Clare. All three sent each other plants as gifts. Although Henderson urged Clare to study plants systematically, his interests were anchored in local species and plant groups that his friend's greater systematic knowledge might help him parse.[53] Their friendship may have urged a modicum of compromise, but Clare did not hesitate to invoke the pettiness of taxonomic distinctions to comment on a letter addressed to him "under a Frank" that required him to pay an extra one-penny charge. He notes tartly that "'knaves in office' watch chances as the cat watches mice and are of that species of animals that catch their prey by supprise" (*JC*, 211). Clare's diction here recasts taxonomic distinction as the work of figure: postal handlers were by turns elevated and decommissioned as "knaves in office."

As he shared books and activities with Henderson and Artis, Clare began to keep experimental records on plant questions that interested him. Like many other botanists, amateur and professional, during the early nineteenth century, Clare investigated the directions in which different vines climb as they grow. The question engaged botanists because the motility of twining plants implied that they were in some sense animate. Troubled throughout his career by versions of this claim, Charles Darwin subsequently published his experimental observations on the subject.[54] Clare became so convinced by the "quinary" hypothesis, which suggested that plants were organized numerologically in terms of fives, that he wrote ecstatically about finding evidence to support this possibility. He began an essay on the sexes of plants and made notes on which trees and plants he had found to be hermaphroditic (*NHPW*, 83, 101–2, 115).

As Clare grew more confident of his taxonomic skill, he engaged Henderson in friendly, competitive taxonomic dispute. After seeing Henderson's ferns, Clare noted in his journal that his friend's collection was "far from compleat tho some of them are beautiful" (*JC*, 202; *NHPW*, 210). Later entries chronicle his efforts to match or outstrip Henderson's fern collection: Clare obtained more soil for his own fern collection, hoped to go to Bristol in the spring to get ferns, and found a new species of maidenhair fern. In late December, 1825, he noted his disagreement with Henderson about local species of bramble. Henderson claimed there were just two, but Clare insists in the letter that he has seen four: "the common one that grows in the hedges—the larger sort that grows on commons bring larger fruit—calld by childern 'black berry' the small creeping 'dew berry' that runs along the ground in the land furrows & on the brinks of brooks & a much larger one of the same kind growing in the woods." He adds, "botanists may say what they will—for tho these are all of a family they are distinctly different" (*NHPW*, 213).

What Clare rejected, so vehemently that he tried to isolate his identity as a poet from his sustained botanical and taxonomic interests, was the Linnaean systematic, on nearly the same grounds that he rejected grammatical rules. The problem was not only Linnaeus's Latin names, which Clare neither understood nor wished to understand, preferring the "vulgar" glossary of common names. He took issue with the Linnaean system itself, as an almost machinic reduction of the variability of plants and their names. In "A Ploughmans Skill at Classification after the Lineian Arrangement," written between 1808 and 1819 but never published, Clare skewered the linear plot of the Linnaean sexual system by recasting it as a sharp-tongued exchange of marital insults between a cranky plowman and his wife. In a nice display of figurative name-calling, she reduces her husband Ploughman Hodge to a "hog," to which he replies, "For surely there is no disgrace / For *hogs* to herd wi' *Sows*" (*EP* 1: 211, ll. 6–7). From its ponderous title to its conclusion, the poem riffs on Linnaeus's claim that the male and female parts of plants determine their taxonomic location and identity to suggest that this "arrangement" produces narrow categories and names, even as the sparring husband and wife do the same. Whereas Erasmus Darwin's elaboration of Linnaeus's plant marriages was playful, admiring, and explanatory, Clare's insisted on its reductiveness to a set of line items in a taxonomic hierarchy that had little systematic interest in determining plant families and none in variations below the level of species.

In the first of the letters on natural history intended for a natural history of Helpston modeled on Gilbert White's epistolary *Natural History of Selbourne*, Clare commented, plant by plant, on Elizabeth Kent's *Flora Domestica*, correcting her names by supplying his own or local names. Here is Clare on Kent's lesser celandine:

> 'Celandine' page 77 this is my 'crow flower' & 'buttercup' the childen often call them 'golden daiseys' some of the common people know them like wise by the name of 'pile wort' but none by the name of 'little celadine' the large one \or major/ is calld 'wart weed' & the yellow juice emitted when the stalk is broken is applyd to warts as a certain cure & I myself have known it suceed often—my mother has a poetical superstion about them she calls (I mean the major Celadine \still/) them Dane weed as they grow plentifully in a field at her native place were it is said the danes & redshanks fought a desperate battle & on that day which is said to be Whit sunday she assures me for certain tho she never tryd it herself that they emit a red juice instead of a yellow I have never had the curosity to travel so far to contradict her & she will not be pesuaded by the contrary at home for she is certain they are not the same flower & that her Dane weed owns its wonderful property still—what belief is stronger then superstion (*NHPW*, 17)

Clare's digression strategically buried Kent's greater or little or lesser celandine underneath a pile of common or, as he put it in this letter, "vulgar" names, including "pile wort," then testimonials for herbal medicine and flat-out superstition, which permited yet another foray back in time to the ancient history of Dane and Viking invasions. He reacted against Kent's common but perhaps more elegant "marsh marigold" by substituting "horse blob" or "water blob" (*NHPW*, 21). Never one common or vulgar name when two or more offer themselves. He responds to Kent's dismissal of the lowly "cowslip" by defending it: "this is a very favourite flower with us among all classes" (*NHPW*, 18). Kent's letters to Clare present them both as concerned with variations in species that other writers ignore.[55]

In an era when a place name like Buckden could be as easily rendered Bugden and still be understood, plant names could and did morph into adjacent names (*JC*, 338n). Orthographic variability was only the beginning. A plant often had many common names, some of which suggested a figurative relationship to something or someone else—a bird, a human, a foot, a fly, and on and on—and thus the possibility of more than one set of taxonomic relationships. Clare's recognition is attentive to the key presupposition of the Natural System that the morphology of a plant species may suggest several taxonomic relationships, some of which will cross genera and family classifications. Although Clare probably knew something about the Natural System, since Henderson was using it to classify his ferns, its Latin nomenclature would not have attracted him to learn more. Yet his handling of common plants and their names is similarly attentive to the way both might be complexly affiliated in local speech and botanical practice during his own time and in a longer history of usage that survived enclosure. Clare's insistence on preserving different common names from all informants echoes his readiness to find and identify different species or varieties. He notes for example that the name used by children for "sallow palms" was "geese and goslings" or "Cats and Kittens" (*JC*, 220). Both as names and plants, synonyms implicitly challenge the early nineteenth-century disposition to preserve all known Latin and Greek synonyms, but report just one common name.

Clare's fascination with wild varieties that defy fixed taxonomic judgments about species and his use of poetic figures that create figurative hybrids constitute a practical and poetic intervention in the modern debate about what words can and cannot do and how they name and organize knowledge. In a sonnet he wrote in the 1830s, he recalls a botanical excursion past a suspicious farmer whose workers were burning straw ("shoaf" is Clare's word). The last couplet reports, "We found the columbine so black/& dug & took a bundle back" (*MP* 5: 386, ll.13–14). Arranged in stanzas that emphasize its Shakespearean rhyme and form, the sonnet plots its outcome in much the way that Clare and his companions plotted theirs. This resemblance insists on discovery as a rhetorical device, in the older sense of discovery as the invention stage of an argument and as a botanical argument about species and difference. A "columbine so black" stands out against Clare's list of different species of columbine, which mentions night brown, stone blue, and ruby red (*NHPW*, 18). The difference in color could be simply a variation produced by the soil (one of the many reasons that Linnaeus refused to consider color in defining species), but what matters here is the uncertainty this plant introduces. In Clare's conversable botanical world, where species stability since creation was the default position, this small variation works against taxonomic stability.[56]

The group of English orchises that Clare so often describes as his "cuckoos" involve both crossover common names and a hint of species variability that he insists on, against most contemporary authorities. As boys, he and Tom Porter had looked for "curious wild plants . . . such as orchises." As an adult, Clare tried to classify orchises at least twice, once in a list of those he found "from the privet hedge" and again as he annotated a list Henderson developed, based on the Latin names and descriptions in Smith/Sowerby's *English Botany*, then sent to Clare in April 1828. Vardy has suggested that Clare had no real taxonomic interest in Henderson's list because it is organized according to Linnaean species and Latin names, whereas the list of twenty-two orchises that he wrote in the inside cover of a book as "from privet hedge" identifies eighteen different species by their common names (*NHPW*, 300–302).[57] Yet Clare collaborated on this project with Henderson beyond what friendship required. The two began thinking about orchises at least a year before, when Henderson wrote to invite him over to look at a copy of Smith/Sowerby's *English Botany*.[58] Apparently Clare came, began looking for local orchises, and then sent them to Henderson. When Henderson sent him the handlist of English orchises, it included all but one of the Latin plant names:—"N. Sepens," for *Neottia Sepens*, one of two Neottia on the Orchis list. Clare put the name in and annotated the list further to indicate where specimens could be found.[59]

Vardy is right to note that Henderson's list of orchises does not match Clare's list of those he found "from the privet hedge" and those he calls his "cuckoos" in a letter on natural history he probably wrote in 1823. Although he could have added Latin names to his own list, he did not. Clare's insistence that he knows his cuckoos may well be right, although his editors and earlier lexicons of plant names in Northamptonshire either do not mention Clare's favorite "pouchd lip cuckoo bud" or explain that it must have been either *Orchis morio* or *Orchis mascula*. Several orchises bloom in early spring and resemble each other; moreover, orchids in general, including English orchises, are rampant natural hybridizers and thus good candidates for creating cultivars. For both reasons, it is sometimes difficult to tell where one species ends and another begins. The early purple orchis may be on the pink side of pale purple; it may also be one of its companion species, *Orchis morio* or *Orchis maculata*.[60]

Here is Clare on the subject, responding to Kent's treatment of *Cardamine pratensis*, also called ladysmock and cuckooflower:

> I never heard it calld cuckoo in my life otherwise then by books—the wood anemone is also often calld 'lady smock' by childern what the common people call 'cuckoo' with us is one which is a species of the 'Orchis' as Henderson tells me: there is a vast many varietys of them with us such as the 'bee Orchis' the 'pigeon Orchis' the fly Orchis' & 'butterfly Orchis' &c &c namd so from the supposd resemblance the flowers bear to those things that are my cuckoos & the one that is found in Spring with the blue bells is the 'pouch lipd cuckoo bud' I have so often mentioned its flowers are purple & freckld with paler spots inside & its leaves are spotted with jet like the arum they come & go with the cuckoo & in my opinion are the only cuckoo flowers in England let the commentators

> of Shakspear say what they will nay shakspear himself has no authority for me in this particular the vulgar wereever I have been know them by this name only & the vulgar are always the best glossary to such things—there is one sort that comes on the pastures a little later then the wood sort & blossoms in a variety of colours such as pink lilac & dark purple there is another sort later still that lifts on a tall stem a wreath of flowers in the form of a sugar loaf of a pale freckld colour—now you know by 'cuckoo buds' & the small english shepherds cuckoo buds england over the botanist gives their vulgar name 'foolstones' what a silly name for such beautiful flowers & what a fool must the botanist have been to disgrace them with it—so much for cuckoos (*NHPW*, 15–16)

Clare's cuckoos do not require a fixed name or species because they are distinguished not by specific species names but by what they look like and, as Mabey notes, by where and when they grow, whether in a wood or in the open, early or late in the spring. One might grow elsewhere some year and "become" something else. As a poet and a naturalist, Clare willingly entertains what Mabey rightly calls a sustained congress between plants, their common names, and the work of metaphor. That congress arises, as Clare was well aware, from the natural mimicry that occurs across the kingdoms of nature, but that plants appear to do with special fervor. Among the specimens that the staff of the Huntington Botanical Gardens recently gathered for an exhibition of "Natural Mimicry" were two orchids. One of them, *Trichoceros tupoipi*, looked like a small, fuzzy, female bee. It is not, but the male bee of the species it resembles mistakenly tries to mate with it and so pollinates the orchid. This orchid belongs to a genus that includes many species that mimic in ways that invite pollination by unsuspecting insects. Another plant, *Lepanthes calodictyon*, has a flower that looks like a tiny gnat resting on the leaf.[61] Clare seems to have understood that plant mimicry invites the poet to join in the dance, by way of poetic figure.

BOTANICAL POETICS

At their most allusive and inventive, Clare's botanical figures nearly always have mimicry, whether natural or nominal, in view. If his poems offer an afterlife of plants, to paraphrase Dan Chiasson, it is because Clare understood, watching the enclosure acts take material shape around him, that plants would have to have an afterlife of an other than material kind. This recognition is in the end commonable as much as it is personal, both because he insisted that his plant names were (mostly) those he held in common with his neighbors and because his attunement to botany was shot through with the sociality, a sense of common regard, that Ngai suggests the mood of speakers can be.

To suggest how commonability, mood, and figure are collectively at work in Clare's poems about plants that mimic other species, I want to read across several poems and then conclude with a reading of a sonnet that replies, in its own key, to the problem of figuration and personification in "The Lament of Swordy Well." The first is "Reccolections after a Ramble," published in *The Village Minstrel* (1821), where the speaker chronicles human and natural history, including the poet's. One stanza considers the greed of those who go on botanical excursions looking for rare plants:

> Some went searching by the wood
> Peeping neath the weaving thorn
> Where the pouch lip'd cuckoo bud
> From its snug retreat was torn
> Where the ragged robbin grew
> With its pipd stem streakd wi jet
> & the crow flowers golden hue
> Carless plenty easier met.
> (*EP* 2: 195, ll. 225–32)

The cuckoo bud, the ragged robin, and the crow flower teasingly suggest a species cross from plants to birds, a category "error" a first-time reader not as botanically adept as Clare would be likely to make, at least until arriving at the last flower name in the stanza, crow flowers, which gives the game away. As a poet, Clare found his "carless plenty easier met" in those common plant names that create opportunities for poetic figure.

Clare's bird names include similar crosses: "Fern Owl" or "stone chat" (*NHPW*, 34, 109). In a poem on that most romantic of birds, the skylark, he considers how its common name provides effective camouflage. Thinking that a skylark must live near the sky, boys look for its nest high up and so miss it on the ground. Clare's sense of the array of poetic opportunities for figurative display and concealment begins early and continues late: he wrote "Reccolections after a Ramble" before 9 February 1820 (*EY* 2: 786n.), and in "The Botanist's Walk," printed in 1841, he reuses the name "dog mercury" [*Mercurialis perennis*] that had appeared in *The Rural Muse* (1835) (*LP* 1: 36n.). The formal reciprocity suggested by plant shape and English name across Clare's poetry shelters a figurative possibility that may also extend beyond plants to their surroundings, as in "ox-eye" daisy and "Oxeye Wood"—and behind these two, perhaps the eye of an ox as well.

Clare mentions at least one other wood with a more pronounced (and more slippery) plant affine. "Royce wood," where he finds many plants, looks very much like "foulroyce twigs" in the poem "Pleasures of Spring" (*MP* 3: 53, ll. 95–98).

The editors of his poems suggest one version of the implied kinship; their glossary of his language suggests another. On maps, Clare's "Royce Wood" is usually spelled "Rice Wood," but Clare's spelling may recall either the name of the farmer after whom the wood was named or the Old English word *hris*, meaning brushwood or underwood, that is a wood managed by coppicing, its brush thus available for commoners to gather (*MP* 3: 616n.). The glossary is still more botanically precise, offering "foul rice" or "foulroyce" as the common name for either the dogwood (*Cornus sanguinea*) or the spindle-tree (*Euonymus europeaus*). Both, the editors note, are used to make musical instruments (*MY* 5: 667). Grainger and Druce avoid Clare's "foulroyce twigs" entirely. It is impossible to know by what logic Clare uses "foulroyce" to name twigs, but the verbal resemblance between those twigs and that wood conveys a relationship that is oddly poised between the possibility of taxonomic kinship (are foulroyce twigs found in Royce Woods or so named because they appear there?) and something less fully declared that may go like this: "royce" names two aspects of Clare's nature, and in the world of Clare's language this doubling spins a verbal thread between a place and a kind of brush. In the interval between name and referent, an imagined and flexible taxonomy emerges in which human figurality and plant variability work like a sea of markers for a world in which several affinities rather than a single species or attribute mark the location of plants and birds.

In an early poem, "The Wild Flower Nosegay," dated 1820, Clare presents a "childish calendar" of the year:

From the sweet time that springs young joys are born
& golden catkins deck the sallow tree
Till summers blue caps blossom mid the corn
& autumns rag wort yellows oer the lea

. . . .

Crimp frilled daisey bright bronze butter cup
Freckt cows lip peeps gilt wins of mornings dew
& hooded aron early sprouting up
Ere the white thorn bud half unfolds to view

. . . .

The jaundic'd tinctured primrose sickly sere
Mid its broad curdl'd leaves of mellow green
Hemmd in with relics of the parted year
The mournfull wrecks of summers that has been

(*EP* 2: 9–10, ll. 9–12, 17–20, and 25–28)

The seemingly nostalgic gesture of this floral calendar, which reappears in brief snapshots elsewhere in Clare's poetry and prose, is in part, at least half, camouflage. For its long view across the seasons of the year and backward in time reinstate, in a different, specifically figurative key, the commonability lost through parliamentary enclosure.

The common plant names that structure these lines assist this figurative project, among them "cows lip," with its companionable manuscript variant "cows lap," and the "hooded aron," Clare's usual but also singular spelling for arum, or *Arum maculatum*. The other, more common name for this plant, which he used elsewhere, is "lords and ladies." Its omission here may be strategic in a list of wildflowers found on common lands that once belonged to commoners, not aristocrats.[62] By breaking up plant names that are now spelled as single words—among them cowslip, buttercup, ragwort, and bluecap, for either "cornflower" or the bird "blue tit"—he points to their resemblances to other natural things and, in some cases, a specific resemblance between plants and birds. If anything, Clare's poetry magnifies such resemblances. In the June section of *The Shepherd's Calendar,* he writes of "the wheat that swells into ear," punting toward one of his favorite bird names, "wheat-ear." The verbal echo glancingly suggests actual ears, human and animal. In the section for the month of May, the line "& cowslip [ms A 20 "cows lip"] cucking balls to toss" refers to a game that children play in which they toss balls of flowers, usually cowslips, although also used is the variant cowslap, which suggests a muckier projectile (*MP* 1: 73, l. 440).

The poem "The Flitting," which Clare may have begun in Helpston, then finished after his move to Northborough, conveys what happened when he found himself outside the taxonomic membrane of nature and human habitation at Helpston. The speaker chronicles details of the landscape and life he has left behind, marking their difference from the "foreign things" that now surround him, which the poem names, or rather repeats and unnames, calling them "vague unpersonifying things." "Unpersonifying things" telegraphs a seismic shift in his way of writing about his surroundings. For the phrase claims that personification is also what nature does, or did until he arrived at Northborough. The unpunctuated adjectival sequence "vague unpersonifying" insists that these coupled attributes mirror in reverse the way particularity, the bedrock of Clare's writing about nature, assists figuration.

Mabey can, as it were, channel Clare, because they are both attentive to available points of crossover between nature and figure. Consider, for example, Mabey's notice of Clare's "ramping dialect" in the poem "To the Snipe."[63] The word *ramping* refers to the way that some plants creep or crawl or climb, wantonly, riotously, up a support (OED). Mabey's use of the term to describe Clare's dialect invites notice of how the poetic line of this poem moves with a ramping metrical swing:

> the clump
> Of hugh flag forrest that thy haunts invest
> Or some old sallow stump
>
> Thriving on seams
> That tiney island[s] swell
> Just hilling from the mud & rancid streams
> Suiting thy nature well.[64]

The "flag forrest," a collective name for a swampy cluster of flag iris, transforms natural description into poetic figure. "Ramping" lines of poetry transform an adjective into a verbal, much as Clare himself does in these lines when he turns "hill" into "hilling" to describe the way that the movement of mud and near stagnant (and stinking) streams make "tiney island[s] swell."

The way Mabey reads Clare, much as Seamus Heaney reads Clare's use of the verb *prog*, mirrors the way Clare reads nature for figure.[65] The hybridizing impulse of his prose and verse writing on natural history looks very much like the work of figure because it is. Whereas plants called "cuckoo" flowers of various persuasions trod roughshod over classificatory distinctions, Latin nomenclature insure one name but insist on distinct species. Or when a plant name looks like a bird but is not or vice versa—such as the bird "wheat ear" and the wheat that Clare often notices growing "into ear"—it offers an opportunity to imagine a cross between one natural kingdom and another, or others. As such, common plant names and, Clare insists, the plants to which they refer, invite an ongoing figurality that cannot be legislated out of existence because such figures evade regulation by transforming themselves into something else, in something like nature's version of the god Proteus, who changes shape to resist capture. This figurative project extends to personification and allegory, the kinds of figure that most risk a loss of particularity to the work of abstraction, itself a poetic version of the troubled relation, for Clare, between particular things in name and taxonomic category.

Writing about John Clare's "Lament of Swordy Well," Simpson has argued that Clare's personification in that poem raises many of the questions this figurative gesture can and should raise about the efficacy of speaking for or of nature.[66] So situated within the literary itineraries of ekphrasis, elegy, personification, and ecopoetry, Clare is, as Simpson observes, more invested in poetic tradition than he has sometimes been understood to have been. Characteristically, that investment is signaled by Clare's resistance to traditional personification. Here the figure of a speaking voice is a diminished shadow of its former self, a voice that speaks as a place that is not a *locus amoenus*, nor even the place it once was. How can one speak for nature when it is presented as a place that is under erasure? How can we explain the relation between Clare's figures and individuals, species, and larger categories that can be said to be taxonomic? At times, Clare celebrates singularity, much as Wordsworth does in the *Intimations Ode*: a single tree, flower, or, in the case of Clare's "Lament of Swordy Well," a singular place, marks a collective history or ecological niche. Yet despite Clare's edginess in this poem about the work of embodiment that traditional personification performs, as a poet and essayist he frequently, even casually, makes use of personification and other poetic figures. I note a few instances.

Among his autobiographical fragments is a pair of potentially allegorical figures: charity and avarice. Of the latter, he remarks, "avarice is a cruel beast—it would throw water on a drowned mouse—cheat an adam out of his fig leaved apron and paint the very devil with lamp black—with a visage all the while as sincere and sanctified as if it was preaching a charity sermon" (*JC*, 161). With its quick relay between literal and figurative components, this vignette conveys the characteristic achievement of romantic allegory, which trawls for meaning on both sides of the border between figurative avarice and its realist activities. In a journal entry for 1824, Clare remarks: "Neglect is the rust of life that eateth it away and layeth the best of minds fallow and maketh them desert." Clare's adage echoes and revises the biblical text "Lay not up for yourselves treasures upon earth, where moth and rust doth corrupt" (Matthew 6:20, King James Bible; *JC*, 183). As he elaborates the figure of rust's corrosion of mental and spiritual potential, its referent becomes agricultural wasteland that begins with fallow, untilled ground and ends in desert. A third example: at the beginning of the story of his escape from Dr. Allen's asylum in Epping Forest, Clare invents an allegorical companion to keep him (figurative) company, and then a plural "troops" to protect them both: "having only honest courage and myself in my army I Led the way and my troops soon followed" (*JC*, 257–58).

Clare was evidently not wary of personification or allegory. In one way this is surprising, since such figures might be said to perform a poetic version of the abstraction required to constitute a taxonomic category and assign individual species and plants to it. Whether the subject is a plant or a poetic figure, the particular individual risks being outnumbered, outclassed by the category, as in "this species belongs to this genus," or "this poetic figure articulates this idea." There are hazards in the other direction as well: being without a category means being without place, unrepresented, living beyond and outside known categories. Poetic figures constitute the limit case of this hazard, for without abstract—in the sense of figural—content or meaning, a figure or metaphor simply cannot exist. To insist, as the American imagist poet William Carlos Williams does, that there are "no ideas but in things" makes a claim about where figurative ideas begin, not about where they end.[67] For the work of figuration is not spontaneous or instantaneous, but developed, mediated, however quickly or invisibly that process occurs, as Wordsworth suggests in his definition of poetry in the preface to *Lyrical Ballads* and as Clare's commonable plant poetics insists.

In the sonnet "Swordy Well," published in the *Stamford Champion* in 1830, Clare presents a scene of pillage that recalls Wordsworth's 1798 poem "Nutting." Yet the sonnet also imagines, by way of orchises, another relation to the poet-speaker that conveys Clare's sense that botanical particulars might ground figures and the abstracting from particulars that figures risk:

Ive loved thee swordy well & love thee still
Long was I with thee tenting sheep & cow
In boyhood ramping up each steepy hill
To play at 'roly poly' down—& now
A man I trifle oer thee cares to kill
Haunting thy mossy steeps to botanize
& hunt the orchis tribes where natures skill
Doth like my thoughts run into phantasys
Spider & Bee all mimicking at will
Displaying powers that fools the proudly wise
Showing the wonders of great natures plan
In trifles insignificant and small
Puzzling the power of that great trifle man
Who finds no reason to be proud at all.

(*MP* 4: 145)

Prompted by the capacity of orchis species to imitate insects—here the Spider and Bee Orchises, to thoughts that "run into phantasys"—Clare's speaker constructs a vision of nature's plan in which plant trifles suggest in the end "that great trifle man" who does not fare well in the implied closing comparison among the "wonders" of such trifles. As members of a genus that prompted a good deal of "phantasy" and at least as much botanical theorizing among nineteenth-century botanists and writers, these orchis or ophrys species present in miniature the relay between plants and common names that Clare uses to do the work of poetic figure on, and then off, nature's hook. It is tempting to suggest that the phrase "cares to kill," is meant to read "careless to kill," as carelessly killing plants by uprooting them. Yet what the poem declares here is worth keeping in a punning yet oppositional relation to what came earlier: here the speaker does have "cares to kill" as he looks at a place he has long loved and trampled. One way to kill cares or at least assuage them is to make a poem from them, in this instance a poem whose figurality takes its cue from a plant whose name and appearance invite this poetic work. The poetic argument runs toward chastisement, but its punning relay between plants that trifle with categories and the small trifle that man is link these two kingdoms of nature in a hierarchical array that refigures the power otherwise accorded the human.

Writing of Clare's other Swordy Well poem, the earlier "Lament," Simpson describes a poetic voice of that place that is wary of its poetic task—speaking for and as a place—and wary on compelling grounds. As the Swordy Well that Clare and his contemporaries knew fades, its identity and purpose scraped to bare ground, how can it coherently be said to speak or for what can it speak? Clare's sonnet "Swordy Well" sketches a reply to this question that works in the folds of an irony that is material and poetic. The "man" who uproots orchises at Swordy Well assists its gradual destruction as the place it once was. This trifling is met by another poetic trifling that works, if not in opposition to this destruction, at least toward other, figurative, ends, by imagining on the particular hook orchis species and names provide. The "man" in the case is at once generic man and Clare himself. Here and elsewhere in Clare's poetry and prose, commonability, plant taxonomy, and figuration attach particular, specific man to his kind, and poet to the work of figure.

Mala's Garden: A Caribbean Interlude

> Mala's companions were the garden's birds, insects, snails and reptiles. She and they and the abundant foliage gossiped among themselves. She listened intently. With an ear pressed to the ground she heard ant communities building, transporting food and breeding. She listened to worms coiling arduously from place to place. She knelt on the ground and whispered to the grass and other young plants, encouraging them to grow, and then she listened as they stretched up to her. She did not intervene in nature's business. When it came time for one creature to succumb to another, she retreated. Flora and fauna left her to her own devices and in return she left them to theirs.
>
> At first Aves, Hexapoda, Gastropoda and Reptilia burrowed instinctively into nooks and crevices. They realized eventually that they had no cause to hide.[1]

Set on a Caribbean island modeled on Trinidad, sometime near the end of the colonial era, Shani Mootoo's novel *Cereus Blooms at Night*, quoted above, imagines a Caribbean garden where the taxonomic names used to classify exotic plants for European readers have become oddly redundant. Mala Ramchandin is a descendant of Indian (that is, East Indian) indentured workers transported to the Caribbean to work in its fields in the postslavery colonial era, yet another wave in the global transplantation of peoples that is mirrored in the transplantation of plant species between the East and West Indies, the two poles of colonial enterprise that structure Abbé Raynal's *Histoire philosophique et politique des établissements & du commerce des Européeans dans les deux Indes*, or, more familiarly, *L'Histoire des deux Indes* (1777).[2]

In Mootoo's novel, Mala's garden of delights is the reiterated emblematic scene for queer partnering that goes back two generations and sets in place the scene and events for another couple when Tyler, a shy, homosexual nurse who cares for Mala late in life, meets the crossgendered Otah. The plant figure that gathers the symbolic energy of the narrative is the demipersonified cereus plant, the night-blooming cactus that others prefer to "pass-along" to someone else.[3] Precisely because it blooms only once a year, its night-blooming habit had long attracted European notice; in *The Temple of Flora*, Thornton alternated between two different engravings of "the Night-Blowing Cereus." Seductive, hidden, and heavy with scent when it blossoms—it attracts a generation of lovers as they stroll by Mala's garden—the Cereus embodies exotic profusion, be it botanical or sexual (the Latinate name insures that Linnaeus's sexualized categories are ever in the background) or both, that outpaces its taxonomic name.

Elsewhere in Mala's garden, the family names of the mid- nineteenth century Natural System only just hang on as artificed, quasi-abstract names ill-matched to the movement of actual creatures ("Aves, Hexapoda, Gastropoda and Reptilia burrowed instinctively").[4] The taxonomic categories that served the colonial imaginary so well across most of the globe no longer count for much in Mala's garden and Mootoo's novel, where life and plant forms proliferate on their own hook and human sexual difference is similarly resistant to known or permitted categories. In Mala's garden no one tries to manage plants and animals. Mala herself, who long ago "stopped using words," has no interest in giving any of them names, not even indigenous ones.[5] In the pas de deux between Mootoo and the taxonomic categories that materially assisted the work of empire, there are no compromises and nothing remotely collaborative.[6]

Mootoo's novel pushes back against a long history in which taxonomical description and colonial botany in Latin America worked in concert, even though evidence of local informants and slave gardeners sketches a more intricate history of knowledge gathering about botany in the New World. Jill Casid has argued that slave gardens, their unsanctioned use of planters' prized specimens and hidden cultivation by Maroons, escaped slaves, in the mountains of Caribbean colonies, offer unacknowledged evidence of resistance to the plantation agricultural economy. Londa Schiebinger and other scholars provide evidence of local informants and knowledge that disturbs the picture offered by official British and French accounts of colonial plantation and Caribbean natural history, whose notice of Indian medicinal expertise is uneasily divided between their knowledge of poisons and cures.[7]

British natural histories of the Caribbean before and after 1800 were, to put it mildly, reticent about who collected plants and drew them. They were not interested in local plant names, and most preferred to anchor their accounts in the Linnaean system and nomenclature. Patrick Browne's 1789 *Civil and Natural History of Jamaica*, illustrated with copperplate engravings by George Ehret, said a good deal about commercially useful Caribbean plants, notably sugarcane, cotton, coffee, indigo, pepper, and the medicinal benefits of aloe, native use of cassava (sanctioned by plantation owners eager to have slaves feed themselves), and the poisonous properties of Indian arrowroot. But Browne rarely mentioned local names. His explanation that the Indian arrowroot is so-called "because it was thought to extract the poison from the wounds inflicted by poisoned arrows of the *Indians*" concisely indicates which names matter and why. M. E. Descourtilz, who had been Rousseau's physician and edited his letters on botany, published an account of Caribbean natural history in *Voyages d'un naturaliste* (1809) that is only intermittently interested in local plant names like the synonym for *Mimosa urens*, the Baie-à-ondes, or bayaonde, and pictorial traditions. Some of its engravings look un-European: the engraving of bayaonde has enormous spines, rather fiercely out of proportion to its size; another presents pre-Columbian Taino figures carved in stone and perhaps on wood.[8] In these works, notice of local informants and assistance is typically indirect, if declared at all.

In the next chapter I turn to the other pole of Abbé Raynal's treatise, *L'histoire des deux indes*, or *The History of the Two Indies*, the Indian subcontinent, to examine a different history of indigenous natural history and European taxonomies. As the West-East polarity of Raynal's treatise assumed, India was likewise colonized and botanized in the name of empire. European governments assiduously pursued these ends under the aegis of various East Indian companies that were organized for trade and conquest. Yet European writers, including the British, could neither avoid nor fail to record the indigenous names and knowledge they found in ancient Indian texts and received from Indian informants. Because the British "discovery" of the Indian flora, arguably the most strenuously examined of all the colonial botanies, is so crosscut by indigenous knowledge and practices, it is not a story of single and determined purpose but instead one perforated and diffused by cross-purposes and unexpected doublings. It is at once a story that includes strong imperialist rhetoric and one in which Indian contributions are unevenly marked. Together these elements convey the friction of colonial contact, a friction also at work in Raynal's discussion of colonial power in West Indies, but less frequently recognized by colonial botanists in the slave colonies, where the political contract to imagine slaves as lesser and less able may have guided the rhetoric of natural histories.

CHAPTER 6

Reading Matter and Paint

INDIAN BOTANY AND THE BRITISH

> As *Archimedes*, who was happily master of his time, had not *space* enough to move the greatest weight with the smallest force, thus we, who have ample space for our inquiries, really want *time* for the pursuit of them. "Give me a place to stand on, said the great mathematician, and I will move the whole earth:" *Give us time*, we may say, *for our investigations, and we will transfer to* Europe *all the sciences, arts, and literature of* Asia. . . . Some hundreds of plants, which are yet imperfectly known to *European* botanists, and with the virtues of which they are wholly unacquainted, grow wild on the plains and in the forests of *India*.[1]

In these lines from "Design of a Treatise on the Plants of India," Sir William Jones declared what Indian and English readers in Calcutta and those back in England and on the Continent knew to be the case: the British who botanized in India did so for imperial gain, both scientific and economic. This bifurcation appears everywhere in Western knowledge production about and in the Indian subcontinent: spice trading, textile production, mapping, and massive surveys in which all aspects of Indian culture, cultivation, and social organization were detailed, state by state.[2] Jones made explicit how this double goal propelled the British effort to "discover," name, and classify all the plants of the vast Indian subcontinent, from its southern states to the Himalayas, for knowledge and profit. By the mid- eighteenth century, the business of the East India Company (EIC) was wide-ranging: it included indigo plantations; cotton manufactories; investigating Indian dyeing techniques and dyes; experimenting with planting cotton species in India that might, it was hoped, outproduce American long-staple cottons; and trying to set up tea plantations in India that could rival those in China and thus restore the loss of gold bullion to the far East.

My argument in this chapter puts this imperial and commercial history beside another story that haunts its folds and shadows. The products of British botanizing on the subcontinent—massive, illustrated books on the India flora, new systematic accounts of genera and species and plants that were mostly new to Europeans, and thousands of botanical drawings by Indian artists for British and some Indian employers—imply an intriguing quid pro quo: in exchange for an exotic, inexhaustible array of Indian plants, the British gave back in the kind that was theirs to give, with a thoroughly Derridean supplement—thousands of Indian drawings that recorded and repeated, in the mirror of art, the array of intricate, often exotic, botanical differences that constitute the appeal of Indian flora. Looking for all the world like gorgeous decoration, the Indian drawings were essential to the taxonomic project in which the British were engaged: in part because herbaria often did not survive in India or in transport back to Europe, but also because these hand-colored drawings were more than, or other than, documentary evidence of the exotic flora of the Indian subcontinent that the British cataloged and described for science, art, and profit.

Although the Indian drawings were commissioned to document plants with precision and in the style of botanical illustration that the British practiced and preferred, they were in fact different: more stylized and less absolutely faithful to documentary protocols. I surmise that it is precisely their difference that Westerners surreptitiously prized, a pictorial record that was as overabundant and exotic and material as Indian plants themselves. The British absorption in the Indian flora (so many plants, so little time) conveys a delight in its formal and taxonomic diversity that contradicts Western disdain for an allied baroque complexity in Indian art and ritual. G. W. F. Hegel's influential version of that imperial disdain lingered, tellingly, with what he called the perverse and degenerate forms of Indian art.[3] Among British botanists working in India, its flora prompted a competitive mimesis: in exchange for exotic plant structures and intense colors and scents, they responded with the best they had to offer: investigative zeal that stopped at nothing, neither mountains nor deserts, nor sickness; exhaustive classification of plant traits and indigenous plant names; and extended debates about names and categories. But there is more. By employing Indian artists to paint plant specimens, the British also gave back, as in a painted Indian mirror, the preeminent aesthetic representation of a scientific achievement that was in the end neither wholly British nor wholly Indian.

For British botanists in India, difference and differentiation, the twin engines of plant classification, required thinking about India minutely and particularly. Yet this impulse, more or less generic to botanists, was in fundamental friction with a concurrent European and British effort to imagine India as a civilization that could be understood and mapped in ways that would support imperial conquest.[4] Thus whereas European and, more insistently, British efforts to define Indian castes would appear to have been concerned to specify cultural differentiation, British arguments about caste quickly simplified its manifold complexity as an idea and cultural practice. The outcome was a view of caste that was limited to four fixed, religious groups whose imagined religious and cultural character was largely a matter of imperial convenience. The untouchables were quite literally outcasts—*out-castes*—because they belonged to none of those groups. This view of caste vastly simplified and rigidified an array of social and hierarchical categories that fluctuated locally and nationally across time.[5] In marked contrast, the British and European discovery of Indian flora was riveted by the variety of plants and their names and uses. Such plenitude invited the imposition of taxonomic order, but it also a sustained recognition of a complexity (morphological but also linguistic) that botanists tirelessly drove themselves (and others) to record.

These two perspectives on the British and Indian botany—the one imperial, the other something else living inside that imperial history—recall Dipesh Chakrabarty's analysis of the two histories that live uncomfortably beside each other in Marx's *Capital*. Much as the historicism of an earlier generation of European philosophers understood history as progressive and cumulative, however unevenly delivered at different times and in different places, so, argues Chakrabarty, did Marx understand capital and labor as terms that gradually assumed abstract, universalizing properties, whereby "abstract labor" acquires something like universal bargaining rights precisely because it can lift up, reinterpret all the local differences of labor and its products that an equally abstract Capital suppresses. For this reason, even Marx's history of capital paradoxically suppressed different labors and even the material products of labor in order to give them collective, abstract status to rival that of capital and money. Undeveloped yet implicit in Marx's analysis is what Chakrabarty calls History 2, which reported the particular and thus differentiated reality of work that in turn conveyed specificities and possibilities that Capital would prefer to foreclose.[6]

In the British encounter with Indian botany, History 1 is the story advanced in the name of imperial ambition with an eye to profitability and mastery, a story that subsumed cultural and labor differences such that profit and scientific mastery transformed botanical differences among Indian plants into a common, abstracting denominator whose name was profit or science or both. History 2, the story of Indian involvement in the apparently British, apparently colonial, mapping of the Indian flora transformed the usual binaries of colonial discovery and conquest—colonial/colonized; dominant/oppressed, native informant vs. colonial or metropole—as well as their postcolonial inversions, hybridity and mimicry. None of these terms and pairings captures the fungible and widening "border" inhabited, at key moments, by the British and the Indians who worked on Indian plants and illustrated them for the British. The character of this border was less that of a secure, fortified frontier, at least on its insides, than a permeable membrane through which British and Indian claims or features briefly passed.

Contemporary scholars have begun to describe a finely tuned network of practices and exchange between British and Indians a few decades before and after 1800. Examining Indian legal practices before Jones arrived to codify Indian laws and others arrived to survey other aspects of Indian culture, Bernard Cohn argues that Indian informants and teachers performed more significant roles than the British were typically willing to grant.[7] In his investigation of Indian manufacturing communities in the eighteenth century and broader financial developments that assisted British efforts to control markets and then whole states, C. A. Bayly shows that Indian artisans and financial traders continued to operate with some degree of independence throughout the century, even as Indian art and culture expansively redirected traditional techniques and styles.[8]

Much has been written about how the British and Europeans saw botanical India—which Himalayan plants seemed to them "European," how they invoked the language of the picturesque, the sublime, and the phytogeographical perspective of Alexander von Humboldt, transplanted from South America to India.[9] Arguing against this Eurocentric tendency, Richard Drayton predicts that "we may eventually discover that the modern world was produced through the collaboration of the labour, wit, and learning of all the world. Neither the Scientific nor the Industrial Revolutions, no merchant bank or university would have arisen in the West had they depended on the material or cultural resources of the people who lived closest to their headquarters."[10]

Drayton's suggestion points to a home truth about the European discovery of Indian flora. For more than two millennia before Europeans arrived in India, a vast and consistent body of knowledge about Indian plants, called *Vanaspati*, organized medical and ritual knowledge. In the early seventeenth century, when the Dutch consulted Erhaza physicians and toddy tappers to create the first European flora of Indian plants along the Malabar coast, what the Dutch received was existing botanical knowledge about one region of the Indian subcontinent. Had they consulted other Indians elsewhere who were similarly knowledgeable, the Dutch could have gathered similar information about the rest of India.

Viewed from the vantage point of what the Dutch learned in India and then communicated to other Europeans, the term *native informant* is remarkably inept. It is also stubbornly insistent that what could be learned from Indians would be explained only by using Western systematic protocols. This is not precisely what happened. The Dutch recorded plant names and systematic relations that Indian physicians provided. Linnaeus then adopted plant identifications, names, and systematic protocols for more than two hundred Malabar plants from Hendrik Adriaan Van Rheede's *Hortus Indicus Malabaricus* (1678) and published them in *Species Plantarum*.[11] The transmission of botanical knowledge from Indians to the Dutch and then to Linnaeus invited allied notice of how, two centuries later, Indians provided botanical information to British botanists who hired them to collect, organize, and depict plants. Henry J. Noltie's insistence that Indian plant drawings have distinct taxonomic value marks an important step. Savithri Preetha Nair's groundbreaking analysis of the career of the Raja Serfoji II of Tanjore, South India (1777–1832) and ongoing scholarship on botanical knowledge preserved among textile communities in Masulipatam convey what may lie ahead as scholars step back or to the side of British imperial thinking about Indian plants and plant knowledge. Serfoji hired Indian artists to paint plants, including one identified as "Thomas from Masulipatam," who was employed to make drawings of the "curious plants, creepers, trees, and birds."[12]

Eighteenth-century British botanists relied on Indian gardeners, pandits, and artists to a degree that the imperial rhetoric of Jones's provocative call to map Indian flora obscures. Those who sent Indian drawings home to be reproduced in expensive folio volumes hoped both to profit from those publications and to persuade their EIC employers to let them botanize on company time. Their carefully worded requests that company officials authorize a cotton plantation here or more botanical exploration there mask a botanical ambition, even desire, that could be nearly unstoppable. The British, or more precisely the Scottish, botanists who dominated scientific inquiry among members of the EIC, moved from one rare, new, or even strangely familiar Indian plant to another, from South India to high in the Himalayas, beyond the Indian border, as though a necessary engine had been set in motion, with no prospect of an end. If this is in one sense the repeating history of taxonomy, it takes a fascinating turn in India, where EIC employees tried to survey all of India, in all its human and natural variety. That all of this effort focused on a vast subcontinent whose plenitude exceeded the ability of travelers to map or classify it is, I surmise, one reason botanists were fascinated by the flora of India. Here were glorious, exotic Indian plants—or, more precisely, glorious, exotic drawings of those plants. Give us more time, more support, the British (as well as the Scottish, Irish, and Welsh) urged, and we will send back more drawings, more plants, from a subcontinent that appears to offer an endless supply.

On both sides of my argument, I am attentive to efforts to recover subaltern, or as Chakrabarty puts it, "subalterned," informants. Gyan Prakash argues that it has always been impossible to recover "the subaltern as a full-blooded subject-agent," in colonial documents, not because the subaltern has no voice, as some have understood Gayatri Spivak to have claimed, but rather because the colonial archive either does not give the subaltern a place from which to speak or it allows only as much speech as it wishes to preserve. For this reason, Spivak contends, the category of the indigenous or native informant is foreclosed, closed over by and in an intervening colonial record.[13]

Noting that the Greek for *archive*, αϱχωνα, refers to the domicile of the archon, to whom all important documents were addressed, Jacques Derrida presented the archive as an "uncommon place, this place of election where law and singularity [that is, the singularity of documents so arranged and addressed] intersect."[14] The archive/archon insisted as its rule of law on organizing and safeguarding records according to its own protocols. Whatever it decided was not needed was thrown away, or perhaps it never arrived at the archive because others had already decided what was important enough to send in the first instance. The magisterial archon of Derrida's analysis of what he calls "archive fever" brilliantly captures the principle of disciplined judgment and collection that marked the colonial archive, where other archives might be more casually assembled, without much if any supervision of their contents. In the regime of the colonial archive, a native informant turns up only if the space of colonial production required his or her presence.

As Spivak's ventriloquism of the logic of the colonial archive makes clear, such archives aspire to totality, a complete transcription of the world and peoples they govern. Yet even here there may be glimpses of something beyond the frame of colonial speech and purpose, precisely because colonial narratives and facts are, Sara Suleri has suggested, "vertiginous: they lack a recognizable cultural plot; they frequently fail to cohere around the master-myth that proclaims static lines of demarcation between imperial power and disempowered culture," moving instead "with a ghostly mobility."[15] For precisely because, as Michel Foucault and Giorgio Agamben have argued, archives display incompletely the cultures they purport to chronicle, their incompleteness invites another route in, another way of reading what is preserved that attends to the possibility of speaking behind what is actually said.[16]

My project here is not to recover that informant as an originary other but to read/address what Spivak calls a ghosted or implied other that might be glimpsed in a narrative that confuses, rather than solidifies, the distinction between colonial self and native other.[17] I do not claim that there existed "a commonwealth of letters" in which Indians and Europeans were equal partners; nor do I attempt to create what Rey Chow calls "a phantom history in which natives appear as our equals."[18] Rather, I revisit the British investigation of India and its flora to consider how fixing the British on one side and the Indians on the other misses subtle and defining interactions.[19] Even the writing of those who elsewhere speak for the British colonial presence and interests offers glimpses of a moment when Indian and

British claims to knowledge existed side by side in the fifty years between the arrival of British botanists in the 1780s and the Indian Mutiny, or War of Independence, in 1857. Here is a clandestine marriage of a different sort, hidden in the apparently tangential presence of Indian knowledge and tradition in an apparently British and European inquiry.

This chapter next surveys private and public British views on India from the last decades of the eighteenth to the early nineteenth century—both those that were emphatic about the cultural superiority of white Europeans over oriental darkness and depravity and some that were more critical of British imperialism. In different ways, each of the romantic writers I discuss in this part, Thomas Moore, Sydney Owenson, and Percy Shelley, is immersed in the larger British argument about India that makes even a radical like Shelley sound like an imperialist. In a later section ("British Botanists in India"), I argue that as the British wrote about Indian plants, they did so by reading plant materials through various lenses, their own and those provided by Indian informants and artists. The two botanists I consider (one amateur, the other professional, by late eighteenth-century standards) are William Jones (the passionate amateur) and William Roxburgh, the EIC medical officer and botanist whose massive effort to codify the Indian flora included hiring Indian artists to depict thousands of plant specimens. Another section ("Indian Botanicals and Pictorial Traditions") turns to the Indian artists whom Roxburgh hired. Their botanical drawings, which mix European practices with elements that echo widely dispersed traditions in Indian pictorialism across several media as well as time and place. Finally I return to the British colonial archive to glimpse Indian artists as they declare who they are and what they know.

BRITISH READING MATTER

Despite Robert Clive's brutal campaigns for British ascendancy in India and Warren Hastings's subsequent use of his position as the first governor-general of India (1776–1784) to profit personally from trading in Indian goods (first among equals in this regard), in the last decades of the eighteenth century the British who went to India as EIC officials were, as William Dalrymple has put it, not "a small alien minority locked away in their Presidency towns, forts and cantonments," but participants in "unexpected and unplanned minglings of peoples and cultures and ideas."[20] At times mingling involved cultural disguise. Arriving in Bombay in 1782, the Scot John Gilchrist grew a long black beard and put on Indian clothes to travel through northern India, hiring "learned Hindoostanees" to help him write a grammar of "Hindustani."[21] Satirizing English mores at the end of the eighteenth century, Elizabeth Hamilton adopted the voice and perspective of a "Hindoo Rajah" who tours England in shock and horror (and occasional nips of pleasure) over the strange ways of the English. The sultan in Isaac Bickerstaff's play *The Sultan* turns the tables by behaving more like the stereotype of an Enlightened European monarch than the Oriental despot that his English visitors expect.[22]

Elsewhere across the metropole, progovernment rhetoric tried to drown out opposition to the official (and unofficial) British rush to colonial power in India. The noise began even before the Hastings impeachment trial began in 1786. The point more widely debated was not whether the British should be an imperial power in India but how they might (profitably or ethically) rule India like a feudal demesne.[23] Although clearly in the minority, anticolonial commentary was nonetheless acute. Writing privately to Horace Mann during the 1760s and 1770s (Mann was in France at the time, assigned to keep track of the Jacobite Pretender), Horace Walpole skewered British maneuvers to consolidate and extend imperial control over India, beginning with the self-serving claim that the East India Company was formally distinct from the British government. Walpole rolls both institutions into his accusatory *we*: "We have outdone the Spaniards in Peru. . . . We have murdered, deposed, plundered, usurped. . . . Well! I wish we had conquered the world and had done! I think we were full as happy, when we were a peaceable quiet set of tradesfolks, as now we are heirs apparent to the Romans, and overrunning East and West Indies."[24] For its part, the *Gentleman's Magazine* refused to prettify British complicity in the Bengal famine of 1769: "Nay, what think you of the famine in Bengal, in which three millions perished, being caused by a monopoly of the provisions by the servants of the East India Company? All this is come out, is coming out—unless the gold that inspired these horrors, can quash them."[25] Walpole comments bitterly: "Recollect what I have said to you, that '*this world is a comedy to those who think, a tragedy to those who feel.*'"[26]

Commenting on Fox's India Bill of 1783, which went down in the ensuing brouhaha that led almost immediately to Fox's defeat and the installation of William Pitt as prime minister, Walpole bitterly assessed the contrast between Spanish and Portuguese imperial manners in the New World and those of the British in old India:

> Our nabobs do not plunder the Indies under the banners of piety like the old Spaniards and Portuguese. I call Man *an aurivorous* [gold devouring] *animal*. We pretend just now to condemn our own excesses, which are shocking indeed—*sed quis custodiet ipsos custodes* [but who will guard the guardians]? a Parliament is a fine court of correction! The Lord Advocate of Scotland, who has sold himself

> over and over, is prosecuting Sir Thomas Rumbold for corruption at Madras! . . . We talk and write of liberty, and plunder the property of the Indies.[27]

Two years later, Walpole accurately predicted that Hastings would eventually be acquitted, inasmuch as his private predations merely echoed an illegality that the British government would never challenge: "I suppose Mr Hastings will be honourably acquitted. In fact, who but Machiavel can pretend that we have a shadow of title to a foot of land in India; unless as our law deems that what is done extra-parochially, is deemed to have happened in the parish of St. Martin's in the Fields, India must in course belong to the crown of Great Britain."[28]

Some novelists and poets were nearly as sharp. The protagonist of Henry Mackenzie's *Man of Feeling* (1771) asks, "What title have the subjects of another kingdom to establish an empire in India?" William Cowper asks the same question with a poetic syntax that begins, or drawls, with an air of vapid colonial doublespeak that is sustained for two lines, but snapped away in the third: "Is India free? And does she wear her plum'd / And jewell'd turban with a smile of peace, / Or do we grind her still?"[29]

Thomas Moore, whose *Lalla Rookh* (1817) profits from its heavily annotated and spicey Orientalist exoticism, not incidentally furnished with a great many Indian plants and trees, snarls in a note:

> Objections may be made to my use of the word liberty, in this, and more especially in the story that follows it, as totally inapplicable to any state of things that has ever existed in the East: but though I cannot, of course, mean to employ it in that enlarged and noble sense which is so well understood at the present day, and, I grieve to say, so little acted upon, yet it is no disparagement to the word to apply it to that national independence, that freedom from the interference and dictation of foreigners, without which, indeed, no liberty of any kind can exist; and for which both Hindoos and Persians fought against their Mussulman invaders with, in many cases, a bravery that deserved much better success.[30]

Here and elsewhere, Mohammed Sharafuddin notes, Moore's highly polished contribution to the Oriental romance offers an implicitly allegorical argument about Ireland. Defending the Indian desire for liberty despite being defeated by invading powers hints, as such allegories do, that a precisely parallel argument could be made about the Irish desire for liberty despite British conquest.[31] Moore's doubly charged insistence that Indians (like the Irish) desire liberty even after being conquered sharply rebuts the claim, often repeated, that Indians had no desire for liberty, having been, it was argued, spinelessly passive before invading forces.

One version of that claim appeared in Alexander Dow's translation of a standard history of India by the Persian historian Firistah (who would have had his own political reasons for dismissing Hindu India), typically identified as Dow's *History of Hindostan* (1768–72). Ventriloquizing his source, Dow declares that the "mild, humane, obedient, and industrious" Hindus "are of all nations the most easily conquered and ruled."[32] Oliver Goldsmith reprocesses Dow's description to construct this darkened image of the Oriental subaltern (Goldsmith chose not to invoke its Orientalist counterpart, the Mughal despot who rules over his subservient Hindu subjects, perhaps because his will to power over the Indian subcontinent might remind readers of British efforts to do the same):

> The Indians have long been remarkable for their cowardice and effeminacy, every conqueror that has attempted the invasion of their country, having succeeded. The warmth of the climate entirely influences their manners; they are slothful, submissive and luxurious: satisfied with sensual happiness alone, they find no pleasure in thinking; and contented with slavery, they are ready to obey any master.[33]

Balachandra Rajan and Nigel Leask have noted that Percy Shelley, otherwise on the liberal if not radical side of English politics, accepted the view Goldsmith presented, at least in broad outline, as the reason the Indians needed the British. In the *Philosophical View of Reform*, Shelley argued that the British introduction of missionaries there introduced "an outworn incumbrance" inasmuch as Indians had long been "cramped in the most severe and paralyzing forms" of belief. Indeed, Shelley assumes here that the only real knowledge Indians have is Western: "Many native Indians have acquired, it is said, a competent knowledge in the arts and philosophy of Europe, and Locke and Hume and Rousseau are familiarly talked of in Brahminical society. But the thing to be sought is that they should, as they would if they were free, attain to a system of arts and literature of their own."[34] The modal qualifiers of Shelley's second sentence, to which Leask calls attention, insist that Indians are not free (cramped by belief and by Mughal rule) and lack "arts and literature" of their own.[35]

This last claim is remarkable given the wide dissemination of British writing on India by 1800, a virtual cottage industry that Sir William Jones began in Calcutta with *Asiatic Researches*, then quickly disseminated to British readers in successive editions of *The Asiatic Miscellany* and *The New Asiatic Miscellany*.[36] Balachandra Rajan drily notes that "Romantics, even radical romantics, were not particularly sensitive to the expansion of empire in India under their noses at a time when paradigms of liberty were in the air."[37] Leask says of this strange and widespread romantic blindness that by 1818 liberal British writers viewed India as a kind of laboratory for modernity where, if outmoded beliefs could be untaught, there would be a wide cultural and commercial space in which to craft a modern state. Shelley's unfinished Oriental tale of 1814, *The Assassins*, anticipated this view by imagining India as an empty space (a textbook example of a "condition contrary to fact"), thereby avoiding, as Leask astringently puts it, "the perennial problem of *real* colonialism, namely what to do with the natives."[38]

In 1817, the same year that Moore first published *Lalla Rookh*, the colonial drumbeat intensified with James Mill's *History of British India*, which concludes by insisting, forty years before Indian officially became "British India" in 1858, on the advantages of British rule for India:

> For, although the country has suffered, and must ever suffer, many and great disadvantages from the substitution of strangers for its own functionaries, its own chiefs, its own sovereigns, it has been, in some degree, compensated for their loss, by exemption from the fatal consequences of native mis-rule—by protection against external enemies—by the perpetuation of internal tranquility—by the assured security of person and of property—by the growth of trade—the increase of cultivation—and the progressive introduction of the arts and sciences, the intelligence and civilisation of Europe.[39]

Five years after Thomas Macauley published the *Minute on Education* (1835), with its insistence that Indian knowledges amounted to nothing more than "History, abounding with kings thirty feet high, and reigns thirty thousand years long—and Geography, made up of seas of treacle and seas of butter," he presented an argument for British imperial rule over India that was by then familiar, if couched in a strange mix of bluntness and evasiveness: "There never, perhaps existed a people so thoroughly fitted by nature and by habit for a foreign yoke."[40] Blunt, that is, about Indian fittedness for foreign domination, but evasive about whose foreign yoke is at issue, a point that had not been in doubt since 1777, when Abbé Raynal considered whether the British might rule India ethically or rapaciously.[41] Macaulay's recalibration of Wordsworth's notice of the "fittedness of man and nature" seeks to make the imposition of a British yoke both inevitable and in some sense therefore (putatively) natural.

Perhaps the most damaging early nineteenth-century account of India is G. W. F. Hegel's in his *Philosophy of History*, published in 1837 but based on lectures he gave in 1830 and 1831. Recycling and redirecting fifty years of writing in which India had been characterized as weak, irrational, and easily conquered, Hegel elevates this argument to a philosophical program whose revisionary, dialectical energy appears to serve Spirit at India's expense. The botanical language that permeates Hegel's strategically "soft" formulation of his position emphatically declares the figurative suggestibility made available by British and European discussions of Indian flora over the same period. Put briefly, India is said to be in the line of succession that begins with female beauty and ends with the female soul and the garden of love. The "flower-life" and "rose-breath" of India inhabit a state of pure emotion that is, like the Garden of Adonis in Spenser's *Faerie Queene*, delusory:

> There is a beauty of a peculiar kind in women, in which their countenance presents a transparency of skin, a light and lovely roseate hue, which is unlike the complexion of mere health and vital vigor—a more refined bloom, breathed, as it were, by the soul within—and which the features, the light of the eye, the position of the mouth, appear soft, yielding, and relaxed. This almost unearthly beauty is perceived in women in those days which immediately succeed child-birth; when freedom from the burden of pregnancy and the pains of travail is added to the joy of soul that welcomes the gift of a beloved infant. A similar tone of beauty is seen also in women during the magical somnambulic sleep, connecting them with a world of superterrestrial beauty. . . . [such as that presented in] Schoreel's painting of the dying Mary . . . Such a beauty we find in the loveliest form of the Indian world; a beauty of enervation in which all that is rough, rigid, and contradictory is dissolved, and we have only the soul in a state of emotion—a soul, however, in which the death of free and self-reliant Spirit is perceptible. For should we approach the charm of this Flower-life . . . in which the whole environment and all its relations are permeated by the rose-breath of the Soul, and the World is transformed into a Garden of Love—should we look at it more closely, examine it in the light of Human Dignity and Freedom, we might the more its first sight impressed [or "bribed"] us, find all the more depravity in every direction.[42]

Once Hegel makes the soft, yielding spiritual beauty of women analogous to "the loveliest form of the Indian world" (he has more to say elsewhere about monstrous Indian forms), that beauty is swiftly devalued. Its enervation reveals a soul in a state of emotion that turns the "dying Mary" of Hegel's intervening analogy into the death of the free Spirit.

Instead of agreeing with Friedrich Schlegel's 1803 exclamation that "absolutely everything, is of Indian origin," an opinion fostered by Jones and other syncretists who found in India evidence of original literature and knowledge and points of correspondence to Western myth and religion, Hegel concluded that India represented a historical dead-end for the work of Spirit, an unchanging culture immune to progress, caught in the folds of the rigid caste system that, to be sure, Europeans had already worked hard to rigidify. From the perspective suggested by Balachandra Rajan's long view of what Western writers have said about India, Hegel completes a line of figurative critique that began with Milton's *Paradise Lost*, where Satan's appearance in Eden beside the "tree of life" demonized the banyan tree (*Ficus bengalensis*) and the serpent sacred to Hindus and Buddhists.[43]

Despite the swelling tide of British rhetoric, punctuated as it is by stereotypes that Edward Said described in *Orientalism*, romantic writing about India is double-voiced, divided between a "soft" imperial cadence and one that is more radically attentive to the possibility that India has knowledge as well as cultural and even religious practices that might be superior to some in the West. Moore's *Lalla Rookh* offers a slight, but suggestive first instance. Its female protagonist is named after a flower ("Tulip Cheek") and the poem's annotations frequently discuss Indian flowers and trees. The repeated notice of Indian plants and trees in the verse and the notes thickens the exotic texture of the poem. "Tulip-beds" gets this note: "The name of tulip is said to be of Turkish extraction and given to the flower on account of its resembling a turban"; the sweet-night flower is identified in its note as "The sorrowful nyctanthes, which begins to spread its rich odour after sunset" (no source given but it is likely Jones, who discusses this plant); Moore's note describing the Cámalatá quotes Jones, who in turn quotes another (unidentified) source: "The Cámalatá (called by Linnaeus, *Ipomaea*) is the most beautiful of its order . . . its elegant blossoms are 'celestial rosy red, love's proper hue,' and have justly procured it the name of Cámalatá, or Love's Creeper."[44] Read against Hegel's use of botanical figures like "flower-life" and "rose-breath" to present India as feminine, exotic, but not much else except deliciously dangerous to the survival of Spirit, these annotations invite a different view of India and its flora, one that supposes that knowing these plants and trees and their role in Indian belief and story matters, and further that learning about India in this way could be expected to engage the British reading public, as indeed it did.

Percy Shelley's *Prometheus Unbound* (1820) and Sydney Owenson's novel *The Missionary* (1811) also refer to the sensuous and amorous appeal of Indian flora. In both works, female characters—Shelley's Asia and Owenson's Luxima—are presented allied with way of botanical figures without the usual British rhetoric about weak, effeminate India. In all three works, but particularly these last two, exotic Indian flora register ethically valuable figures. At the beginning of *Prometheus Unbound*, Asia lives in a valley of the Caucasus, as does her lover Prometheus, from whom she is separated by his punishment or perhaps by his curse. As Rajan notes, before she arrives there, Asia's valley was as desolate and isolated as the gorge where Prometheus is imprisoned. Thanks to Asia's "transforming presence" her valley is now verdant, full of flowers and herbs.[45] The allegorical figure of Love in the play, Asia compares her task to that of a vine:

> and Love he sent to bind
> The disunited tendrils of that vine
> Which bears the wine of life, the human heart.[46]

As the one who creates or preserves botanical life in the same region of the Indian subcontinent where Prometheus's rebellious presence, as Earth sorrowfully notes in the play, introduces a scorched earth policy instead of the rhythm of the seasons, Asia is nothing like the supine female figure that appears so often in contemporary British accounts of India as an effeminate, superstitious people. When she confronts Demogorgon in act 2 of Shelley's drama, she asks him some tough questions about the origin of evil that move closer and closer to identifying Prometheus as the bringer of fire, intelligence, thought, but also war and pestilence to the world. Her line of questioning is so uncomfortable that Demogorgon (and perhaps Shelley the poet, who has vested a great deal in Prometheus's "moral excellence") shuts it down, booming from the depths of his cavern beneath the earth, "the deep truth is imageless."[47]

Shelley's Asia is modeled, as several scholars have noted, on Luxima, the heroine of Owenson's *Missionary*. Leask suggests that Shelley's admiration for Owenson's heroine, a Kashmiri priestess of a Hindu religious cult, prompted him to mute his radical skepticism about Hinduism.[48] Owenson's plot, briefly put: the Portuguese Jesuit Hilarion travels to India to convert the natives and is advised to begin with Luxima, inasmuch as her conversion would be likely to bring in many more. Instead, the two fall in love, marry, then go into hiding. Eventually she dies, and the novel's ambivalent stance on Hilarion's rigid Catholicism leans toward Luxima as the embodiment of spiritual as well as human eroticism. Owenson sidesteps British colonial notice of the sexual initiations of temple priestesses by priests or landlords by identifying Luxima as the granddaughter rather than the ecstatic sexual partner of the Brahmin priest, Rah-Singh (88).[49] Although the marriage of Hilarion and Luxima seems to offer an instance of a "conciliatory marriage between colonizer and colonized,"[50] the pressures on this marriage are far too great for the partners to bear.

What interests me is the narrator's split and shifting point of view on these characters. Because she is identified as "languid" (37), Luxima in name and temperament seems to embody the European view of effeminate India. She is so evidently near kin to Luxury—the province of Oriental silks, soft paisleys, perfumed flowers, and spices—and to Lakshmi, the Hindu goddess of wealth and prosperity, who Jones identifies as the "Goddess of Abundance," "the Ceres of India, the preserving power of nature."[51] Yet beside Hilarion she is the richer moral and ethical figure precisely because her temperament is illuminated both by the divine and by her capacity for human attachment. By contrast, Hilarion is a problematic figure of sublime isolation and, as such, not of the same "species" as ordinary men (74). The narrator's reserve in earlier volumes of the novel about both characters disappears in the scene that immediately follows the account of their union, positioned at the beginning of the third volume:

> These rude fragments [of rocks] collected by time and chance, cemented by the river Slime, and intermixed by creeping plants, and parasite grasses, become small but lovely islets, covered with flowers, sowed by the vagrant winds, and skirted by the leaves and blossoms of the crimson lotos, the water-loving flower of Indian groves. This scene, so luxuriant and yet so animating, where all was light, and harmony, and odour, gave a new sensation to the nerves, and a new tone to the feelings of the wanderers, and their spirits were fed with the balmier airs, and their eyes greeted with lovelier objects, than hope or fancy had ever imaged to their minds.—Sometimes they stood together on the edge of the silvery flood, watching the motion of the arbours which floated on its bosom, or pursuing the twinings of the harmless green serpent, which, shining amidst masses of kindred hues, raised gracefully his brilliant crest above the edges of the river bank. Sometimes from beneath the shade of umbrageous trees, they beheld the sacred animal of India breaking the stubborn flood with his broad white breast, and gaining the fragrant islet, where he reposed his heated limbs; his mild countenance shaded by his crooked horns, crowned by the foliage in which he had entangled them; thus reposing in tranquil majesty, he looked like some river-deity of ancient fable.
>
> Flights of many-coloured perroquets, of lorys, and of peacocks, reflected on the bosom of the river the bright and various tints of their splendid plumage; while the cozel, the nightingale of Hindoo bards, poured forth its song of love from the summit of the loftiest *mergosa*, the eastern lilac. It was here they found the *Jama*, or rose apple-tree, bearing ambrosial fruit—it was here that the sweet sumbal, the spikenard of the ancients, spread its tresses of dusky gold over the clumps of granite, which sparkled like coloured gems amidst the sapphire of the mossy soil—it was here that, at the decline of a lovely day, the wanderers reached the shade of a natural arbour, formed by the union of a tamarind-tree with the branches of a *covidara* [kovidara, *Bauhinia variegate*],[52] whose purple and rose-coloured blossoms mingled with the golden fruit which, to the Indian palate, affords so delicious a refreshment. (201–2)

Modern readers of this long passage might agree with the caustic reviewer of Moore's *Lalla Rookh*, who declared him or herself ready to "die of a rose in aromatic pain."[53] The secondary target of this remark is Keats's poetry, but it might as easily have been Shelley's *Alastor*, where a neoplatonic diction of "odours" and "many-coloured" natural objects governs the figurative landscape.[54] Apart from the "river Slime," remarkable in its own right, Owenson's prose insists that this Indian landscape is quasi-human and, to the extent that it is human, it is female: "the enchanting glen . . . reposes," "the bosom of the Behat," and, most explicitly, in the declaration that Luxima looked "like the emblem of that lovely region" (202). An iconic Indian paradise, this place has a serpent too, not the toxic satanic form of Milton's poem and Genesis but a "harmless green serpent." The only troubled occupant of the place seems to be Hilarion, whose "diseased mind discovered a lurking crime in the most innocent enjoyments" (202). Unlike Luxima, whose being expands in this place, leaving that Oriental languidness behind, her lover's cramped "moral economy" poisons his temper.

The narrator's decision to abandon here the novel's earlier ambivalence about both characters in favor of Luxima is learnedly botanic. Indian plants and trees that specify the texture of this paradise are among those that Jones repeatedly discusses, in particular the "spikenard of the ancients," a phrase the Owenson lifts from Jones's essay on this topic, and the *Jama* or rose-apple, which Moore had mentioned in his oriental tale. Between 1857 and 1858, in the last months of her life that coincided with the Indian Mutiny/first Indian War of Independence, Owenson retitled the novel *Luxima, the Prophetess: A Tale of India*, adjusting its presentation of Indian religion, adding a "converted Hindoo" teacher to the portfolio of Hilarion's early religious training and a new preface that comments on "the recent melancholy occurrences" in India.[55] Revoiced or translated into botanical figures, the idea of India is transformed in Owenson's novel. Guided by Jones, when Shelley and Owenson imagine Asia and its women as flowers or embedded in a densely floral Indian setting, they convey a view of India emptied of critique and reservation. Its counterargument is suggested by earlier British accounts of the banyan tree.

The aerial root system of the banyan (*Ficus bengalensis*) allows branches to root well beyond the central trunk, which often dies off as the rest of the tree survives for centuries, spreading out and around to provide space and shelter beneath for worshippers, traders, and even whole villages. For Hindus, the banyan signifies immortality and sacred, ritual fertility, hence the frequent appearance at its base of the emblems of Shiva worship, the yoni and lingam, to signify the fertility promised by their union as stylized male (a lotus stalk) and female members. Believed by Hindus and Buddhists to be the male version of the female bodhi tree (*Ficus religiosa*), the banyan is revered equally with the bodhi. Early European travelers who fixed on the yoni-lingam accounts tended to identify the banyan with a promiscuous sexuality, no doubt encouraged by the fantastic appearance of those aerial roots.[56] European and British depictions of banyan trees, a virtual set piece in picturesque travel accounts, typically emphasize their vastness and the social, commercial, and religious activities they shelter. Maria Graham's 1810 journal account of a banyan near Bombay, and her drawing of it, which appears as an engraving in the published work, present it as the site of religious activity. A small relief of

Ganesh, the god who protects Hindu temples in India, is tucked into the base of what appears to be main trunk of the tree. Barely discernible beside the Ganesh are two smaller sculpted versions of the lingam and the yoni, separate and thus not in the suggestive alignment that earlier visitors had found alluring or shocking or both. Noting the sacredness of the tree and its typical proximity to "*Pagodas*, as the Europeans call the Hindoo temples," Graham records what she saw: "The natives walk around it in token of respect, with their hands joined, and their eyes fixed on the ground; they also sprinkle it with red and yellow dust, and strew flowers before it; and it is common to see at its root stone sculpture with the figures of some of the minor Hindoo gods."[57]

Forty years earlier, the British artist James Forbes had sketched what may have been the same tree, identified as "A View of Cubber Burr, the celebrated banyan tree on the banks of the Nerbudda." In 1787, James Wales exhibited several paintings based on Forbes's drawings, among them *Cubeer Bur, the Great Banyan Tree.* When William Hodges included yet another engraving of this iconic tree in his *Travels in India during the Years 1780, 1781, 1782 and 1783* (published 1793), after a painting he had exhibited in 1786, he was following an already well established trade route for the Indian picturesque. After him would come Thomas Daniell's 1796 watercolor of another banyan in Bihar, Graham's 1810 drawing, and finally Forbes's, which he had engraved and printed in his *Oriental Memoirs* (1813). The marked pro-British flavor of Hodges's narrative and drawing is, Daniel O'Quinn notes, consonant with the occasion, for Hodges traveled in Mysore with his patron Warren Hastings during one of his campaigns against Tipu Sultan. Hodges's iconographic rendering of the banyan as a refashioned emblem of British security (the prow of a British ship is visible on the shore to the left) and stability for Indians who gather under this banyan tree near the Malabar coast (and near Tipu Sultan) involved a degree of delicate adjustment.[58] In his 1813 *Memoirs*, Forbes describes a gathering of EIC members who camped beneath the banyan tree in the scene he had depicted in his earlier drawing:

> The chief was extremely fond of field diversions, and used to encamp under [the celebrated Cubber Burr tree] in a magnificent style; having a saloon, dining-room, drawing-room, bed chambers, bath, kitchen, and every other accommodation, all in separate tents; yet did this noble tree cover the whole; together with his carriages, horses, camels, guards, and attendants. While its spreading branches afforded shady spots for the tents of his friends, with their servants and cattle.[59]

Forbes now noted the military advantage of using banyan trees for cover. Fighting in Gujarat against stubborn Indian rulers who preferred to remain sovereign powers, the British found that a single banyan tree could protect between seven and

Fig. 28. *Banian Tree*, Maria Graham, engraving, *Journal of a Residence in India*, Edinburgh, A. Constable and Company, 1812. 12.7 × 19.4 cm. (5 × 7⁹⁄₁₆ in.). Courtesy of the Department of Special Collections, General Library System, University of Wisconsin–Madison.

Fig. 29. *The Indian Burr, or, Banian Tree*, James Forbes (1749–1819), pen and ink and watercolor, 21.5 × 31cm. (8.46 × 12.2 in.). In A *Natural Description of the Island of Bombay; with its Trees, Fruits, & Vegetable Productions*, vol. 2 (1767) of Forbes's A *Voyage from England to Bombay, with Descriptions in Asia, Africa, and South America*, 1776–1784 (vols. 1–13 of his *Descriptive Letters and Drawings Presented to Elizabeth Forbes*, 1800). Yale Center for British Art, Paul Mellon Collection.

***Fig.* 30.** *Convolvulus gangeticus R.*, Roxburgh Indian artist, watercolor on paper. 49 × 35.5 cm. (19.29 × 13.98 in.). Royal Botanic Gardens, Kew (2006). Roxburgh No. 1793, *Flora Indica* 1, 467.

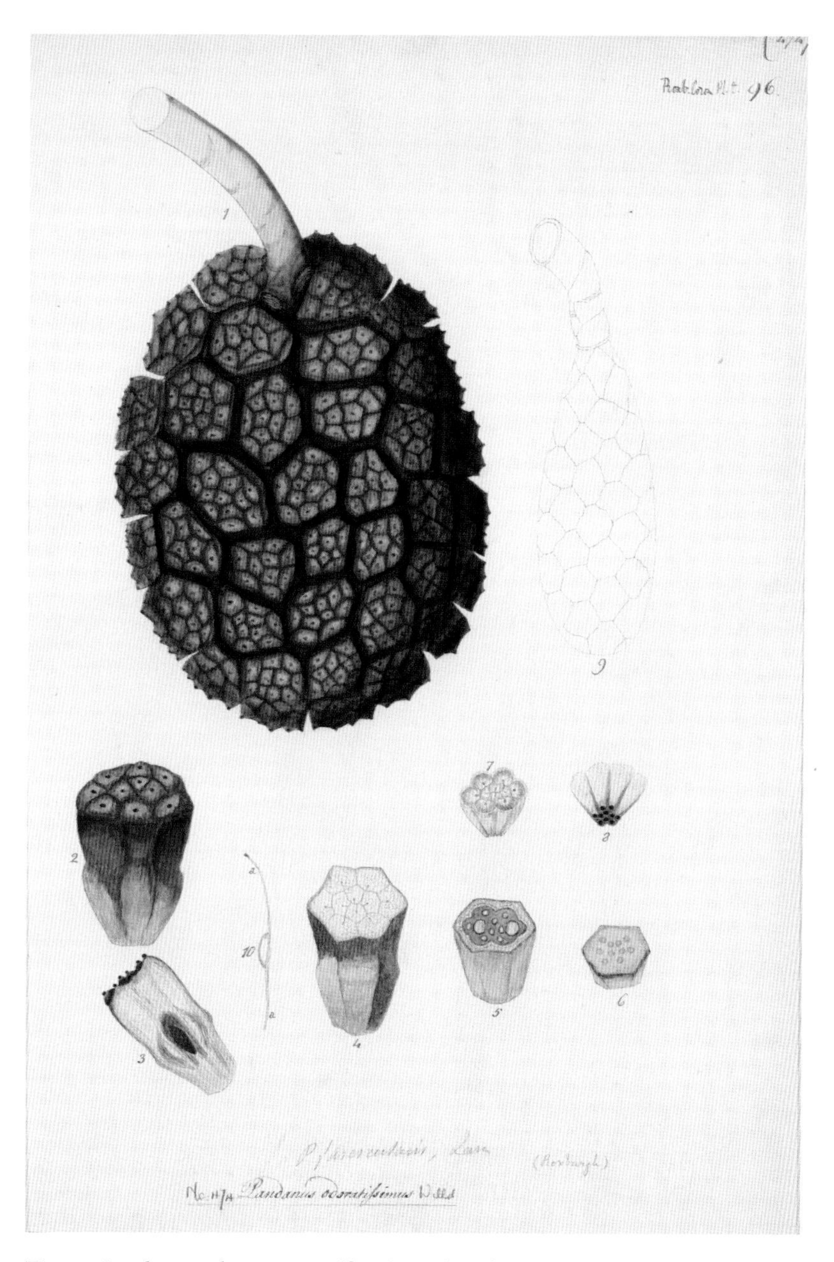

Fig. 31. *Pandanus odoratissimus* [fruit], Roxburgh artist, watercolor on paper, 49 × 35.5 cm. (19.29 × 13.98 in.). Royal Botanic Gardens, Kew. Roxburgh No. 474, *Flora Indica* 3(1832), 738.

Fig. 32. *Pandanus odoratissimus* [whole plant], watercolor on paper, 47 × 26.9 cm. (18.5 × 10.59 in.). Copyright Natural History Museum, Botany Library, London. Fleming Collection, plate 598. Negative no. 019649, accession no. 368740-1001.

***Fig.* 33.** *Borassus flabelliformis*, watercolor on paper, 46.8 × 32.7 cm. (18.43 × 12.87 in.). Copyright Natural History Museum, Botany Library, London. Fleming Collection, plate 882. Negative no. 031946, accession no. 368740-1001.

Fig. 34. Detail, *Trapa bispinosa*. Royal Botanic Gardens, Kew. Roxburgh No. 1345, *Flora Indica* 1 (1832), 428.

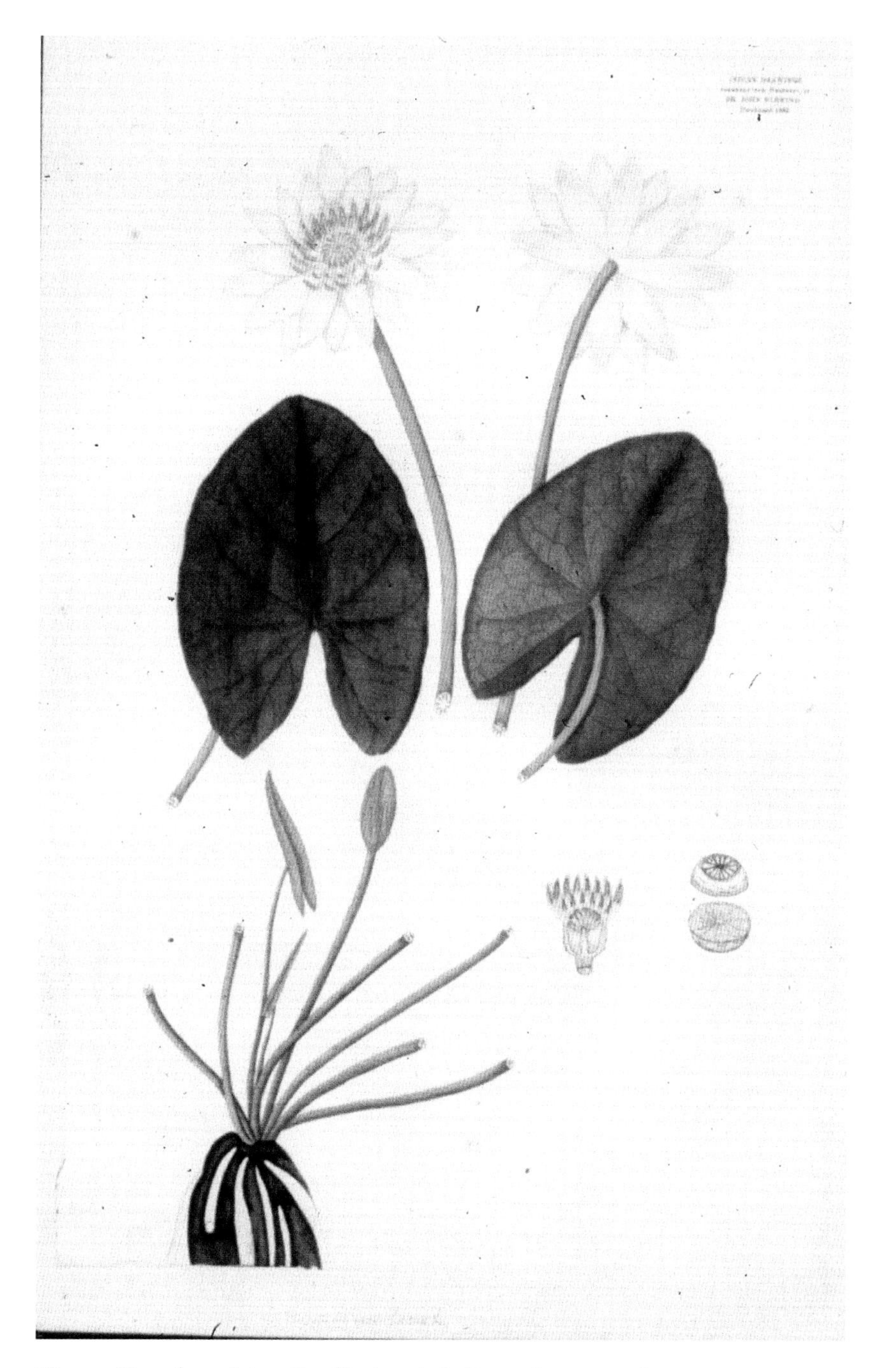

***Fig.* 35.** *Nymphaea lotus alba*, Roxburgh Indian artist, watercolor on paper, 50.2 × 32.6 cm. (19.76 × 12.83 in.). Copyright Natural History Museum, Botany Library, London. Fleming Collection, plate 6, accession no. 368740-1001.

Fig. 36. *Nelumbium speciosum*, Roxburgh Indian artist, watercolor on paper, 49 × 35.5 cm. (19.29 × 13.98 in.). Royal Botanic Gardens, Kew. Roxburgh No. 664, *Flora Indica* 2(1832), 647.

***Fig.* 37.** *Nymphea Lotos, Blue Water-Lily of Guzerat,* James Forbes, watercolor on paper, 54 × 37.5 cm. (21¼ × 14¾ in.). Vol. 12, *A Voyage from England to Bombay,* 291. Yale Center for British Art, Paul Mellon Collection.

***Fig.* 38.** *Leea macrophylla,* Roxburgh Indian artist, watercolor on paper, 49 × 35.5 cm. (19.29 × 13.98 in.). Royal Botanic Gardens, Kew. Roxburgh No. 925, *Flora Indica* 1(1832), 653.

Fig. 39. Bombax Heptaphyllum, Roxburgh Indian artist, watercolor on paper, 50.2 × 32.6 cm. (19.76 × 12.83 in.). Copyright Natural History Museum, Botany Library, London. Fleming Collection, plate 21, accession no. 368740-1001, negative no. 040224.

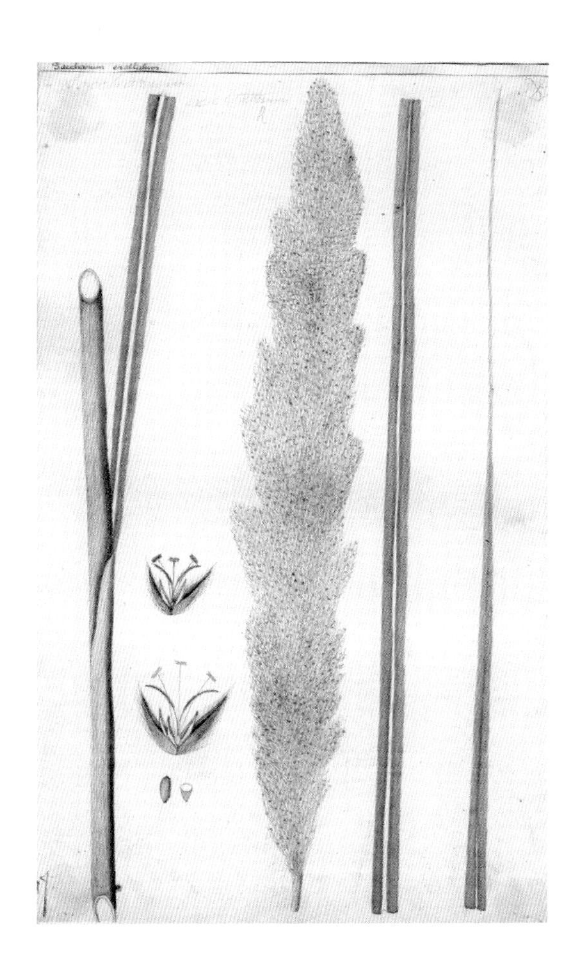

Fig. 40. *Saccharum exaltatum*, Roxburgh Indian Artist, c. 1790, watercolor on paper, 26.1 × 41.4 cm. (10.28 × 16.34 in.). Reproduced with permission of the Royal Botanic Garden Edinburgh. Roxburgh Collection, no. 551.

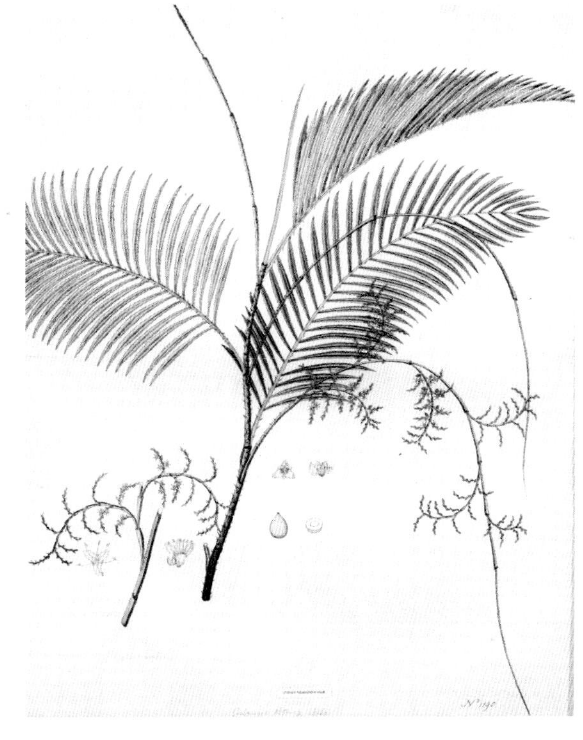

Fig. 41. *Calamus Rotang*, Roxburgh Indian Artist, watercolor on paper, 49 × 35.5 cm. (19.29 × 13.98 in.). Royal Botanic Gardens, Kew. Roxburgh No. 1190, *Flora Indica* 3(1832), 777.

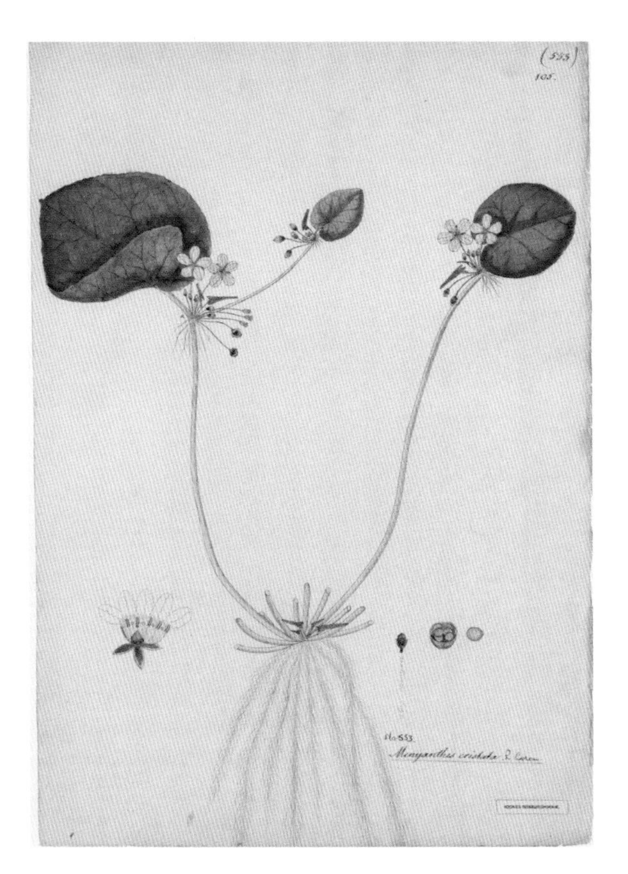

Fig. 42. *Menyanthes cristata*, Roxburgh Indian artist, watercolor on paper, 49 × 35.5 cm. (19.29 × 13.98 in.). Royal Botanic Gardens, Kew. Roxburgh No. 553, *Flora Indica* 1(1832), 459.

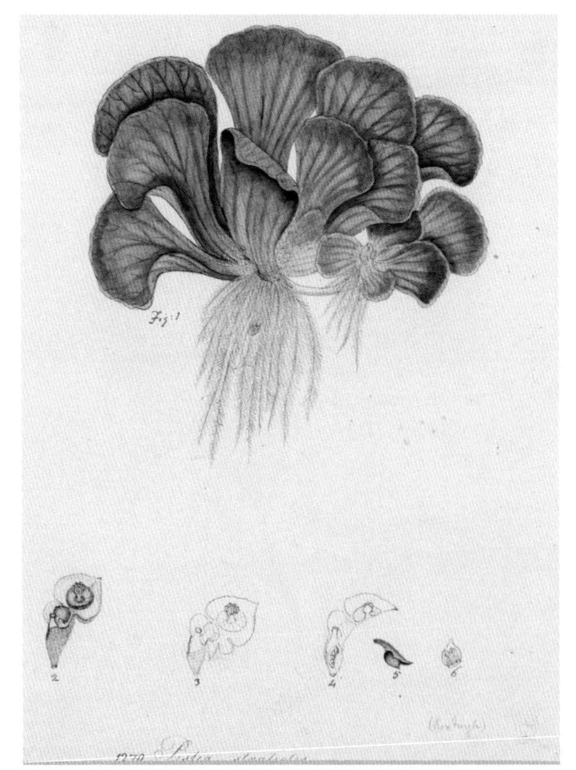

Fig. 43. *Pistia stratiotes*, Roxburgh Indian artist, watercolor on paper, 49 × 35.5 cm (19.29 × 13.98 in.). Royal Botanic Gardens, Kew. Roxburgh No. 1270, *Flora Indica* 3(1832), 131.

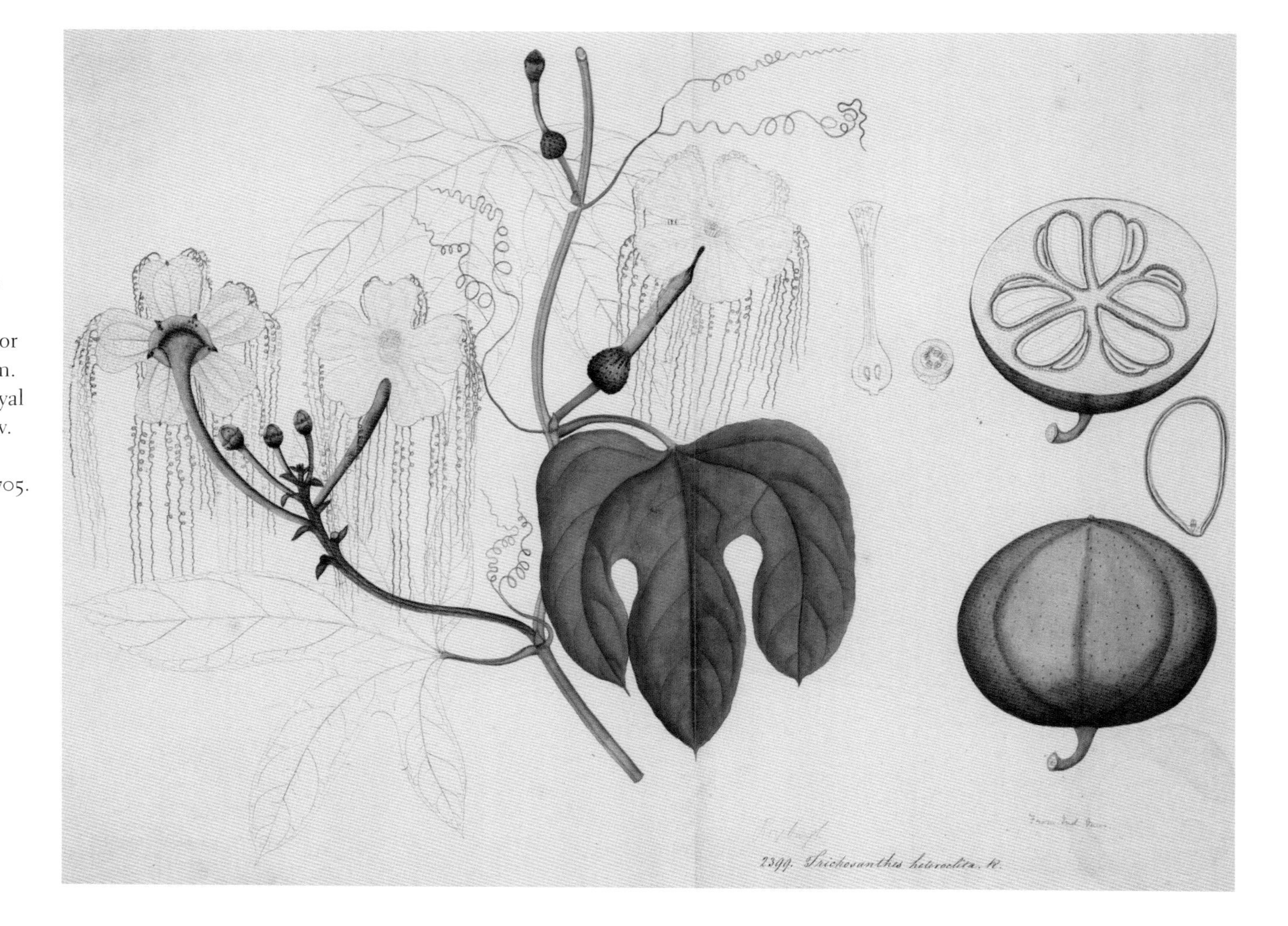

Fig. 44. *Trichosanthes heteroclita*, Roxburgh Indian artist, watercolor on paper, 49 × 35.5 cm. (19.29 × 13.98 in.). Royal Botanic Gardens, Kew. Roxburgh No. 2399, *Flora Indica* 3(1832), 705.

Fig. 45. (*Above*) [*Orchidaceae of Mexico and Guatemala*], G. Cruikshank, engraving, 7.8 × 10.7 cm. (3.1 × 4.2 in.), James Bateman, *The Orchidaceae of Mexico and Guatemala*, p. 8. Reproduced by permission of The Huntington Library, San Marino, California.

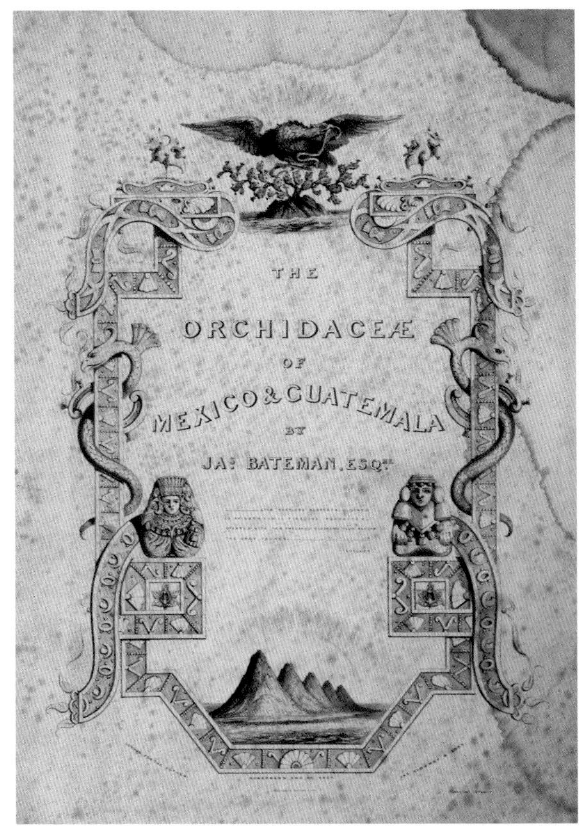

Fig. 46. Title Page, engraving, 34 × 23.5 cm. (13¼ × 9⅜ in.), Bateman, *The Orchidaceae of Mexico and Guatemala*. Reproduced by permission of The Huntington Library, San Marino, California.

Fig. 47. (*Above*) *Vignette*, after *Schomburgkia Tibicinis* orchid plate, engraving, 13.5 × 16.6 cm. (5.3 × 6.5 in.). Bateman, *The Orchidaceae of Mexico and Guatemala*. Reproduced by permission of The Huntington Library, San Marino, California.

Fig. 48. *Cycnoches Ventricosum*, Jane Edwards, hand-colored engraving, 34 × 23.5 cm. (13¼ × 9⅜ in.), Bateman, *The Orchidaceae of Mexico and Guatemala*, Plate 5. Reproduced by permission of The Huntington Library, San Marino, California.

For the tail-piece Lady GREY of Groby has kindly contributed a most ingenious device, compounded of divers Orchidaceous flowers, which, with very gentle violence, have been induced to assume the attitudes in which they appear below.*

—— "Nature breeds
Perverse, all monstrous, all prodigious things,
Abominable, unutterable, and worse
Than fables yet have feigned, or fear conceived,
Gorgons, and hydras, and chimeras dire."

MILTON.

* The hag came forth, broom and all, from a flower of *Cypripedium insigne;* her attendant spirits are composed of *Brassia Lauceana, Angræcum caudatum, Oncidium papilio,* &c. &c.; two specimens of *Cycnoches* sail majestically on the globe below, on the right of which crawls *Megaclinium falcatum.* In the centre stands a desponding *Monachanthus;* on the left a pair of *Masdevallias* are dancing a minuet, while sundry *Epidendra,* not unlike the "walking leaves" of Australia, complete the group.

Fig. 49. *Tail-piece*, after Katherine Charteris Grey, engraving, 12.8 × 15.7 cm. (5 × 6.2 in.). Bateman, *The Orchidaceae of Mexico and Guatemala.* Reproduced by permission of The Huntington Library, San Marino, California; RB 325034.

ten thousand EIC soldiers. Forbes's title for his engraved drawing, *A Banyan Tree, consecrated for Worship in a Guzerat Village*, suggests the political enormity of the British encampment. In war and pictures, the British takeover of the banyan tree conveys the imperial prehistory for the British exploration of Indian flora.

BRITISH BOTANISTS IN INDIA: SIR WILLIAM JONES AND WILLIAM ROXBURGH

Although Sir William Jones spoke for British imperial interests when he called for a mastery of Indian flora in his "Design of a Treatise on the Plants of India" (1790), quoted at the beginning of this chapter, Jones became an amateur botanist only after he arrived in India, where he expanded, more or less, his knowledge of Oriental languages and worked for the EIC and the Crown as a judge charged with codifying Indian law. Almost immediately after he arrived in Calcutta, he launched the Asiatic Society of Calcutta and published essays, his own and those of others, in the journal *Asiatic Researches*.[60] Botany, "the loveliest and most copious division in the history of nature," was the project Jones shared with his wife, Anna Maria, when they lived in a cottage in Khrishna-nagar, outside Calcutta, during four months of the year.[61] In a 1791 letter to an old friend back in England, Jones writes:

> As to botany, it is my greatest delight in our vacations, partly because it is the most agreeable and interesting branch of natural history, but principally because it is the favourite amusement of my darling Anna, who will have the pleasure of showing your Ladyship her botanical drawings of Indian plants, which we have examined together. Though we have read . . . many . . . excellent theological discourses, yet we find a more exquisite lecture, on the being and attributes of God, in every flower, every leaf, and every berry, than can be produced by the mere wisdom or eloquence of man: the sublime doctrine of final causes is no where so beautifully proved and illustrated as in the plants of the lakes and forests, when their different parts and the uses of them are minutely and attentively observed.[62]

Jones's thoroughly deist description of the work and pleasure of studying plant specimens joins activities that might, at another time and in another life, have been kept separate: theology, domestic pleasure, and botanical minuteness. In his last letters before he died in 1794, written to Samuel Davis, a new friend with whom he shared botanical specimens after Anna Maria's return to England, Jones teases Davis about whether he might arrive for a visit, like "Crishna, decked with holy Tamāla blossoms."[63] In ten years, Jones assembled a considerable knowledge of Indian plants from books and from the specimens that collectors and Anna Maria provided and she drew. Botany was also, as his 1790 "Design" makes clear, a passion that could be folded into an imperial vision of what might be done with the Indian flora.

The imperial gesture of Jones's public statements about India after his arrival register one side of a bifurcated political identity. Throughout his adult life Jones had radical and liberal friends, among them Joseph Priestley, those who frequented meetings organized by his father-in-law, Bishop Shipley, Rev. Richard Price and other Lunar Society confederates, including Benjamin Franklin. Jones disliked Burke, spoke openly about the need for parliamentary reform to preserve, as he once put it, the British constitution, and urged that a union be formed by "the young supporters of liberty."[64] After reading Robert Orme's 1772 *Extract*—the extract of a letter in which Hastings argued that the children of Indian criminals might be enslaved on the grounds that doing so would break the cycle of criminality because "slaves are treated as the children of the families to which they belong, and often acquire a happier state by their slavery than they could have hoped for by the enjoyment of liberty"—Jones wrote in the margin: "This is the most spurious argument for despotism which all despots use."[65] Knighted just before he left for Calcutta, Sir William was in the main a liberal Welshman with radical sympathies. He was also poor and worked hard to secure his Indian judgeship.

In India, Jones and Hastings became friends, living near each other in Alipore, then a suburb of Calcutta.[66] Perhaps because he recalled the difficulty his political views had caused him in England, Jones's letters from India urged friends to avoid politics. In private, though, he could be critical of British conduct in India. Writing about the disastrous end of the third Mysore War against Tipu Sultan, he remarked, "Our nation, in the name of the king, has *twenty three millions* of black subjects in these two provinces; but nine tenths of their property are taken from them, and it has even been publickly insisted, that they have no landed property at all: if my Digest of Indian Law should give stability to their property, real and personal, and security to their persons, it will be the greatest benefit they ever received from us."[67]

Even so, the space Jones imagines himself mastering in his "Design for a Treatise on the Plants of India" is at once India and, following his Archimedean template, a full systematic map of its plants. Yet his understanding of the "virtues" of those plants is light years away from that of his EIC contemporary Robert Kyd, whose successful 1786 pitch to the (British) Bengal government for the establishment of the botanic garden near Calcutta sternly (and strategically) refused to admit that such a garden would be anything but useful:

> I take this opportunity of suggesting to the Board the propriety of establishing a Botanical Garden not for the purpose of collecting rare plants (although they

> also have their use) as things of mere curiosity or furnishing articles for the gratification of luxury, but for establishing a stock for disseminating such articles as may prove beneficial to the inhabitants, as well as to the natives of Great Britain, and which ultimately may tend to the extension of the national commerce and riches.[68]

Kyd's dismissal of "luxury" and "curiosity" as motives for botanical exploration in India lets that cat out of the bag. Certainly, all the EIC officials who pursued botanical interests worked hard to identify commercial and medicinal uses for Indian plants, among them indigo, coir, cotton, coffee, tobacco, opium, and tea, as well as key minerals used in the production of textiles traded to the West. With a quick slip of category, Kyd makes Indian "inhabitants" and British "natives" twin recipients of somebody's "national commerce and riches."

Yet the botanists who traversed and planted across the Indian subcontinent were also compelled to do so by their curiosity, often tinged with anticipation of the glorious botanical profusion they expected to, and did, find. Jones's "some hundreds of plants" soon amounted to several thousand, and the scientific pleasure in this botanical array was always, as such pleasures tend to be, deeply aesthetic as well. William Roxburgh reported, after attending a Dutch physician and botanist who had become an EIC employee, Johann Gerhard Koenig, as he lay dying, that when he showed his colleague the plant *Roxburghia gloriosoides* (now called *Stemona tuberosa*), "He desired that I would describe it particularly, for he thought it was new, and uncommonly curious and beautiful."[69]

Jones gave the British public its first extended account of the Indian flora in, so to speak, their native linguistic habitats—the Sanskrit, Persian, and Arabic sources and local informants that had provided Jones with the names Linnaeus had used (relying on Hendrik van Rheede's informants). Even before Roxburgh published his *Plants of the Coast of Coromandel* in 1795, Jones had begun to produce works that blended his botanical and literary fascination with India, including a catalog of Indian plants; a "Design" for a massive treatise on Indian plants; his subsequent "Botanical Observations" on select Indian plants, two tracts in which he argued that the Indian plant that matched ancient descriptions of the spikenard was *Valeriana jatamansi*;[70] a number of verse hymns; *Sacontalá*, his translation of Kālidasā's *Śakuntalā*, a play in which a plant creeper plays a cameo role; and a verse narrative, *The Enchanted Fruit, or the Hindu Wife*. All of these were printed in Calcutta and reprinted soon afterward in England, either in reprints of *Asiatic Researches*, which Jones began publishing in Calcutta in 1787, with Warren Hastings's commendatory letter to introduce the first volume, or *Asiatic Miscellany*, first published in two volumes in 1785 and reissued by divers editorial hands in 1787, 1789, and 1818.

Some have argued that Jones was a jacobin sympathizer for Indian religion and culture, yet imperial language peppers his official remarks in *Asiatic Researches*, In the same anniversary discourse that celebrates Indian botany, Jones repeats the well-worn claim that any reader of the history of India "could not but remark constant effect of despotism [by the Mughal emperors] in benumbing and debasing all those faculties which distinguish men from the herd that grazes; and to that cause, he would impute the decided inferiority of most Asiatic nations, ancient and modern, to those in Europe who are blest with happier governments." Although Jones goes on to suggest that since those happier governments rarely remain so, the same wise reader will recognize that "the British constitution (I mean our public law, not the actual state of things in any given period)" is "the best form ever established," and, although "we can only make distant approaches to its theoretical perfection," it is fortunate that in "these Indian territories, which Providence has thrown into the arms of Britain for their protection and welfare, the religion, manners, and laws of the natives preclude even the idea of political freedom." True, Jones's manifesto for providential British rule in India is hedged with awkward demurrals: most singularly, that the British, the best of all constitutions, is not ideal. Eight years earlier, Jones had insisted in his second "Anniversary Discourse" (1785) that Asia might be appropriately characterized as the "Handmaid" to Europe, a conceit recycled from Alexander, that earlier, would-be conqueror of the subcontinent. Asia's form of service, says Jones, derives from her "momentous objects of research," available now to Britain as "an *imperial* but, which is a character of equal dignity, a *commercial*, people." In the next paragraph, Jones offers botany as the first such object, emphasizing its utility.[71]

Yet Jones regularly consulted Indian scholars, ancient texts, local gardeners, and plant collectors whom he respected as trusted Indian pandits.[72] Writing to Patrick Russell from Krishna-nagar in September 1788 about which Indian plant might be what ancient writers had called the spikenard, Jones mentions that "two native physicians in my family" have seen only dried specimens of the plant.[73] Jones's botanical remarks are split between respect for Linnaean taxonomy and utter frustration with its nomenclature, in part because Linnaeus and his informants did not know Sanskrit, but more fundamentally because Jones found the names ludicrous. In a letter to Banks written from Krishna-nagar in 1791, Jones describes studying plant specimens there:

> I have neither eyes nor time for a Botanist; yet, with Lady Jones's assistance, I am continually advancing; & we have examined about 170 Linnean genera. She brought home a morning or two ago, the most lovely Epidendrum, that ever was seen; but the description of it would take up too much room in a letter: it grew

> on a lofty Amra; but it is an air-plant, & puts forth its fragrant enameled blossoms in a pot without earth or water. None of the many species in Linnaeus correspond exactly with it. You must not imagine that because I am & shall be saucy about the Linnean language, that I have not the highest veneration for its Author; but I think his diction barbarous & pedantic, especially in his *Philosophia Botanica*, which I have a right to criticize, having read it three times with equal attention & pleasure.[74]

Jones was neither the only nor the first to be "saucy" about Linnaean names, yet he remained attentive to Linnaean arguments. Although he keenly missed Koenig's botanical expertise, Jones acknowledged when Linnaeus, not Koenig, appeared to have been right. Jones writes Russell in 1786:

> I have carefully examined a plant, which Koenig mentioned to me, and called *pentapethes protea*, from the singular variety of leaves on the same tree. The natives call it *Mascamchand*; and one of its fragrant fleshy blossoms, infused for a night in a glass of water, forms a mucilage of very cooling quality. The pentapethes phoenicia, which now beautifies this plain, produces a similar mucilage. . . . But I mention this plant, because Koenig told me that Linnaeus had inverted nature in his description of it, by assigning to it *five* castrated filaments, to each of which were annexed three prolific ones; whereas, said he, (I am sure I did not mistake him) the flower has fifteen castrated, and five prolific; so that in truth it would have been *pentandrian*. Now I have examined all the flowers of this species that I could get, and I find the description of Linnaeus to be correct; but there is no accounting for the variety of a protean plant.[75]

The botanical terminology is Englished Linnaeus—castrated versus prolific filaments ("filament" was the name given to the stalk that may extend from the male stamen of a flowering plant), yet Jones's concession at the end to difference as the norm for protea or, perhaps more generally, any plant with protea-like variability, moves away from the logic of Linnaean classification, in which the number of stamens determines the class to which a plant belongs. Four year later, Jones wrote to Banks (who had all of Koenig's manuscripts) to disagree with Koenig again:

> The Madhuca is beyond a doubt the Bassia; but I can safely assert, that not one of fifty blossoms which I have examined, had *16 filaments, 8 above the throat & 8 within the tube*: that Koenig whom I knew to be very accurate, had seen such a character I doubt not, but he should not have set it down as constant. I frequently saw 26 & 28 filaments, sometimes 12; & the average was about 20 or 22.[76]

Jones then asks Banks for copies of Koenig's manuscripts as a return for sending back "accurate" Sanskrit names, plant specimens, or seeds. The ensuing botanical summary of evidence gleaned by examining several madhuca/bassia specimens shows that Jones had by this time sufficient botanical knowledge to make good use of Koenig's unpublished plant notes. It also shows that Jones has found fault with Linneaus's claim that precise rather than variable numbers should determine plant classification.

In his "Design of a Treatise on the Plants of India," Jones objects to Linnaean plant names on grounds that were familiar and popular and would become even more so in the decade ahead, when botany, women, and propriety become a triple threat in *The Anti-Jacobin* and other venues similarly disposed. He says of Linnaean plant names in his "Design" that their "allegory of *sexes* and *nuptials*" is both "unbecoming the gravity of men" and so "wantonly indecent" that "no well-born and well-educated woman can be advised to amuse herself with botany."[77] He perhaps hoped that his readers would not recall his own Linnaean allegory in the 1784 poem *The Enchanted Fruit or the Hindu Wife*. In later essays, including some that appeared posthumously, Jones specifies the deeper ground of his objection to Linnaean nomenclature.

Jones insists that neither Linnaeus nor his many informants on Indian plants had knowledge of the Indian names presented in ancient Sanskrit texts, particularly poems. Instead, he frequently remarks that Linnaeus had introduced names that had little apparent relation to the specific Indian plant or tree. Of the tamarind, Jones says first that its stamens argue for it being removed to another Linnaean class; then more damningly, that "it were to be wished that so barbarous a word as Tamarindus, corrupted from an *Arabic* phrase absurd in itself, since the plant has no sort of resemblance to a date-tree, could, without inconvenience, be rejected, and its *Indian* appellation [that is, the Sanskrit name Jones gives this tree, *Amlicá*] admitted in its room."[78] More insistently than many of his contemporaries, Jones preferred Sanskrit names and sources, presenting them ahead of and in preference to plant names derived from other Indian languages. Apparently in reply to Patrick Russell's caveats about the reliability of Sanskrit dictionaries for plant identification and names, Jones only half-agrees: "Those books do not carry full conviction; but they lead to useful experiments."[79]

Jones was blunt about the perceived linguistic failings of those on whom Linnaeus had relied. Looking back to the *Hortus Malabaricus* in his tenth anniversary discourse, Jones remarked:

> It is much to be deplored, that the illustrious Van Rheede had no acquaintance with *Sanscrit*, which even his three *Bràhmens* . . . appeared to have understood very imperfectly, and certainly wrote with disgraceful inaccuracy. In all his twelve volumes I recollect only *Bunarnavà*, in which the *Nágari* letters are

> tolerably right; the *Hindu* words in *Arabian* characters are shamefully incorrect; and the *Malabar*, I am credibly informed, is as bad as the rest.[80]

Jones in fact knew no Malabar and was not credibly informed. Grove argues that the Ezhava names that appear in van Rheede's *Hortus* convey a systematic grasp of plant affinities that was important for later botanical investigations of tropical plants.[81]

Published posthumously, Jones's "Botanical Observations on Select Indian Plants" and "Catalogue of Indian Plants" are, along with the spikenard treatise, his most extensive commentaries on Indian botany. Plant by plant, they specify when and where he refuses Linnaean names and relies on Indian informants or his own (usually Sanskritic) expertise. They also specify more fully than elsewhere the logic that informs these judgments. His "Catalogue" takes no prisoners. Although it begins with a brief explanation that both Sanskrit and Linnaean names are included where these can be "ascertained," the list is given exclusively in Sanskrit names, followed in some cases by Linnaean names. At a time when the indices of botanical works would frequently include two alphabetical lists of plant names, one Linnaean, the other in the vernacular, Jones's decision to emphasize Sanskrit names is both remarkable and consistent with his unshakeable conviction that Sanskrit is essential for recording (and preserving) Indian knowledge.

As the protocols of botanical description required, Jones's "Botanical Observations" lists plant names and synonyms in other languages, noting the informants and published works that he had consulted. Among those informants he lists Indian gardeners and plant collectors whom he employed.[82] But whereas van Rheede had adopted local Indian plant names, Jones uses Sanskrit names first, as in the case of *Bhúchampaca*, a Sanskrit plant name for the Indian spice plant "round-rooted galangal," its Hindi name transcribed into English.[83] Other synonyms for this plant include a "vulgar" form *Bhúchampac*, which seems to be transcribed from spoken Hindi or Bengalese, and then its Linnaean name, Round-rooted *Kaempferia*. Jones objects to the name given in van Rheede's *Hortus*: "This plant is clearly the *Benchâpo of Rheede*, whose native assistant had written Bhu on the drawing, and intended to follow it with *Champá*: the spicy odour and elegance of the flowers, induced me to place this *Kaempferia* (though generally known) in a series of select *Indian* plants; but the name *Ground Champac* is very improper, since the true *Champaca* belongs to a different order and class; nor is there any resemblance between the two flowers, except that both have a rich aromatic scent." He adds that it would be impossible to determine the genera of the natural order of plants that includes the *Bhúchampaca*—that is, plants Jones calls "*scitamineous*"—until all Indian plants belonging to that order had been "perfectly described."[84]

Jones relies on Indian knowledge even more emphatically to distinguish different species of the night-blooming tree that Linnaeus had named *Nyctanthes tristis*. Jones prefers the Indian name *Sép'halicá*, but he includes several Indian synonyms (*Swvahá, Nirgudí, Nílicá, Niváoicá*) and two "vulgar" Indian names (*Singaha, Nibári*), as well as Linnaeus's half-Englished "Sorrowful *Nyctanthes*." Of these, *Nirgudí* is a plant name in Marathi and *Nílicá* a variant of *Sép'halicá*, the name Jones prefers for plants that have flowers in which bees sleep. His description of this tree makes quick work of the Linnaean epithet: "This *gay* tree (for nothing *sorrowful* appears in its nature) spreads its rich odour to a considerable distance every evening." Explaining that that his pandits call this tree *Sép'halicá*, he notes that other Indians call it *Párijáticá* or *Párijáta*.[85] To describe other varieties of Linnaeus's *Nyctanthes*, distinguished by Englished epithets ("Many-flowered," "Narrow-leaved"), he consults Persian and Arabian poets, van Rheede on the Brahmins in western India (Jones was in Bengal), and Indians and Brahmins among his acquaintance.

The most arresting accounts of local knowledge in Jones's "Botanical Observations" concern plant names that specify differences between species. On these occasions, he willingly pits local Indian informants against Linnaeus, who remained for British botanists of this period, including those working on the Indian flora, the primary authority for names and taxonomic placement. Challenging Linnaeus as a European was one thing; presenting indigenous knowledge as superior to Linnaeus's was quite another, to no small degree because doing so conveyed a powerful turn in the status of subaltern or indigenous knowledge. In the first of two examples, Jones identifies Linnaeus's "Great flowered Jasmin" by the name *Málatí*, then notes that among the string of synonyms for this plant is one, *Játi*, that may be more defensible as the name of this species of jasmine. Jones puts it this way: "Though *Málatí* and *Játi* be synonymous, yet some of the native gardeners distinguish them; and it is only the *Játi* that I have examined."[86] Within a sentence, Jones moves from the categorical subjunctive "be synonymous," which carries something like the force of a rule (as in "these are held, known, said to be synonymous") to adopt his informants' distinction between the *Játi*, the (species or variety) he had examined, and the other one, which he had not. He goes on to note (though I will not) still more names discussed by other Europeans, who got them from still other native informants.

In his botanical description of the plant he calls *La'ngali*, Jones presents two misidentifications in succession. The honor of the first mistake goes to Linnaeus, who named it *Nama of Sila'n* [Ceylon], although it also grows in India. The second mistake Jones finds is the vulgar synonym *Canchrà* of *Bengal*. Here his informant is his gardener, who, Jones said, "insists that *Canchrà* is a very different plant, which, on examination, appears to be the *Ascending Jussieua* of *Linnaeus*," a plant he describes as "leaves inverse-egged, smooth."[87] This sentence is suggestively unclear about which of them, Jones or his gardener, supplied the correct Linnaean name and which then examined and described the plant as Linnaeus and his descendants would have done. The gardener would have done so only if he was well acquainted with Linnaean protocols, by some means that might or might not have included being able to read Latin or even English (he could have learned these protocols from speaking and looking at plants extensively with English botanists, for example). This first scenario is perhaps the less likely one. Yet even if Jones is the explainer of the case, his sentence slips from his gardener's knowledge to his own. Here subaltern knowledge is not subsumed by British authority. Rather the gardener knows what he says he knows and Jones assents to that authority.

Who speaks, in this instance, may be less critical for thinking about colonial knowledge and power than the way Jones's text melds two authoritative voices, with the Indian gardener's voice in the lead. At textual moments like this, something more subtle than the hardening of colonial knowledge occurs. If these moments in Jones's essay are akin to the "capillaries" that Foucault insisted would need to be scrutinized to get a grip on operative relations between knowledge and power, what they suggest cannot be easily described as "power sharing" or "hybrid knowledge."[88] Yet even the phrase *hybrid knowledge*, misses the narrative development Jones implies: a dynamic plot that begins with Linnaeus then puts that authority aside in the name of indigenous knowledge, which Jones then takes up as if he had produced it. The *as if* is just that. Even if we grant that Jones's local informants did not read or know Linnaeus and so were not in a position to recognize that their knowledge challenged Linnaean authority, Jones clearly asserts that they knew the plants in question as Linnaeus did not, a recognition that echoes all the way back to Rheede's and Linnaeus's reliance on Malabar physicians and toddy tappers to name and classify plants of the Malabar coast.

In botanical essays and in the literary translations he produced or imaginatively recast, Jones exploits the figurative potential of plant epithets, those descriptive terms that frequently attach to the species part of a Linnaean binomial or its English equivalent. Of the Indian tree *As'óca*, for which there is no Linnaean name, Jones writes:

> The vegetable world scarce exhibits a richer sight than an *Asóca*-tree in full bloom: it is about as high as an ordinary cherry-tree. A Bráhmin informs me, that one species of the *Asóca* is a creeper; and *Jayadéva* [a twelfth-century CE court poet in Orissa] gives it the epithet *voluble*: the *Sanscrit* name will, I hope, be retained by botanists, as it perpetually occurs in the old *Indian* poems, and in treatises on religious rites.[89]

For Jones, keeping the history of Indian botanical names alive allows other figures to catch hold, to be voiced as it were, even as the *Asóca* creeper becomes "voluble." Jones's experiments in literary translation insist that epithets like this that give identity and character to plants are what their names encourage. His popular translation of Kālidasā's fifth-century CE play *Śakuntalā*, or *Shakuntala* (in Jones's translation, *Sacontalá*), excerpted from the Indian epic *Mahábhárata*, is a case in point. Jones's translation, rather than the original Sanskrit/Pakrit text, was the basis for Georg Foster's widely read German translation. For several German romantic writers, *Śakuntalā* became key evidence that ancient Indian literature and culture were superior to that of classical Greece.[90]

In "Botanical Observations," Jones notes that the plant he identifies as *Atimucta*, for which the Linnaean name is the rather generic "Bengal Banisteria" (that is, a Bengali creeper that climbs banisters), was the favorite plant of the eponymous heroine of Kālidasā's play *Sacontalá*. The synonymic relay of names and identities in the play extends from this plant to the heroine to forecast her eventual marriage to the king, Dushmanta, who happens by in the opening scene, then hides behind a tree so that he can spy on Sacontalá and her attendants as they tend the plants of a sacred grove.

When one of Sacontalá's attendants compares her to "the fresh-blown Mallicá," she replies, "I really feel the affection of a sister for these young plants." In an aside, Dushmanta picks up the figurative thread, noting that the robe of fibers thrown over her shoulders is "like a veil of yellow leaves enfolding a radiant flower." One botanical figure leads to another, this one implied (although it could hardly be missed): "The water lily, though dark moss may settle on its head, is nevertheless beautiful . . . Many are the rough stalks which support the water lily; but many and exquisite are the blossoms which hang on them." An attendant expands the chain of analogy to the Amra tree, a mango, that seems to point "with the finger of its leaves" and, she tells Sacontalá, "seems wedded to you, who are graceful as the blooming creeper which twines round it." As it happens, the *Amra'taca* is listed just after the Mallicá creeper in Jones's "Botanical Observations."[91] The besotted Dushmanta goes on: "Her lip glows like the tender leaflet; her arms resemble two flexible stalks; and youthful beauty shines, like a blossom, in all her lineaments." Another attendant amends this unfortunate drift toward generality: "See, my Sacontalá, how yon fresh Mallicá, which you have surnamed Vanàdósini, or Delight of the Grove, has chosen the sweet Amra for her bridegroom."[92]

Reluctant to leave the sacred grove to marry Dushmanta, Sacontalá asks her foster father, who is its priest, to "suffer me to address this Mádhavi creeper, whose red blossoms inflame the grove" (Inevitably, Jones's "Botanical Observations" lists "Mádhavi" as the "vulgar" synonym for the Mallicá plant).[93] And address this plant she does: "[*Embracing the plant*] O most radiant of twining plants, receive my embraces, and return them with thy flexible arms, from this day, though removed to a fatal distance, I shall for ever be thine. —O beloved father, consider this creeper as myself." Which he in turn does: "Now, since my solicitude for thy marriage is at an end, I will marry thy favourite plant to the bridegroom Amra [aka Dushmanta], who sheds fragrance near her."[94] The course of their marital love does not, of course, run smooth in the rest of the play, which still has three acts to go.

Writing to Banks, Jones defends his translation from the imputation of salaciousness by taking shelter in numbing botanical detail. He explains that the epithet Vanàdósini is from the Pakrit (the common language used along with the Sanskrit spoken by aristocratic speakers in the play) term for the Banisteria, *Vana dosini*"; the "filaments" entwined about Sakuntala's wrists in act 3 are separated "threads" of the lotus stalk—a practice he has often seen—hence not in this instance a botanical term; that the blue lotus is found only in upper India (that is, the lower Himalayas); and that local species are white and rose-colored. Of the masque-like scene in act 6, in which women who are allegorical aspects of the "God of Love" pick a flower bud that is sacred to the god and a court Chamberlain scolds them, saying that it has not yet shed its "prolifick dust," Jones explains this line as only a Linnaean botanist with nomenclature on his mind would do: "The use of pollen in flowers is I believe well known to the Brahmens; but I am not sure that I have not added the epithet *prolifick* to distinguish it from common dust, which would have been the exact version of *renu*; but they also call it *rajas*, which means I believe a seminal substance, they even apply the word to animal seed." Jones also identifies the other plants and trees of the play, giving information about local specimens and Linnaean equivalents for local names:

> The *Amra* is Mangifera; the Mallica (I believe) is Nyctanthes Zambak; the Madhavi creeper, Banisteria; the Cusa I cannot see in blossom; the Sirisha is Mimosa odoratissima: the *Pippala, Ficus religiosa*: if I recollect Lacsha is not a plant, but *Lac*. . . . As to *Nard*, I know not what to say: if the Greeks meant only a *fragrant Grass*, we have Nards in abundance, Acorus, Schoenus, Andropogon, Cyperus, &c, &c; but I have no evidence that they meant any such thing.[95]

In Jones's translation, botanical figures conduct the mind inevitably to Linnaean arguments about plant marriage, although here the marriage between creeper and tree (with the usual gender roles safely in place) or Sacontalá and Dushmanta is not salacious but blessed. Above all, it is a chaste and monogamous relation between one male and one female.

No such defense is available for Jones's poem *The Enchanted Fruit; or, the Hindu Wife*, published in 1784, six years before he charged Linnaean nomenclature and taxonomy with being unfit for ladies to read. The copious annotation of the published poem established a model for the Oriental verse tale, including Moore's *Lalla Rookh*.[96] When Erasmus Darwin settled in to write "The Loves of the Plants" a decade later, he unquestionably had Jones's scandalous Linnaean "marriage" between one Hindu wife and her five husbands in mind.

In *The Enchanted Fruit*, Jones rewrote the story of the Hindu princess Draupadi in the *Mahábhárata*, letting her keep five husbands there but shedding their dalliances with other women. The moral of the story as Jones tells it turns on the truth telling required of each husband and then Draupadi after one of them, Erjun, shoots an arrow at an enticing fruit high up on the tallest tree in a temple grove. As luck would have it, the tree is sacred to Krishna, who explains that its fruit is reserved for a pious "Muni," or inspired poet, who will no longer sing if he does not receive his monthly supply of the fruit in question. If all involved tell the truth about their individual failings, Krishna explains, the fruit will fly safely back to the tree. They mostly do and it almost does. The figurative texture Jones gives to this plot involves many Indian plants—some personified and others not. He identifies them by coding each with a typographic symbol that corresponds to a note at the bottom of each page. When this pattern of annotation is particularly thick, the symbols in the text and in the corresponding notes expand from a single asterisk or cross or double addition sign to parallel lines to multiples of each of these. At this point Jones's text begins to look rather mad, as though the poem's various stories of male prowess with spears and other engines of war had somehow leaked into the typesetting.

Jones's "Antediluvian Tale Written in the Province of Bahar," as the poem is subtitled, illustrates, we are told, the ancient Brahmin motto, "what pleaseth, hath no law forbidden." What pleases in this instance is the happy marital arrangement in which Draupadi has "sev'ral husbands, free from strife, / Link'd fairly to a single wife!" This arrangement is slightly less scandalous than the Linnaean class that the poem offers as a botanist's riddle:

> Thus Botanists, with eyes acute
> To see prolifick dust minute,
> Taught by their learned northern *Brahmen* [Linnaeus, among others]
> To class by *pistil* and by *stamen*,
> Produce from nature's rich dominion,
> Flow'rs *Polyandrian Monogynian*,
> Where embryon blossoms, fruits, and leaves
> *Twenty* prepare, and ONE receives.[97]

The term *polyandrian* refers to a class of hermaphroditic plants that has many stamens or any number above twenty. Jones's "five-mal'd single-femal'd flow'r" is by comparison a slightly more modest version of sex and marriage in the plant kingdom.

Whereas each of the husband/brothers tells a tale that illustrates his characteristic fault (pride, revenge, rage, intemperance, and avarice), and with each telling the fruit flies another ten cubits back up toward its original location on the tree, Draupadi's truth telling falls slightly short, though she has, and at some length revealed her (inevitable) female vanity. Apparently she has not told the whole truth, since the fruit stops its flight just two cubits from the tree. All the husbands look at each other and at Draupadi, who shamefacedly confesses to them that while she was studying Sanskrit with a learned Brahmin (hired or sent by her five husbands, she notes), the way he told stories about Krishna and his gopis made the scholar himself seem "Bright as a God." Trembling at her husbands' shock, she insists, in capital letters and apparently in truth, he "ONLY KISS'D MY CHEEK."[98] The fruit flies back to its branch.

Jones draws a comparative lesson from the tale: would a Christian wife fully confess her failings "like this black wife," and so restore us to "primeval life,/and bid that apple, pluck'd for Eve/By him, who might all wives deceive/hang from its parent bough once more/divine and perfect as before?" The lesson of Pope's *Rape of the Lock* suggests not, or that she will need Pope or Jones to set her right. The allegorical texture thickens as Britannia arrives to vanquish filthy *Scandal* and save the reputation of British women. The poem ends edgily, with a judgment scene in which Britannia, that "victorious Fair," reigns equally "In *British* or in *Indian*, air!" Franklin suggests that this imperial scenario is not the one Jones planned in his proposed epic *Britain Discovered*, which would have charged Britain with commercial plunder and disrespect for the ancient cultures of the Indian subcontinent.[99] Yet Jones's "Hymn to Gangá" concludes its history of the course of the river by praising the British for having brought peace and prosperity to the Gangetic plain.[100] Jones's handling of Indian botany is split between moments like the conclusion of *The Enchanted Fruit; or, the Hindu Wife*, where a sense of colonial possession by way of botany settles back in place, and moments that can be glimpsed in the folds of the plant descriptions of Jones's "Botanical Observations" and the dramatic role that plants and their names play in his translation of *Sacontalá*). In these works Jones grants considerable authority to Indian botanical knowledge, never questioning it as he questioned the reliability of Indian informants with whom he discussed Hindu law.[101] True, it is the Hindu wife in Jones's poem who does not at first tell the whole truth, a failing that the poem suggests is typical of women, whose minor faults ought to be easier to confess (the comparison implies) than the big male ones about which her husbands are so admirably forthright. Nobody's perfect.

William Roxburgh was the first British botanist to make the investigation of the Indian flora a quasi-official project of the EIC (this was not easy to do). Like other EIC botanists who investigated commercial crops and plants, he took special note of plants he knew to be economically valuable to the British and made detailed notes about plants that Indians valued as sources of food, drink, writing material, structural materials, and household items. Trained in medicine in Edinburgh, where he studied botany with John Hope, the professor of botany who taught an entire generation of Scottish botanists, Roxburgh joined the EIC first as a surgeon's mate on one of its ships, then in 1776 as assistant surgeon to the Madras General Hospital. From the Dutch doctor and botanist Koenig, who came into the full employ of the EIC in 1778, Roxburgh learned a good deal about Indian botany. Initially he spent most of his time in the Northern Circars, at a garrison station about two hundred miles north of Madras. There he discovered a native Indian pepper that he planted, along with many other plants that he hoped might be commercially successful: coffee, breadfruit, cinnamon, indigo, and *Opuntia*, the cactus that is host for the cochineal insect that exudes a red dye, long the subject of intense European interest in the "East" as well as "West" Indies, eventually extending his plantations to one hundred thousand plants.

Much of this early activity, which Roxburgh did with his own funds and for which he sought EIC support, was reported back to England in Alexander Dalrymple's 1793 *Oriental Repertory*, clearly part of a campaign at home as well as in India to get Roxburgh the funds to continue his venture. By 1789 (this, too, Dalrymple reports), Roxburgh wrote that he had hired two native artists who had produced between four hundred and five hundred drawings. Appointed to the botanic garden at Howrah, Calcutta, in 1793, he set up a team of Indian artists to draw more Indian plants. The final tally of drawings at the time of Roxburgh's death in 1815 varies slightly, from 2,542 to 2,595.[102] Hand-drawn copies of the Calcutta originals were sent to Kew, except for those shipments that were on ships that sank. Friends and colleagues who saw the drawings also asked for copies, so that the original set of under three thousand expanded with duplication.

From about 1789 until he left India on furlough for Scotland and England in 1813, where he died in 1815, Roxburgh studied plants and produced thousands of botanical descriptions that were to be accompanied by engraved replicas of the Indian artists' drawings. The only publication in which this goal was partly realized during his lifetime was *Plants of the Coast of Coromandel* (1795–1820), which reproduces three hundred drawings of plants he had examined in the Coromandel region before he moved up the coast to Bengal. Although he completed the manuscript (in several fair copies) for *Flora Indica*, this work appeared only after his death, in two formats: an incomplete edition of two volumes edited by Nathaniel

Wallich (1820–24) and a verbatim but unillustrated three-volume edition published in 1832 at Serampore, near Calcutta, at the request of Roxburgh's sons. Another EIC botanist, William Griffith, published an account of Roxburgh's cryptogamic plants in 1844.[103]

In the large-folio format used for *Plants of the Coast of Coromandel* and in the manuscript he prepared for *Flora Indica*, Roxburgh comments at length on the commercial potential of numerous plants. This pointed notice of the profitability that might follow from botanizing India is part of the logic of the *Coromandel* publication, which gave its purchasers a large, illustrated preview of botanical opportunity nearer their grasp than the spice markets of Indonesia and all the tea in China. If these plants could be grown commercially in India, where the British had established a strong military and commercial presence, the three-way trade to markets east of India and the drain of gold bullion from west to east might be reversed.[104] Roxburgh scrupulously noted the medicinal properties of plants, vegetable dyes, and mordants, all kinds of local uses, and explained planting and extraction experiments. This practice is better marked in the *Coromandel* volumes than it is in the published editions of *Flora Indica*, even the 1832 edition that reproduces his notes without omission.

In the folds of this commercial venture are at least two other impulses: to provide a pictorial and scientific record of Indian plants and to make that record aesthetically pleasing. Roxburgh's plan for a fully illustrated Indian flora probably owed a good deal to John Hope, whose remarkable hand-drawn illustrations for his Edinburgh lectures on botany instructed more than a generation of botanists who studied with him, including Roxburgh. The aesthetic impact of this affiliation is conveyed in the large-folio *Coromandel* volumes, each of which presents one hundred Indian plants and illustrations. Like Thornton's *Temple of Flora*, which began production three years later, in 1798, *Plants of the Coast of Coromandel* is a big, expensive book in which botanical and commercial explanations and gorgeous plants are printed side by side, in a far bigger format than even the weakest eyes might require. Exotic, gorgeous, or bizarre, the Indian plants presented in this large format were meant to entice a public that could not get enough images of plants, especially exotic plants.

The *Coromandel* volumes point toward the still more massive botanical wealth of the Indian subcontinent. Plant by plant, Roxburgh's descriptions amass details that collectively specify this remarkable botanical array. The arguably endless relay set in motion by the tight symmetry of number in the *Coromandel* volumes (one hundred plants illustrated and discussed in each volume, for a total of three hundred) is explicitly synecdochic: these plants represent only a portion of the botany of the Coromandel coast, and the Coromandel plants are themselves only a portion of the flora of India. The prospective expansiveness of thinking about the Indian flora in this way is the cumulative enterprise of British botanists from Roxburgh and his contemporaries to Joseph Dalton Hooker near the end of the nineteenth century, exploring Sikkim and the Himalayas, piling up evidence of Himalaya orchid species and genera that far outstrips Roxburgh's discussion of this plant family in *Flora Indica*.

The second story embedded in the commercial enterprise of Roxburgh's Indian botany is suggested by his attention to Indian botanical knowledge and artistic practices. The *Coromandel* entry on dyeing and painting with the root of *Oldenlandia umbellata* (omitted in *Flora Indica*) explains that it is "called by the Indians chay and by the English madder root." Madder was a highly valued, strong, red dye that Indian artisans used extensively to dye cottons known as chintz to the European market. Roxburgh's lengthy description of the Indian process for using madder to dye fabric is shorter than earlier European accounts of Indian cloth dyeing and painting; it may also omit a couple of steps.[105] He begins with an admission that is apparently unique among the several manuscripts on this topic: "I have tried various experiments to enable me to dye red with this root (I may say two or three hundred), in a more expeditious and less troublesome way than what the natives follow, but all with no satisfactory success . . . all that is necessary for me to say at present, is to give the process for dying [sic], painting, or printing red with this root, as practised by the natives."[106]

Another experimentalist might have summarized unsuccessful trials (as do entries in *Cook's* magazine on failed culinary experiments), but Roxburgh lets them go, preferring to observe that the recipe he gives is the one used in the Masulipatam district on the Coromandel coast, famous at the time for the quality of its painted cottons and particularly their reds. The British had established a trading presence there in 1611, and by 1769 this had expanded into military control of the region where French, Dutch, and English trading interests had long vied for Indian loyalties and textiles. Prompted by an evidently commercial motive—to document a superior Indian dyeing method that Europeans had long wished to copy—Roxburgh departed from the usual format for botanical description to include Indian practices and an Indian plant name (chay root). This example is not unique in the *Coromandel* volumes and the *Flora Indica* manuscript, which often fold in Indian plant names and uses.

Roxburgh's transcription of Indian plant names in several languages required another, more subtle adjustment in the usual format for taxonomic description, in which plant names and synonyms were given with the source or authority for each. At least since van Rheede's *Hortus Malabaricus*, this format had included local plant names. The reason for this was eminently practical. Working, as British botanists seem always to have worked, with Indian plant collectors and gardeners, exchanging information about plants required knowing what Indians called them. Yet Roxburgh's care with Indian names, emphatically presented in occasional manuscript corrections of his clerk's transcriptions, convey his repeated efforts to get right an array of names and plants that over time fan out to include more languages and more learned sources. The *Coromandel* entry for the large palm tree *Pandanus odoratissimus* provides something like a baseline example. Because Linnaeus included and named this plant in his *Species Plantarum*, its Linnaean name is listed first. The next two are transcriptions of local names for the male and female versions of this species: "Mugalie is the Telinga [Telegu] name of the male plant, and Ghaezanghee that of the female." In last place Roxburgh mentions that "Caldera" is the name used for both male and female plants "amongst Europeans on this coast." For taxonomic purposes, the Telugu names are more informative because they distinguish the two versions of this plant.

Whereas Jones regards Sanskrit names as the only true Indian names, and others as synonyms, Roxburgh lists every Indian name reported to him. In *Flora Indica*, which he began on the Coromandel coast but expanded in scope when he moved to Calcutta to take up the directorship of the botanic garden there, he adds names used in Bengal, among Hindus, by Sanskrit and Persian speakers (including Jones) and others. Plants that had long been significant to Indian ritual and belief across the subcontinent, like the lotus and the mango tree, invited this proliferation of names. For the smaller white lotus, *Nymphaea alba*, the names Roxburgh gives in his manuscript are: "Kamada the Sanscrit name. Salak of the Bengalese. Koe the Hindu name. Tella-calwa of the Telingas."[107] For the larger lotus *Nelumbium speciosum*, a plant whose joints may with age become "as large as a man's fist," the list of Indian names is longer still.

For *Mangifera indica*, the mango tree, identified in Hindu art and culture with the fecundity of nature and women, especially Krishna's consort Radha, Roxburgh lists several authorities for its Latin name, followed by its Sanskrit, Hindi, and Bengali names, its English name, and finally its Telinga [Telegu] and Tamul [Tamil] names. If there is an implicit hierarchy in this list, it would appear to be, from top to bottom—that names with the widest currency precede more local names, with English less local that Telugu and Tamil, but less widely used than Bengali, Hindi, Sanskrit, and Latin names.[108] Roxburgh does not set out to discount English names; he simply lists the "Mango tree of the English" as he does because this name is not one used by the greatest number of speakers. This sequence is odd given that the intended reading audience for *Flora Indica* was English-speaking. By foregrounding Indian instead of English or Linnaean names, Roxburgh implies that to study the Indian flora one must use Indian names and knowledge. Even when he describes an Indian plant that he does not know to be either medicinal or sacred to Indians, he lists its Indian names. He writes of *Leea macrophylla*, a species he was the first European to identify, that "Samudraca is the Sanskrit name. Dhal-samuds of the Bengalese." He adds that its root "promises to yield a colour fit for dyeing," but is apparently unaware of the medicinal value accorded this plant in ancient Sanskrit texts.[109]

Roxburgh's account of *Ocimum sanctum*—literally "holy basil" and so regarded by Hindus for its ayurvedic properties and spiritual powers—and related species is still more emphatic about the priority of Indian names.[110] In the *Flora Indica* manuscript he calls this species "a very grateful smelling plant," a wonderful inversion of its Latin name *O. gratissimum* (it is rather humans that ought to be grateful to species of *Ocimum* for their scent or curative powers).[111] Roxburgh's Indian drawings of this species and *O. caryophyllatum* are inscribed with names that are Latin (for the genus) and Indian: the drawing for *O. gratissimum* is inscribed "Ocymum Ram-tulasi," and that for *O. caryophyllatum*, "Ocymum Goolal-tulasi."[112] The manuscript entry for *Ocimum sanctum* includes these Indian names and published sources: "Nalla-Tirtava. Rheed. Mal. 10. T. 85. Parnasa the Sanscrit name, see *Asiatick Researches. 4. P. 288*. Kalla, or Crishna Tulasi, or Tulsi of the Bengalese, *and* Hindoos [Roxburgh adds this last phrase]. Christna [Roxburgh corrects to *Crishna*]-tulasi of the Telingas [Telugu]."[113]

With at least three fair-copy manuscripts of *Flora Indica* to review, he proofread selectively. To head off attributing agency to a plant, he substitutes "drooping" for "nodding" to describe the palm leaves of the *Calamus rotang*. In one manuscript copy of the entry for *Trapans bisnosa*, water chestnut, he crosses out some phrases and changes "Hindoos" to "natives" in the sentence "The Hindoos admire their kernels." He adjusts the clerk's spellings of Indian plant names, but misses other spelling errors in the entry for *Poa Cynosuroides*, which he identifies, after Jones, by its Sanskrit name, "Cusa or Cusha": "If so, (& I have no doubt of it) we have here a very ancient president [*sic*, precedent] for the present very convenient practice of giving arbitrary names to plants, & to their mileis [*sic*, milieu]." The homophonic substitutions of "president" for "precedent" and "mileis" for "milieu" look more like errors that a clerk might make in taking dictation. Whenever Indian plant names are at issue, Roxburgh is far more vigilant about spelling.[114]

Roxburgh's correspondence registers the commercial ambition of his botanical work: he supervised the planting of thousands of Indian pepper plants in the Northern Circars, even before the EIC board of directors agreed to support this venture; he described many species of cotton and some twenty-five species of indigo, with their various dye possibilities painstakingly recorded. The aesthetic ambition of Roxburgh's botanizing in India is at least as prodigious. If time and circumstance had permitted, he would have produced a flora of India as copiously illustrated as the *Plants of the Coast of Coromandel Coast*, but one that would have included many more species from all over the Indian subcontinent. He would have, one imagines, botanized without end.

INDIAN BOTANICALS AND PICTORIAL TRADITIONS

There may never be a book like *The Body of the Artisan*, the subject of Pamela H. Smith's study of the role of artisans in the scientific revolution, written about the Indian artists who painted plants for Europeans, in part because so little is known about the Indians beyond, in a few instances, their names, but more fundamentally because their work practices and histories are only barely and obliquely recorded in colonial records.[115] In the first, extended point of contact between Europeans and Indian informants, when the Dutch consulted with Ezhava physicians and toddy tappers, using their names and systematic to create the first flora of India along the Malabar coast, the first edition of that flora included extensive information about Indian informants and their contributions. Much less is known about the Indian artists who produced thousands of drawings of plant specimens for British botanists and other patrons beginning in the last decades of the eighteenth century, although the names and careers of those who commissioned those artists are recorded. That list was already long by 1800.[116] The artists of the unsigned drawings commissioned by Lady Impey have been identified from archival records as Bhawani Das and Ram Das and Shaikh Zain al-Din. Desmond and Archer provide names of several other Indian artists who may have worked for Roxburgh in Calcutta and for other British botanists stationed in the north: Vishnu Prasad (or Vishnupersaud), Gorachand, Rungiah, Haludar, Mahangu Lal, Gurudayal, and Lakshman Singh, who worked for the British at Saharanpur in Himachal Pradesh.[117] Robert Wight, whose opinion of his artists' work was consistently high, learned lithography to engrave their drawings, identifying many of them as the work of Rungiah or Govindoo.[118]

Not all the artists whom the British hired came from the artistic orbit of Mughal culture. Noltie and others note that painters who were hired in the south came from there, a region that was not dominated by Mughal rulers and style. The two painters who worked for Roxburgh on the Coromandel plants almost certainly came from the Northern Circars in the Coromandel region, and the painter Serfoji hired came from Masulipatam on the east coast of India. In the 1840s, Alexander Gibson's artist may have been of Portuguese extraction.[119] Even those artists whose names survived in colonial records seem to have left no local and familial archives that might tell us more than the little that is recorded in colonial documents. Across India, most of the trades, including artistic trades, were inherited, and those who worked in a specific trade (or specialized task within a trade) were members of a single caste. These castes could and often did vary. Some included Muslim and Hindu members. Indeed, despite the British colonial insistence that the caste system was transhistorical and absolute, most aspects of caste appear to have been subject to change over time and from place to place—from the status of a caste to the work in which its members engaged and handed off to the next generation.[120] Surmising who did what and where therefore requires, as postcolonial theorists and historians have long urged, local knowledge about trades, membership, and material practices.

I emphasize the contribution of Roxburgh's Indian artists in what follows, but all the Indian botanical drawings of this period warrant more investigation.[121] More needs to be known about them as works of art and science created mostly for the British by artists who were not British. These botanical drawings look very different from that of Indian artists of the same period whose work is now identified as Company or Patna painting, a style that recalibrates British and European pictorialism to depict Indian social life and landscape. Put briefly, Indian botanical art of the same period implies other or additional provenances that may include traditions in Indian art that may have informed the way Indian artists painted plants for the British. These claims, necessarily provisional, sketch an approach to the Indian botanicals that attends to their cultural and artistic implications as matter and paint.[122]

Without question, the commissioning of Indian artists was an arrangement between employer and employee, with low salaries and some piece-work payment, especially in the early years. Because Roxburgh typically numbered the Indian artists' drawings as he received them, there is some evidence that their artistic practice improved or changed over the quarter century in which these artists worked for him. While he was the superintendent of the Calcutta Botanic Garden and for an unspecified time after his death, the Calcutta drawings fueled a small spin-off industry of original copies that British travelers requested so that they could take them home. Drawings in the John Fleming collection of the Botany Library of the Natural History Museum, some at the Royal Botanic Garden, Edinburgh, and those that surfaced in a 2006 exhibition and sale at the Colnaghi Gallery in London are almost certainly original copies made by the Calcutta Garden artists for Roxburgh's friends and colleagues.[123] The number of Roxburgh "copies," and the use of Whatman paper to produce them as well as the "original" works (both copies and the original drawings of plant specimens were done by hand; the difference being that the first versions depicted actual plant specimens), makes it clear that those who requested and paid for copies valued them.[124]

Yet in the early years of these commissions, their success as botanical drawings was frequently in question. In 1790, Roxburgh sent Joseph Banks the first of many installments of drawings, selected from his collection at the time of about seven hundred drawings, and asked for a frank judgment of their quality. Apparently Banks expressed some reservations without indicating what he thought inadequate. Roxburgh was irritated and perhaps chagrined: "if you had mentioned what the defects were that my drawings and descriptions . . . I would then, probably have been able to rectify them in those that are still to finish, but you have left me in the dark." Other botanists praised some drawings and complained about others. In 1794, Banks praised the work of Roxburgh's Indian artists to the EIC court of directors, no doubt to encourage their financial support for Roxburgh's *Plants of the Coast of Coromandel*, published the next year. Two years later, Banks agreed with Roxburgh that his artists had improved, noting in particular their careful delineation of fructification parts in drawings of grasses.[125] Amateur naturalists tended to be more admiring. In 1790, William Tennant, an EIC chaplain, wrote of "the laborious exactness with which they imitate every feather of a bird, or the smallest fibre on the leaf of a plant." In November 1810, Maria Graham, sister-in-law of Robert Graham, then the Regius Keeper of the Royal Botanic Garden at Edinburgh, described the work of Roxburgh's Indian artists in Calcutta as "the most beautiful and correct delineations of flowers I ever saw. Indeed, the Hindoos excel in all minute works of this kind."[126]

Over time, the Indian artists learned how to satisfy the demands of their British employers for accuracy, perspectival representation, and a lighter application of color more akin to British watercolorist style of the period, which favored little more than a watercolor wash. Traditional Indian artistic practice favored instead a dense, opaque watercolor, sometimes thickened with gum arabic or applied in thick coats. On occasion, Indian artists burnished the back side of the sheet so that the artist could overpaint successive layers and details—a coloristic practice widely used across the subcontinent for painting on cloth as well as paper.[127] Although overpainting in this way permitted the addition of details like fur or feathers, it worked against the delineation of tiny plant ciliae, or fibers. Evidently, it also took more time to add layers and burnish the back of sheets between the application of layers.

Whereas British artists understood depth of field as part of the work of drawing, Indian artists, particularly those who worked early on for Roxburgh and other British employers, rarely employed perspective, choosing instead to depict plants on a flattened picture plane, also a widespread practice in traditional Indian art. The British emphasis on accurate rendition of details was at times put aside so that Indian artists could arrange the plants or their depictions of them. Noltie notes that in some cases if "two species were drawn on one sheet, then it was ruthlessly bisected so that each half could be put in the 'right' place!"[128] This practice is particularly visible in drawings of Indian plants that climb or twine, among them the Convolvulaceae, depicted so that tendrils and vines wind about in ways that in effect compose the surrounding picture space on the page. The compositional care of this drawing is, from a documentary perspective, unnecessary. In other drawings, different patterns emerge: a plant might be depicted as overreaching the page, with a frond or branch cut off. Some Indian drawings depict plants in ways that command the picture space. In these, the plants look slightly expansive and plumped up even when they do not fill the sheet. Noltie contrasts this compositional practice to the large folio illustrations of Curtis's *Flora Londinensis*, which Roxburgh presented to his artists as a model to follow.[129]

The botanicals themselves make it clear that Roxburgh's hired artists chose a different path. Whereas Curtis's illustrations are centered on the page and surrounded by white space, a compositional strategy J. M. W. Turner used in his mezzotint illustrations of the 1820s and 1830s, Indian plant drawings now called the Roxburgh drawings dominate the picture plane, either by filling it up with botanical information or by composing the plant image so that it becomes the focal point. The closest British parallel to this compositional strategy may be the relatively crowded, much smaller pages of the *Botanical Magazine*. Yet even this similarity points to other differences: rarely do the Roxburgh drawings seem crowded. Instead, they appear to take whatever space they need, even if doing so means going off the edge of the paper.

Some have argued that the stylistic distinctiveness of the Indian botanicals moves decisively away from the ideal of a documentary botanical style. This is not a minor dispute: from a taxonomist's perspective, plant depictions display the species traits of a plant. Yet even this project has cliffs of fall. A plant that illustrates the species type must provide a clear visual reportage of that type, regardless of whether the individual plant specimen conforms wholly to type. Botanical stylization exposes a problem that is inherent in taxonomic representation. As Beth Tobin and Bernard Smith have noted, the depiction of a single plant to represent the traits of a species conveys the typicality of the specimen rather than its uniqueness or individuality.[130] To put this point another way, a good specimen and a good drawing of that specimen are botanically and taxonomically good because they display species traits.

The task of drawing botanical specimens is thus inherently, if instructively, caught between these extremes. The drawing itself, precisely because it is highly colored and aesthetically commanding, emphasizes its singularity and perhaps that of the specimen being drawn. Species specificity remains the scientific project, but the aesthetic character of the drawing pulls toward the particular and the individual. Nor do English botanicals of this period necessarily satisfy the ideal of botanical illustration. Those in Thornton's *Temple of Flora* present an extraordinary but hardly documentary management of scale and perspective. Although Sowerby and Smith corresponded regularly about whether the drawings for *English Botany* looked like the specimens or an intermediate live plant that one or the other would invoke as a corrective, when Sowerby engraved his drawings for serial publication, Smith sometimes complained that the engravings lacked documentary rigor.[131]

Evidence about the paints Indian artists used is scant. Although they may have used commercially produced English watercolors that Reeves began shipping to northern India in the late 1770s, it is also possible that they used some local paints, including some that Europeans prized, like indigo and madder. The Kew website for the Roxburgh drawings suggests that the Indian artists may have used "Indian yellow," a pigment purportedly made from the urine of cows fed primarily on mango leaves, a color source and pigment long identified with Mughal art; "Indian Lake," or Lakh, made from the larvae of the lac insect; "Indian Red," a natural iron oxide; and "Indigo," the plant dye that the English worked hardest to plant and produce for commercial purposes.[132] Scholarly discussion of paints and pigment sources remains to some degree speculative. Although chemical analysis of contemporary Kalighat paintings indicates that these Indian artists of Bengal and Bihar may have used some local paints to produce these works on paper well into the twentieth century, no comparable analysis has been performed on the Indian botanical drawings of the late eighteenth and early nineteenth centuries.[133]

Because some traits of the Indian drawings point toward the Mughal miniature tradition, Archer surmised that the artists were trained in that pictorial tradition, which dominated parts of the Gangetic plain in northern India.[134] Yet here, too, stylistic differences suggest a wider provenance and training. Mughal miniatures typically rely on a heavy application of paint achieved by mixing opaque powdered pigments with egg or gum to depict principal figures and elements. The Indian artists who painted plants for the British appear to have used this technique far more sparingly to indicate the edge of a flower or petal or to suggest volume, a practice that looks more akin to a late eighteenth-century Rajasthani, possibly Mewar, iris painting, attributed to a Rajput painter, and thus someone likely to have been at least as much influenced by Rajput as Mughal pictorial techniques.[135] Although the Indian artists whom Roxburgh employed may have occasionally burnished the reverse of a painted surface to allow the paint to become more luminous (this process makes the surface so smooth that it can receive additional layers of paint), this technique is not exclusively Mughal. It is used nearly everywhere in Indian painting on cloth and on paper.[136]

I begin with five images that can be internally dated around 1794; that is, very early in Roxburgh's commissioning of Indian artists, before he moved up to Calcutta. Given this evidence, the artists could not have come from the Mughal-dominated north. They would instead have been among those Roxburgh hired in Coromandel, along the southern coast of India. The first *Pandanus odoratissimus* is an exception to the general rule that Roxburgh's artists later became more proficient.[137] To depict a specimen frond and blossom from this large palm tree, the artist indicates its size by allowing its large frond to escape the edges of the sheet. The detailed line drawing is refined, no easy task given that the Indian artists used a tiny, sometimes single squirrel hair as a brush to create outline and interior details. The artist may have thickened the green paint with gum arabic to produce the darker shading on the palm fronds. The palm kernel is rough, almost furry. If this is an early drawing by one of Roxburgh's artists, he already knew precisely how to use indigenous techniques (fine brush, gum arabic) for maximum effect and how to gesture toward the actual size of this palm, a convention of botanical illustration but not one easily accommodated in the case of very large Indian plants and trees.

Another Coromandel era drawing of another palm, *Borassus flabelliformis*, a tree that has many uses, especially for food during times of famine, is far less delicate in the application of opaque watercolors.[138] Its large cross section of the palm nut is quirky: instead of cutting the kernel across the middle, the more usual method for presenting a cross section of a seed or fruit, the artist has cut it off near the top, as if to display the edible kernels in a market. The depiction of its contents is thickly painted, from white inner kernels to outer covering, as are several budding versions of the kernel and incipient nodules on the plant itself. The drawing is at once slightly clumsy and arresting: its palette offers the strong primary color contrasts and all those kernels and nodules insist on the exotic character of this Indian plant. *Trapa bisnosa*, the water chestnut, is among the most beautiful drawings in the Roxburgh collection.[139] Its dramatic composition, which fills up the sheet, uses a fairly thick application of paint to mark the edges of the leaves. Although its Roxburgh number suggests that it was painted more than halfway through the series, its appearance as an engraving in the *Coromandel* work also means that it was created around or before 1794. Floating on paper as it might on water, with superb line drawings of the feathery, delicately hairy ciliae of the plant stem, the drawing depicts the plant at an early stage of flowering marked by unfolding leaf buds, a deep blush on the new shoots, and meticulously drawn white flowers. The coloristic definition of the leaves suggests a fairly thick application, or perhaps a layered application of color; neither technique recalls the British preference for light washes of watercolor.

What is perhaps most surprising about the Indian drawing of *Trapa bisnosa* is the delineation of its small white flowers. Archer argued that the Indian artists used European paints, in particular a good deal of china white, also called zinc white.[140]

Yet in this image and in other early Indian Roxburgh drawings of white blossoms, there appears to be no paint or, at most, a slight wash of white color that would seem more appropriate in a late drawing by one of Roxburgh's artists. The apparent difference in the white of the blossom against the white of the paper appears to be managed not by the application of the color white but by fine line brush strokes of black to outline and define the volume of the flowers. This is not to say that the Indian artists never used white paint. The Kew drawings include many in which lead white paint was used to color white flowers, easily determined because the lead oxided and the effect is now dirty, smudged white. China or zinc white, which gradually replaced lead white (for obvious reasons and despite its more brittle surface over time) in Europe in the 1830s, had long been used by Indo-Persian artists along with lead white.[141]

The omission of white paint altogether requires unusual skill, apparent even in drawings of two species of white lotus that Roxburgh's artists probably created in the early Calcutta period. The drawing of the smaller of the two *Nymphaea lotus* drawings includes a cross section that exhibits the stiffness and lack of perspective that is fairly typical among the early drawings, but its depiction of the front and back of the full flower is brilliantly stylized in its presentation of the symmetrical structure of the lotus. Here the drawing is attentive to the blossom itself, using bits of a soft brown to tint the back of the petals, and light striations of black to define the rest of the blossoms, except for the colored stamens and pistils in the center.[142] The drawing of the much larger white *Nelumbium speciosum*, the lotus with a stem with nodules that can grow as large as a man's fist, is still more remarkable in close up.[143] Perhaps in deference to its symbolic importance across India, the drawing lists one of its Indian names first as *Tamara alba*, the Linnaean formulaic for a generic-specific name here recast as an Indian plant name. Its spiny stem and flower bud are painted and detailed with ink brush. Its unpainted petals are indicated with fine black lines that convey considerable depth of field, such that the petal tips do appear to be turned up. The central part of the flower is a crowded, highly symmetrical rondelle of stamens and pistils with red-orange pollen dust. Without question, these are stylized images: the odd way that the seed pod of the actual plant protrudes from the surrounding stamens is not indicated except in the separate drawing of the pod; nor does the drawing show how the thick leaves collect and retain water.

The specificity and delicacy of these Indian lotus drawings mark their difference from James Forbes's somewhat earlier drawing of a plant he identifies as "Nymphea Lotos, Blue Water-Lily of Guzerat," a species found in the lower Himalayas, but not in Gujarat in western India. It does not appear in Roxburgh's *Flora Indica*, which deals relatively little with the botany of the Himalayas. Forbes, who worked in India in the 1760s and 1770s, was trained as a draughtsman when he arrived as a very young man. In India he painted more than two hundred natural history subjects. Inscribed on this drawing is a passage from Jones's translation of *Sacontalá*, a detail Forbes must have added much later, after Jones's translation was published in England.[144] Forbes's lotus may, then, be blue for symbolic reasons: it is the flower identified with Krishna, often represented as blue or blue-black and sometimes seated on a blue lotus, and the focus of an important cult in Rajasthan and Gujarat that produced the tradition of painted cloths called *pichwai*, which are still hung behind the Krishna in the village temple at Nathadwara devoted to his worship. In India, it would appear, even the British engaged in plant symbolism. Forbes's lotus is stiffly painted; its petals are crudely ridged and its center looks like a sunflower.

Other drawings that either appeared in the *Coromandel* work or have early Roxburgh numbers suggest a range of technique and expertise. In the drawing of *Leea macrophylla*, the artist drew a very large, almost certainly magnified and uncolored leaf as a kind of background image.[145] Superimposed is a colored depiction of the plant, its flowers and two leaves, one from the front, another from the back. The drawing is not especially precise: the jagged-edged leaf is painted such that its sections appear to bulge, an effect of the application of a thick green paint; the leaf edge is roughly outlined in ink, and the flowerlets, including some that are ink outlined, look generalized rather than specific. Yet the drawing is compositionally inviting. Its imprecisions not particularly apparent until one looks at it close up. Another early drawing of a species of Bengali cotton, *Bombax heptaphyllum* uses an opaque watercolor to create and a strong dramatic contrast.[146] The red blossoms of the cotton plant are painted with strong color, and the woody green seed pods that surround the puffy cotton look like an inverted bouquet. The separate drawing of a single leaf looks slightly mottled, likely an effect of the passage of time and, when it occurs in some of the Calcutta drawings, probably the effect of humidity on thickly layered paint.

Two Indian drawings of sugar cane and a climbing palm convey the remarkable eye and restraint that some of Roxburgh's artists exercised when dealing with plants whose lightness and movement would be ill-served by heavy application of paint and outline. Many of the drawings of grasses among the Calcutta Roxburgh drawings are exquisitely subtle. One sugar cane that probably dates from early in the history of Roxburgh's commissions, *Saccharum exaltatum*, is represented by a drawing, now in the Library of the Royal Botanic Garden, Edinburgh, that was probably copied by an Indian artist from an earlier Roxburgh drawing.[147] In the Edinburgh collection, this drawing is one of several that are executed with such consistent refinement that they could be the work of a single artist. Roxburgh's handwriting on the drawing of *Calamus rotang*, or rattan palm, a large climbing plant used for making canes and cane furniture, gives its local name, again in the genus-species formula, as *Rhapir rotang*, a name not given in an early dictionary of Indian plant names.[148] The drawing fills the sheet, letting one palm frond exceed the edge of the paper to convey its size, and tiny blossoms trail down in lightly curled stems. In close-up, the technique used to produce an image in which lightness and movement dominate becomes more visible. The spiked and segmented cane and the more closely segmented buds are presented with clean, fine outlines and a very light wash of color: this in a drawing that was, judging by its number, created less than halfway through the long history of the Roxburgh workshop, from his two Coromandel painters of 1789 until well after he returned to England for the last time in 1813.

The last three drawings I discuss here, each from a different time in the history of Roxburgh's commissioned drawings, display a degree of expertise that argues against the claim that the Indian artists who worked for Roxbourgh initially lacked a requisite mastery that their British employer helped them acquire over time. *Menyanthes cristata* is an aquatic plant for which Roxburgh provided Bengali and Telugu names.[149] Whereas some drawings flip plant leaves rather awkwardly to display their underside, this one folds up the edge of just one leaf to indicate the fleshy volume of this leaf and plant. The decision to do this is not impelled by a realist protocol, yet the choice is satisfying because it presents the detailed veining inscribed on the top and underside of the leaf. These white flowers do appear to be painted with white paint that has not oxidized, so the paint may have been china or zinc white. Another Indian water plant, *Pistia stratiotes*, also known as "water lettuce," now identified as a tropical invasive in the waters of the United States, prompted Roxburgh to record several Indian synonyms, which he later corrected in manuscript.[150]

Created somewhere near the middle in the production of the Roxburgh commissions, this drawing combines the opaque body color favored in Indian art to depict the thick, green leaves with a light drawing of the multiple little hairs of the roots. The succulence of the image insists on its edibility. Fairly late in the Roxburgh commissions, one artist created a large drawing (covering a full half sheet of Whatman paper instead of the more typical quarter sheet) of *Tricosanthes heteroclita*, a plant Roxburgh was the first European to identify. Initially he believed that it might be native to Bhutan, but in *Flora Indica* he argued that it is native to Bengal.[151] The drawing is extraordinarily precise in its rendering of white flowers (apparently without white paint) and stunning cascades of fringe threading down from the blossoms. Its large red, slightly speckled fruit is placed alongside, depicted uncut and also in cross section, not—as is more typical in the placement of dissections—below or near the bottom of the sheet. The unusual position gives near equal status to the fruit.

Indian botanical imagery tends then to be stylized and symbolic in ways that appear to be capable of endless expansion and variation, from those that F. D. K. Bosch derived from the early Vedas to the unnaturalistic color symbolism of a blue lotus to match the blue body of Krishna in Rajasthan (echoed in Forbes's blue lotus of Gujarat) and the use of propitious colors like green and red, even on the spotted cows that appear on a mid-nineteenth century *pichwai*.[152] The sacred pipul, *Ficus religiosa*, or its characteristic leaf, is often represented in Indian textiles either in a central design or as a repeated leaf pattern. As distinct pictorial styles developed across India, flowering plants were popular motifs by the second century CE. The Mughal development of naturalistic floral motifs around 1600 arose when the emperor Jehangir, whose taste for natural history would have been extraordinary in any era, employed a Persian artist, Mansur, whose drawings of natural history, including three albums of plant drawings, earned him the honorific Nadir al-Asr, "Wonder of the Age."[153] Although few of those drawings have survived, they are echoed by the stone mosaics of the Taj Mahal, the mausoleum Shan Jahan created to honor his Persian wife.[154]

Although Persian craftsmen settled in India in the late sixteenth century to teach Indians who were working in Mughal and Deccani court workshops, there were also Indian craftsmen outside those workshops who adapted Persian style to their own requirements. In doing so, they preserved what Irwin calls "the more earthy and dynamic qualities of Indian art as a whole."[155] By the mid-seventeenth century, the direction of influence appears to have reversed as Mughal art became less naturalistic and more stylized.[156] The relation between Mughal art and that of the rest of India looks, then, more like a seepage of styles and practices than a battleground in which defined frontiers separated Mughal from non-Mughal art and artists or, for that matter, separated courtly from popular art forms. Stuart Cary Welch insists that there was no sharp separation between courtly artists and patrons and popular village traditions.[157] The Rajasthani use of Mughal motifs, for example, can be explained by factors unrelated to recent or sustained contact with Mughal courts.[158]

It would be difficult to find a pictorial tradition in Indian culture of the last two thousand years that does not include botanical figures. Sculpture alone presents numerous instances: the first-century CE stone incisions on the ruins of the Buddhist site at Sarnath, near Varnasi; another first-century Buddhist stupa in the museum on that site that includes among its sculptural details a floral and vine motif, the lotus and other flowers, and an eleventh-century CE depiction of the bodhisattva Padpani, typically represented with two lotus flowers; a stylized floral motif on the Qu-tab Minar in Delhi, an early Mughal site of the eleventh to thirteenth centuries that predates the era of Persian influence; sculpted stone leaf patterns like those at Fatepur Sikri and the Diwan-i-Am of the Red Fort, in Delhi; and floral "clouds" on the ceiling of the Ajanta caves in what is now Maharashtra state, a Buddhist site created from the second century BC to the sixth CE.[159]

Like Indian artists before them, those who painted plants for the British may have learned from a long tradition of painting on cloth and paper that extended from the Mughal courts and communities of the Gangetic plain to their neighbors to the north and far to the south, a tradition in which coloristic practices and pigments moved from paper to cloth and back again. To cite some instances: *pichwai*, cloth hangings used by devotees to decorate the wall behind the Krishna in Rajasthan, typically depicts lotuses, ripe mango trees, and bunches of flowers; an eighteenth-century Rajasthani miniature in which Krishna leads Radha down a garden path, created under the patronage of a Rajput ruler who sought to re-Hinduize the art of the region after the death of the Mughal emperor Aurangzeb in 1707;[160] floral designs and an occasional tree in the Kalighat paintings of Bengal and the Madubhani painters of Bihar; a Madubhani paper pattern of a design, recreated as a painting on a wall to celebrate marriages, that includes a ring of lotus flowers and the bamboo tree.[161] In modern block prints derived from earlier kalamkari models, two patterns are given names—*kamalam*, a generic term for different forms of the lotus, and *tiggalu*, the term for an interlacing pattern of leaves and flowers.[162] Images related to divinatory practices, genealogies, and astronomy also feature botanical figures.[163]

The Pahari court painters of northern India of the fifteenth through eighteenth centuries CE used stylized flowers and trees (so stylized that modern scholars mostly refuse to commit themselves on the question of the species), including a white lotus that nonetheless looks strikingly like that flower in nature.[164] Finally, the visual resemblance between flowering vines on European chintz fabrics and the depiction of vines in Indian botanical drawings has its own East-West history because Indian textile trade with Europe, like its trade with the Far East, had long adapted to what purchasers wanted.[165] As this survey indicates, stylized botanical representation, compositional pattern, and the use of paints and dyes in different media—all these reappear in the Indian botanical art that the British commissioned.

IN A CALCUTTA GARDEN

Ten months after Roxburgh died, in Edinburgh, February 1815, a folder of 156 colored drawings "found in the possession of Doctor Roxburgh" was delivered to the

EIC, which handed it over to the Calcutta Botanic Garden. These drawings may be those involved in a later dispute about whether Roxburgh's heirs or the EIC owned them.[166] Appended below a brief note that explaines how the folder of drawings made its way back to Calcutta (dated 29 December 1815) is this statement, dated six months later and signed "W. E. Gestner": "The painters who drew them and the Sircar who purchased the paper and charged the expenses in the public accounts are here and can vouch for their being the property of the Government. Botanic Garden 10th June, 1816."[167]

What this statement certifies is self-evident; what it implies is elusive yet compelling. I begin with a fact and an inference: more than a year after his death, Roxburgh's artists remained at the Calcutta botanic garden, where they probably continued to paint specimens under the supervision of the new superintendent, James Hare, who at some point managed to get twelve hundred copies of the original Roxburgh drawings before he was stopped.[168] Beyond inference, somewhere near the edge of intuition, I understand this brief official note to convey the artists' collective self-identification as members of a workshop of artists, not unlike those "painters of Patna" who so self-identified when they worked for the Impeys and others before Roxburgh's arrival. Written after Roxburgh's death in England, the statement refers to a later moment when Indian artists who had worked for Roxburgh remained on site to say that they "are here" and "can vouch" for the circumstances in which they produced the drawings. In 1912, those 156 drawings were dispersed, reassigned to folders and drawers organized to reflect Joseph Hooker's late nineteenth-century systematic for Indian plants. With taxonomy on the wing, as it always is in botanical archives and gardens, you catch the edge of the past where you can.

This unsigned statement survives in the colonial archive because it certifies something that was, as Spivak puts it, "needed in the space of imperial production," yet its resonances extend beyond this colonial intentionality.[169] The issue for that archive was legal ownership of a group of drawings that had become separated from the rest that were archived at Calcutta and found their way back. What the artists testify in addition is their authority, as the artists who made these drawings, that they were created for the Calcutta garden and paid for with funds allocated for their production. It is of course possible that all this is a colonial lie, a set-up designed to forestall a possible claim by Roxburgh's heirs, who later insisted that the EIC had illegally acquired some drawings. I want to sketch what follows if we take the document on something like face value.

Its presence in the colonial archive disturbs the impersonal, discursive character that, Foucault argued, muffles and redirects voices so that they speak only within the protocols of the archive. Yet the artists who were still working at the Calcutta

Botanic Garden after Roxburgh's death offer a testimony that conveys their authority about the conditions in which they worked as artists. Chakrabarty has written about moments like this, when one's scholarly encounter with a subaltern or subalterned past comes to the limit of historical and literary narration, but not, I think, to the end of what might be said. The term he uses to describe such moments is *time-knot*, a Bengali concept that I take to mean those moments when our historicizing and imperializing understanding finds itself surprisingly knotted to the past, unable to put it behind us or to narrate it, our present for a moment captured in a past that is not ours to manage but is nonetheless before us, "speaking" as we may not have expected it to "speak."[170]

This archival speech, like Indian art and botanical knowledge, exceeds the mercantile, colonial, and scientific goals of the British. That Indian art and knowledge work this way inside the colonial archive encourages us to consider whether the space of imperial production may say, or imply, more rather than less or nothing at all about indigenous knowledge. The British and Indian employment of Indian artists from the 1780s until well into the nineteenth century registers in stunning visual terms something beyond mercantile and colonial necessity. Indian drawings themselves convey a surprising, not wholly easy, marriage of taxonomic representation with the aesthetic appeal of an individual work of art, as Indian artists produced drawings that register an array of technical crossovers derived from traditional Indian art forms and media. In this project the British were subtly complicit, and they thereby insured that Indian contributions to the world's knowledge of Indian flora would exceed the terms of the colonial imaginary.

INTERLUDE TWO

A Romantic Garden: Shelley on Vitality and Decay

Two poems Percy Shelley wrote near the end of his life, *The Sensitive-Plant* (1820) and the unfinished *Triumph of Life* (1822), dramatize the embodied life and death of plants.[1] I use the phrase *embodied life* to refer to material forms of life without claiming that those forms can be separated from mechanical or nonvital processes, a distinction I have learned from Amanda Goldstein's work on romantic empiricism.[2] From the perspective of thinking about materiality and life conveyed by the phrase *embodied life*, romantic and postromantic efforts to bifurcate vital and mechanical processes neglect what experimentalists and poets found more plausible after 1800—the possibility that forms of life and nonlife, as well as vital and mechanical operations, coexist in the forms of being we call life.

The question that the speaker of *The Triumph of Life* keeps asking—what is life?—cannot, then, be answered simply by claiming that what is vital is alive and what is mechanical is not. *The Sensitive-Plant* hazards a more tenuous or at least less decided case for asking this question by half-displacing vital agency onto the Lady who tends the garden, then dies. Yet both poems test the operative relation between bodily or material form and life by way of figure and each does so by turning away from Darwin's personifications of plant loves in *The Botanic Garden.* Whereas Darwin invoked decay and death as threats that heighten the peril of plant life, Shelley makes them aspects of the story of life that its captives would prefer to deny. Shelley's poetic staging of these questions is hardly categorical. Although *The Sensitive-Plant* appears to take the notion of plant vitality and run with it, in the end the poem cannot run very far. In *The Triumph of Life*, Shelley's ironic portrait of the dead Rousseau as a gnarled root points to an outcome that those following the chariot of life do not wish to know or accept: that like plants they die, often with failure trailing in their wake. These two poems narrate, then, a poetics of plant life and death that sketches the most problematic terms of the romantic debate about life forms and the kingdoms of nature.

By 1820, analogical thinking across the kingdoms of nature already had a long, albeit disputed, history. Julien Offray de La Mettrie's *L'Homme Machine* (1747/1748) pursued a sustained plant/man analogy in which plant terms define the parts of the human anatomy: the human is only a tree on its head; the abdominal walls of the body the outer bark (*écorce*) that houses the seed of man; the lungs are man's leaves, and so on.[3] Shelley's *The Sensitive-Plant* pushes the logic of plant/human analogy much further. Unlike La Mettrie, who insisted that what makes plants, animals, and humans alike is their common identity as material machines, the plants of Shelley's poem, most spectacularly its Sensitive-plant, are animated and personified and, as such, so full of feeling (spiteful as well as loving) that there is no room left for thinking about them as plantlike or inanimate.

Shelley also expands the reach of Darwinian plant personification by teasing out figurative possibilities that move from a paradise of animated plants to its negation, a garden animated by cranky, noxious weeds whose spleen propels the poem's counternarrative and finally chokes out the Sensitive-plant. The poem modulates its figurative argument by rendering botanical particulars so deftly that their figural life depends on their material presence. I quote just two in a series of Shelleyan plant portraits.

> The snow-drop and then the violet
> Arose from the ground with warm rain wet
> And their breath was mixed with fresh odour, sent
> From the turf, like the voice and the instrument.
> (*S-P*, Pt. 1, ll. 14–16, in *SPP*, 287)
>
> And the wand-like lily, which lifted up,
> As a Maenad, its moonlight-coloured cup
> Till the fiery star, which is its eye,
> Gazed through clear dew on the tender sky;
> (*S-P*, Pt. 1, ll. 33–36, in *SPP*, 287)

Although these and other allied descriptions amplify the poem's Neoplatonic address by linking these flowers to odors, to heavenly light, they remain grounded, to use Shelley's term, in the seasonal succession and character of the flowers the poem presents. Whereas romantic botanists and some philosophers describe some plants, the *Mimosa pudica*, or "sensitive-plant," among them, as "sensitive" or "irritable"—meaning that their leaves turn up with contact or that they turn toward the sun—Shelley separates the two descriptors. Being sensitive is here not a mechanical reflex but a sign of life, as it was for several among earlier experimentalists who debated the relative vitality of sensitive and irritable processes in plants and animals.[4]

In the bifurcated taxonomy of Shelley's poem, sensitivity marks humans (and plants) most capable of giving and receiving love as well as keeping the "undefiled Paradise" of its garden of personified flowers alive. Like the Sensitive-plant, the Lady who arrives, tends the garden, then dies in Part Second is "companionless" (the Lady "has no companion of moral race" yet is to the garden as "God is to the starry scheme" (*S-P*, Pt. 2, ll. 1–4, *SPP*, 290). Being "companionless" is Shelley's most deliberate turn away from Linnaeus, among many others, who identified the "sensitive-plant" as an hermaphrodite because it produces male and female parts on a single blossom; or as Erasmus Darwin put it, "in one house."[5] True, Shelley's Sensitive-plant is an "it," but it is an it who loves what it has not. In the language of Shelley's poem, requiring no sexual companion to reproduce is hardly if at all at issue. What matters is rather its capacity for feeling:

> But none ever trembled and panted with bliss
> In the garden, the field or the wilderness,
> Like a doe in the noontide with love's sweet want
> As the companionless Sensitive-plant.
> (*S-P*, Pt. 1, ll. 9–12, in *SPP*, 287)

Unlike the superficial likeness to animal or human reflexes or movement typically identified with such plants, this Sensitive-plant is aptly named because its longing, its capacity to feel, is suprahuman longing. Its "want," as the poem puns at least twice, exceeds the longing of other flowers and herbs that respond to spring as it arises "like the Spirit of love felt every where" (*S-P*, Pt. 1, l. 5, in *SPP*, 286). Where all flowers grow "in perfect prime," this garden is, like the shy *Mimosa pudica*, an "undefiled Paradise," where union is less Linnaean than Neoplatonic:

> For each was interpenetrated
> With the light and the odour its neighbour shed
> Like young lovers, whom youth and love make dear,
> Wrapt and filled by their mutual atmosphere.
> (*S-P*, Pt. 1, ll. 66–69, in *SPP*, 288)

What nourishes this garden is not water or someone tending tender stalks but a current of strong feeling that the Sensitive-plant embodies. Sensing the winds, the insects, unseen clouds, noontide vapors, all "like ministering angels were/For the Sensitive-plant sweet joy to bear" (*S-P*, Part 1, ll. 94–95, in *SPP*, 289). Although the concluding stanzas of this part muffle the syntax that would appear to link "consciousness," dreams, and the Sensitive-plant, that the extended resemblance between this plant and the Lady who tends the garden then dies implies an allied consciousness grounded in structure of feeling in which desire, longing, and love are mixed:

> Her step seemed to pity the grass it prest;
> You might hear by the heaving of her breast,

> That the coming and going of the wind
> Brought pleasure there and left passion behind.
> (*S-P*, Pt. 2, ll. 24–27, *SPP*, 290)

To this, the narrator suggests (briefly channeling Wordsworth), that the flowers respond in kind:

> I doubt not the flowers of that garden sweet
> Rejoiced in the sound of her gentle feet;
> I doubt not they felt the spirit that came
> From her glowing fingers through all their frame.
> (*S-P*, Pt. 2, ll. 29–32, in *SPP*, 291)

Once she dies, though, lamented as only the Sensitive-plant is left to do, the poem undoes the "sweet joy" of botanical bliss to imagine a garden in which plants still feel, but now with abstracted malice. The result is a living death, or something like nature's murder. Shelley may well be replaying his own Wordsworthian impulse in reverse here. The poet-speaker who now finds revolting those plants "at whose names the verse feels loath" puts the cold death dealing of the weeds into sharpened relief. To feel like this is to kill fellow feeling in a paroxysm of natural malevolence. By personifying the plants that invade the dying garden as irritable (not in so many words, but their botanically marked crankiness is evident), Shelley hijacks the sensitivity/irritability debate by applying both terms to plants: some are sensitive, some are irritable, and all are alive in the world of nature and phenomenal existence that includes human beings. The affirmation at the end of the poem that

> For love, and beauty, and delight
> There is no death nor change: their might
> Exceeds our organs—which endure
> No light—being themselves obscure.
> (*S-P*, Conclusion, ll.21–24, in *SPP*, 295)

takes the high road, but those noisome, embodied plants who die or are killed off, as the Sensitive-plant is when its tears become "frozen glue," suggest another, more earthbound outcome. What if feelings like love, beauty, and delight do not exceed the life of the body and its organs? Granted, this way of putting the argument of these lines understands *organs* viscerally to mean just that and not "sensations or reasoning" as Shelley's modern editors would have it. Perhaps those organs are obscure and endure no light precisely because, in the atmospherics of this poem, they turn away from the ideal light that initially bathes the Neoplatonic and romantic garden of *The Sensitive-Plant*.

The plant portrait of the old, gnarled root that is or once was Rousseau in *The Triumph of Life* stunningly reorients the question "what is life?" away from disembodied toward forms of life that are embodied and as such subject to decay and death. At the end of the poem as we have it, light is obscuring, turbid, anything but the bright white light of Neoplatonic eternity. Goldstein's evocative rendering of this moment as less a matter of defacement, as Paul de Man argued, than one in which particles (minds as well as bodies) dissolve and transform as life goes on in other forms redirects the philosophical irony of this poem by embedding that irony in the material (human and botanical) fact of living and dying. Rousseau's spirit or spark cannot on this view be separated from the fact of his body's quite material, quite natural decay, although this is hardly a point Rousseau is prepared to concede, any more than those whom life drives on, can concede their death. On they totter, unwilling to let go, misunderstanding which end of life they convey. The echo of Shakespeare's *Tempest* that slips into the poet's description of Rousseau cannot return him to life, as Ferdinand is, as if by magic, transformed from death to life in that play ("those are pearls that were his eyes," *Tempest*, 1.2):

> That what I thought was an old root which grew
> To strange distortion out of the hill side
> Was indeed one of that deluded crew,
>
> And that the grass which methought hung so wide
> And white, was but his thin discoloured hair,
> And that the holes it vainly sought to hide
>
> Were or had been eyes.
>
> (*TL*, ll. 182–88, in *SPP*, 489)

This, we eventually learn, is, or was, Rousseau, a deferral of the name that sharply registers an identity and being that exist only in dissolution. He claims that none of this "Corruption" would have happened, "if the spark with which Heaven lit my spirit/Earth had with purer nutriment supplied." What he is now is, he insists, merely a disguise that "stain[s] him who "disdains to wear it" (*TL*, ll. 200–205, in *SPP*, 489).

Wrapped in the delays and adumbrations of successive terza rima stanzas, Rousseau's self-pleading, which claims the autonomous being that his degraded physical being denies him, amplifies this poem's ironic relation to Dante's *Commedia* and the figure of Virgil as poet who guides another poet. No more than a skeletal root, Shelley's Rousseau does not grasp, even now, what life is: a densely particulate mix of material being and mind for which not decaying, not dying, is not an option.

Shelley's poetic explorations of what it means to live like a plant entertain but also question the possibility that the Neoplatonic light and idealist reality that bathe human life might also encompass the material life of plants. *The Sensitive-Plant* shifts its narrative claims on this point between doubt and affirmation, yet always troubled by the dubieties that the poem has so remarkably set in motion. *The Triumph of Life* takes a still more radical step away from Neoplatonic (and Newtonian) celebrations of full, eternal light and being to consider how human life might in the end be said to undergo the cycle of life, death and decay that all living, material forms undergo. For this possibility, plant life (and death) is the model, not the exception. The mordant wit of Shelley's material figure for Rousseau reverses the typical, near-axiomatic, figure for organic vitality—the tree or plant that begins as a rooted entity that grows, develops, unfolds into life. Here that figure is put into reverse: backing down and back from the life of plants and trees, we get their end in their beginning. The apparent negativity of Shelley's reading of the damaged life of plants and organic form might be understood as what these forms of organic life teach romanticism about human life, over against romantic aspirations for forms of life beyond death. What makes such arguments viable is, of course, the possibility that life is not material, not mechanical, a possibility that tunnels under the logic of organic form and takes it over. In the chapter that follows, what counts as life, and whether life is something shared by plants and humans, specify the difference between Goethe's view of plant metamorphosis and Hegel's sustained defense of the barrier between plants and animals.

In imagining his own answers to this ongoing romantic debate in *The Sensitive-Plant* and *The Triumph of Life*, Shelley conveys what is at stake in what Goethe writes about plant metamorphosis and what Hegel says by way of reply, as each illustrates the key moves in a much wider debate about what it means to think of plants as animate beings (or not). For Goethe, plants are at once material and engaged in vital development. For Hegel, how plants are said to be alive or dead bears uncomfortably on the status of the human, a point that is hardly surprising given contemporary claims about the vitality of plants, quite apart from whether or not plants make good figures for life, as Darwin had shown they do. Read as a prelude to this debate, Shelley's poems show what is at stake in Goethe's writing on plant metamorphosis and Hegel's on the philosophy of nature and mind. Because the spectacle of living plant nature creates insoluble problems for Hegel's story of the work of Spirit in and on nature, he takes special care to reread Goethe's 1790 essay *The Metamorphosis of Plants* in ways that undermine its suggestions that plants materially direct a highly complex and shifting self-formation.

CHAPTER 7

Restless Romantic Plants and Philosophers

Round 1, Hegel:

> Every product of the spirit, the very worst of its imaginings, the capriciousness of its most arbitrary moods, a mere word, are all better evidence of God's being than any single object. It is not only that in nature the play of forms has unbounded and unbridled contingency, but that each shape by itself is devoid of the concept of itself. *Life* is the ultimate that nature in its existence drives toward, but as a merely natural life is given over to the irrationality of externality. . . . If spiritual contingency or *caprice* goes forth into *evil*, that which goes astray is still infinitely superior to the regular movement of the stars, or the innocent life of the plant, because that which errs is still spirit.

Round 2, Goethe:

> By chance a passage from the preface to Hegel's *Logic* came into my hands. It goes as follows: "The bud disappears in the bursting forth of the blossom, and one could say that it was refuted by it; in the same way through the fruit the blossom is revealed as a false existence, and as its truth, it takes its place. These forms suppress each other as incompatible with each other, but their fluid nature makes them, at the same time moments of organic unity, in which they not only no longer struggle against each other, but one is as necessary as the other; and this same necessity first creates a life out of the whole." It is really not possible to say anything more monstrous. To want to destroy the eternal reality of nature through a poor sophistic game [*schlechten sophistischen Spass*], seems to me unworthy of a rational man.

Round 3, Hegel:

> A World-historical individual is . . . devoted to the One Aim, regardless of all else . . . so mighty a form must trample down many an innocent flower—crush to pieces many an object in its path.

Read against each other, the above quotations—two from Hegel, one from Goethe—bristle with differences that separate Goethe's view of plant nature from Hegel's claim that it must be subordinated to Spirit.[1] Although the two writers are hardly the only ones in the romantic era to consider the nature of plant nature, their differences and glimpsed convergences specify the stubborn antinomy that runs through debates about plants as the middle kingdom of nature between inanimate minerals and animate animals, including the human. These debates animate the divergent accounts of nature and the human among *Naturphilosophie* writers, and even Immanuel Kant in the *Critique of Judgment*. At issue is the degree to which matter or mind takes up or challenges the possibility of thinking systematically about nature, mind, and being.

These debates, which go back to Aristotle and ancient classical thought about forms of life, acquired new impulses and tensions in the late seventeenth century. As questions crossed the Channel between England and the Continent concerning the difference between (or proximity of) mechanical and vital processes in human bodies and muscles, plants entered around 1740 by a near, or hybrid, back door in the odd form of apparently hybrid plant-animals like the polyp and, near the end of the eighteenth century, sensitive or irritable plants like the Venus flytrap. For Edmund Stahl, Albrecht von Haller, and Théophile de Bordeu, the relative autonomy of sensitive or irritable bodies and parts established the terms of this long inquiry. For German writers on *Naturphilosophie*, including J. W. von Goethe, G. W. F. Hegel, F. W. J. Schelling, Gottfried Herder, Novalis, and Lorenz Oken, plants were at times both figure and ground in arguments that nature and matter were part of or resistant to spirit and mind. This broader debate and its key articulations moved in many directions, mostly the consensus of views that has at times been supposed to define German *Naturphilosophie*.[2]

Because the antagonism between Goethe and Hegel on the question of "vegetable nature" (Hegel's nicely dismissive phrase) puts the philosophical difficulties that plants create for romantic thinking about nature into high focus, this chapter tracks the fault lines that separate these two writers in order to make clear, via their differences, the difficulties that plant or vegetable nature created for romantic philosophical thinking about nature. Where, in short, do plants stand in the series of binaries concerning forms of life that had emerged over the preceding two centuries? Are plants alive or dead, animate or inert, conscious or not, and, via all these binaries, can they be systematized into a vision of nature that maintains a clear hierarchy of created forms and life? Hegel leant one way, Goethe mostly the other: between them they specified the trouble plants cause for romantic theories of life and volition.

In the first of the three rounds, Hegel insists that even if Spirit goes badly astray, it remains superior to nature; in the last Hegel argues that nature is at once material and weak, an innocent flower that is, as the logic of gender virtually required around 1800, inevitably female and so both unlikely and unable to resist being crushed. In the middle round, Goethe declares Hegel's account of plant development monstrous.

For his part, Hegel pursues a covert operation against Goethe's account of plant nature from Hegel's early writing about nature to successive versions of his *Philoso-*

phy of Nature, finally published as part 2 of the *Encyclopedia of the Philosophical Sciences* (1830). In its final iterations, this work restaged the argument of Goethe's *Metamorphosis of Plants*, less to praise it than to undermine its conclusions about plant development. What is telling about Hegel's half-mimetic, half-aggressive rewriting is the degree to which these contradictory operations emerged gradually in earlier versions of his *Philosophy of Nature*, which he delivered as university lectures first in Jena in 1805–6 and again in Berlin for many years, beginning in 1819. In the tripartite division of nature to which Goethe and Hegel differently reply, claims that plants that act or look like animals (the Venus flytrap, its kinder, gentler cousin the sensitive plant, the vegetable lamb, orchids that look like insects) could and did cause trouble. That Hegel addressed this situation by reading Goethe both with and against him conveys just how difficult it was, even for Hegel, to insist that plants had none of the inner directedness and life attributable to spirit.

In the long debate about whether nature is animate or mechanical, and specifically how plant nature brings this question into sharp relief, Goethe and Hegel had more in common with each other than either had with early *Naturphilosophie* writers who believed, as did Friedrich Schlegel in 1798, that nature had nothing mechanical about it but was the world soul, organic and self-organized.[3] Those who have read Goethe on science and nature as though he were in agreement with *Naturphilosophie* have missed the two principles on which he and Hegel agreed to part company with their Jena compatriots. Both were committed, albeit differently, to the role of mechanism in plant processes and both were, as early *Naturphilosophie* writers were not, wary about the use of analogy to claim shared identity for all aspects of nature.[4] It is this measure of agreement that prompted Hegel to declare that Goethe's *Metamorphosis of Plants*, "the beginning of a rational conception of the nature of plant-life."[5] Goethe, as did many his scientific contemporaries, argued that plant development combined mechanical processes as well as something that was not mechanical. This point of disagreement between the older Goethe, on the one hand, and the younger Schlegel, Fichte, and Schelling, on the other, took shape as each granted or rejected Kant's declaration in the *Critique of Judgment* that organisms, minerals as well as plants, are organized beings. This recognition opens several doors that Hegel's mature philosophy of nature attempted to close.

Kant put it this way: organisms develop in ways that, rationally considered, would appear to be inner-directed, not compelled from without: "things, as natural ends, are organized beings" [*organisierte Wesen*].[6] Although *Naturphilosophie* writers took this phrase to mean that organized beings are alive, however different their form of life is from that of animals and plants, Kant was wary of this claim, although he nearly made it himself.[7] He cautioned repeatedly that the notion of the organisms as "organized beings" is a regulative or "as if" supposition, not one that is constitutive in the sense that it might be understood as a claim based on sure knowledge.[8] However warily offered, Kant's claim marked a quite new direction in philosophical thinking about nature, including his own. In *Metaphysical Foundations of Natural Science* he had warned about the dangers of "hylozoism," the claim that life inheres in matter.[9] Yet four years later in the third Critique, he argues that organic development cannot be explained by either mechanical processes or physicotheological belief in divine design and the uses of nature.[10]

Humans, Kant wryly notes, may have ideas of what plants (and animals) are for, but those ideas are unrelated to the purposiveness (*Zweckmäßigkeit*) by which nature's particulars grow as individuals and even as members of species and genera, an orientation toward ends that has nothing intrinsically to do with external design.[11] Like the objects of aesthetic judgments, the teleology of organic beings exhibits a purposiveness without purpose (*Zweckmäßigkeit ohne Zweck*) in the sense that organisms cannot be assigned a purpose beyond that of their own inner-directedness. Briefly attracted to think about nature as an analogue to art, he suggests that the more apt analogue might be to "life" (*Leben*), then decides that he has gone too far, for to claim that this "inscrutable property of nature" is "*an analogue of life*" would require conjoining to matter "an alien principle"; that is, soul. To do so would return to hylozoism by another route. (Kant, *CJ*, lemma 65, 254; *KU*, 735). However qualified and tentative all this is, Kant here deliberates with some care about questions and hypotheses that had been widely discussed for nearly a half century.[12]

As Kant writes about how organisms develop as individuals and as contributing members of species and genera, using the example of a tree, his admiration for human classifications of animals and plants is matched, and at times outrun, by his notice of what he calls "the self-help of nature," whereby a tree may heal itself against external injury, producing in some cases malformation or defects while sustaining the life of the individual and contributing to the reproduction of its species (Kant, *CJ*, lemma 64, 243–44; *KU*, 732–34). He further insists that chance and contingency are logically necessary for claiming that forces beyond "blind mechanism" direct organic development. With impeccable logic, he argues against the claim that nothing happens by chance, that it is chance or contingency all the way down, that if we see in the world organic forms that cannot be controlled by blind mechanism, we must in that event admit the necessity (a nice turn of argument) of contingency. For without it we have nothing to say about teleology as the logic of ends, since this notion is meaningless without the possibility that there might be different ends (Kant, *CJ*, lemma 66, 248; *KU*, 740–41).

By insisting on the role of contingency in nature, Kant returns unequivocally to an issue so worrying to early moderns that they created as its specter or counterimage mathematical probability as the "law" of chance.[13] Granting the role of contingency in the development of biological life meant, then, that plants like animals take what chances come their way and develop from unseen, inner-directed force. Although the implications of this argument for evolutionary thought are not ones that Kant pursued, it offers plants in particular a far greater capacity for development and survival than they had before had, philosophically speaking. Plants may not talk or think, but they can and do take up, at each stage, what is on offer for development, even as they develop from within. So described, plants have an inner-directedness that is in many cases strong enough to survive external accident. In the framework of late Enlightenment thought, this double capacity made plants remarkably resilient; it also made them sharers in the excess Kant ascribes to nature because its being "cannot be deduced from the general [mechanical] laws of nature."[14] If this excess is not "life" itself, it is nonetheless an excess identified with nature and represented in an illustrated print culture that made the strongest possible case for the bounty of material nature.

Naturphilosophie writers transformed Kant's regulative principle that we ought to proceed as if organisms have inner purposiveness into a constitutive claim about what nature is. This swerve from Kant decisively set *Naturphilosophie* on its own philosophical course. Not content to suppose that some organic principle is at work alongside mechanical processes of the kind being discovered in chemistry, physics, and biological life, Friedrich Schelling and then others insisted further that mechanical processes are themselves directed and supervised by an organic spirit and, still further, that the difference between spirit or mind and nature or matter is one of degree not essential kind.[15] Lorenz Oken offered a more specific map of the work of synthesis in *Lehrbuch der Naturphilosophie* (1809–11), which linked organic and inorganic nature via analogies that are either constructed, in the sense that they link objects with those that significantly precede it, or directed by the *naturphilosophische* method, which attends to "correspondences between part and whole in the cosmos."[16]

Neither Goethe nor Hegel adopted the *Naturphilosophie* aspiration or claim of a vast, cosmic unity of organic and inorganic forms without appeal to mechanical processes; both granted Kant's regulative principle of the inner directedness of organic growth. Their differences beyond this point capture the culture of romantic nature around 1800, when the early enthusiasm for *Naturphilosophie* took a more skeptical and scientific turn. For Goethe, Hegel and their contemporaries, at issue were questions about abstraction, figurality, and classification that emerge as botany becomes a philosophical as well as material enterprise. To suggest why, I begin by asking some more pointed questions about Hegel's presentation of plants. Why is it necessary that Spirit "crush many an innocent flower," or even that a flower become a figure and an instance for other objects that spirit will need to crush? Why is the spirit's ascent over the world of nature rendered as struggle if, as Hegel elsewhere contends, this outcome is itself an instance of a lifting up (*Aufhebung*) that preserves what it elevates? Why does Hegel define the verb that is the root of this term *aufheben*) by describing fulcrums or levers (*Hebel*) as if to suggest that the process is mechanical? Should we be suspicious of Hegel's claim that *Aufhebung* always preserves what lifts up? How, to speak from a materialist perspective that challenges the idealist tilt of Hegel's thought, is preservation possible if plants and flowers need to be crushed or made into miniature levers that awkwardly lift plant development from one (discrete) stage to another? How could Hegel, who was once attracted to the *Naturphilosophie* principle of spirit in nature, turn on nature with a rhetorical ferocity that belies his claim to present the development of spirit from its origin in nature to its dominion over nature as a process that nonetheless preserves its point of departure, now elevated and transfigured?

In one sense, it is hardly accidental that plants are the allegorical, flower-figured focus of Hegel's philosophical turning away from *Naturphilosophie* and his calculated rewriting of Goethe on the metamorphosis of plants. For Hegel as for his contemporaries, plants looked categorically unreliable: some seemed to move; most seemed to procreate if one took Linnaeus at his figurative word; and still others looked like animals. Unless plant development could be presented as wholly mechanical, plants could not (as they always had) mark the barrier between life and nonlife. For the same reason, they could not mark the difference that for Hegel divided nature from spirit or mind in its highest animal form, that is, human form. Then, too, by 1800 plants raised more questions than answers, making them, rather than the human beings who studied them, appear to hold some hidden cards. For a philosopher who contended that spirit became spirit by sublating (that is, both incorporating and overcoming) its original relation to nature, a botanical nature with this much hidden power needed to be crushed.

Hegel's developing philosophy of nature minimizes the inner purposiveness of organisms, so that the Spirit takes from nature what it needs, then subordinates it. For the Hegelian Spirit cannot in the end (and for Hegel there is an end of nature and particulars in Spirit) encompass nature as idea if nature itself won't settle down into a stable array of species and characters. He prefers the contingency put in motion by an erring spirit, even if that spirit slips into evil. Nature's restlessness thus becomes the Spirit's, open to the contingency of its becoming.[17] Granting that nature's "play of forms has unbounded and unbridled contingency," Hegel asserts that nature never achieves the life it strives for because natural life is caught up in what Hegel calls the "irrationality of externality"—the contingency of the phenomenal world—to be called life. Nor will he admit into the careful tripod structure of his examples the possibility of astral contingency, of comets or other irregular (and irrational) motions in the heavens. What swirls then beneath Hegel's assertion is the contingency of things, particulars, the natural world, a contingency that he wants very much to give over to Spirit.[18] In marked contrast, Goethe, as do many of his scientific contemporaries, understands contingency or chance (*Zufälligkeit* and *Zufall*) as involved with irregularities in the development of plant and animal morphology that introduce unusual or, biologically speaking, "monstrous" forms. By studying how plants adapt to chance events and accidents, he argues, it is possible to learn how regular as well as irregular or monstrous forms occur.

In what follows I trace Hegel's and Goethe's writing on plant nature in stages, beginning with Hegel's *Phenomenology of Spirit*, then Goethe's 1790 *Metamorphosis of Plants*, his 1798 poem of the same title and related scientific writing from later years. I turn then to Hegel's *Philosophy of Nature* (1830) as it rereads Goethe's *Metamorphosis* essay in the light and shadow of the older Goethe's scientific writing and reputation. I conclude by way of Theodor Adorno, who tells a brief tale in *Minima Moralia* about two rabbits that take what chances for life come their way. As I read it, this tale invites notice of how nature survives contingencies that are anathema to the work and goals of Hegelian Spirit.

BECOMING HEGEL

I return briefly to Hegel's notice of how the world Spirit, who has important work to do, ends up trampling innocent flowers. Whenever plants turn up as figures in Hegel's writing, they do not fare particularly well. Consider the botanical figure that he uses to map the relation between spirit and nature in the beginning of *Phenomenology of Spirit*. Rather than fixating on truth or falsity, he proposes, it is essential to "comprehend the diversity of philosophical systems as the progressive unfolding of truth," a process he compares to the way a plant buds and flowers:

> The bud disappears in the bursting-forth of the blossom, and one might say that the former is refuted by the latter; similarly, when the fruit appears, the blossom is shown up in its turn as a false manifestation of the plant, and the fruit now emerges as the truth of it instead. These forms are not just distinguished from one another, they also supplant one another as mutually incompatible. Yet at the same time their fluid nature makes them moments of an organic unity in which they not only do not conflict, but in which each is as necessary as the other. (Hegel, *PS*, 2; *Werke*, 3: 12)

The first part of this remark reprises the idea that Goethe called monstrous, unworthy of a rational man. Yet because it naturalizes a sequence that involves putting things aside, disappearing them, Hegel finds this plant analogy useful to Spirit. His broad objection to analogy, which he announces in the same work, gives him an out should this analogy, or any other, prove troublesome: "Analogy not only does not give a perfect right, but on account of its nature contradicts itself so often that the inference to be drawn from analogy itself is rather that analogy does not permit an inference to be made" (Hegel, *PS*, 152; *Werke*, 3: 194). With this syntactically tortuous backward leap, Hegel seems to retract the famous analogy between the birth-time of the era and the emergence of Spirit that launches his argument in the *Phenomenology*, along with its botanical coda: "when we wish to see an oak with its massive trunk and spreading branches and foliage, we are not content to be shown an acorn instead." Rightly so, for "the actuality of this simple whole consists in those various shapes and forms which have become its moments, and which will now develop and take shape afresh, this time in their new element, in their newly acquired meaning" (Hegel, *PS*, 6–7; *Werke*, 3: 19–20).

Late in the *Phenomenology* he argues that the "revealed religion" of Spirit succeeds the lesser modes of spirit involved in "natural religion"—first pantheism with its "flower religion," then animal lives that are lived "unconscious of universality." All these are taken up, though not yet understood, by spirit as an artificer who creates or understands in much the way that bees do (or don't) when they build honeycombs to store honey. This "instinctive operation" begins to render the worlds of plants and animals as formally intelligible, without yet performing the work of self-consciousness which can recognize that intelligibility (Hegel, *PS*, 420–21; *Werke*, 3: 507–8). By shrinking pantheism to a "flower religion" (which he elsewhere links to Hindu effeminacy), he binds plant nature and beauty with unthinking worship, a weak theology because grounded in a weak aesthetic in which the beauty of nature trumps the work of minds and hands. Displaced here precisely because it is made over as a weakly, synonymous religion, botany has no purchase on Spirit. Hegel is precise and rather chilling on this point. The movement from plant to animal to Spirit prompts the passiveness of plant existence to invite a counterattack from its "spiritual atoms" that in turn triggers animal destructiveness. All of this must go for the "ensoulment of this kingdom of Spirit" to occur. The plant's "innocent indifference" to all of this produces "a hostile movement in which the hatred which stems from being-for-itself is aroused." Exit plant nature, done in by its own singularity, so that the newly ensouled Spirit can assume its powers. So much for organic analogy in Hegel's grasp.

Hegel supposes that the problem of *differentiae*, identifying which plant characters distinguish one species from another, should address the role of the universal Spirit in the world. His account of the state of taxonomic confusion about categories is a nightmare of concerns that run just beneath the surface (sometimes on the surface) of taxonomic debates of the early nineteenth century: "Observation, which kept them properly apart and believed that in them it had something firm and settled, sees principles overlapping one another, transitions and confusions developing; what it at first took to be absolutely separate, it sees combined with something else, and what it reckoned to be in combination, it sees apart and separate" (Hegel, *PS*, 150; *Werke*, 3: 190–91). Hegel's syntactic irony is sharp: as the twice-delayed referent for *them—principles*—gets entangled, and their universal claims are put aside precisely because observation observes its own protocols to the letter: the objects it identifies it also treats as merely or only objects, whose "passive, unbroken selfsameness of being," in Hegel's phrase, neglects the universal idea. For this reason, Hegel argues, the *differentiae* of taxonomic practice are "vanishing moments of a movement which returns back into itself," moments that Reason would instead scrutinize to discover the "*law* and the *Notion* of such determinations" to arrive at an adequate understanding of what is Real, not simply observed.

Although Hegel's description could be said to predict the frustrations conveyed by later nineteenth-century efforts to map the array of affinities between plants in particular in ever increasingly complex and multitiered models, his point is not that another systematic might work but that objectifying systems fail because they get caught in their own either/or distinctions, warring over differences and contradictions that Hegel says should instead be understood as developing moments of Spirit. Thus all taxonomic systems that are not open-ended and provisional and thus in literal terms unsystematic are likely to stumble in similar ways, except if they end, as Hegel's does, in Spirit.

Yet precisely because the need to systematize runs evidently in tandem with Hegel's project, when he returns to the task of identifying species and genera later in the *Phenomenology of Spirit* he does so by considering why the Earth (*Erde*) as a "universal individual," undoes taxonomic distinctions with near abandon. What he finds is an incipient version of the enabling negativity the Spirit will eventually use to bring the phenomenal world into its own orbit. What interests me here is the violence in Hegel's scenario:

> The genus, which divides itself into species on the basis of *general determinateness* of number, or which may adopt as its principle of division particular features of its existence, e.g., shape, colour, etc., while peacefully engaged in this activity, suffers violence from the universal individual, *the Earth*, which as the universal negativity preserves the differences as they exist within itself—their nature, on account of the substance to which they belong, being different from the nature of those of the genus—and in face of the systematization of the genus. This action of the genus comes to be a quite restricted affair which it is permitted to carry on

> only inside those powerful elements, and which is interrupted, incomplete and curtailed on all sides by their unchecked violence. (Hegel, *PS*, 178; *Werke*, 3: 224)

So, Earth keeps in reserve a set of *differentiae* whose negativity is of a different kind from those of the genus constructed on the basis of observable traits. Nomenclature matters here: "nature" is the lexical sign of incommensurable natures as well as differences between a negativity whose "powerful elements" restrict the observable differentiae used to create a genus in all manner of ways, with a violence that is "unchecked."

In Hegel's account of the Earth's negativity, its *differentiae* may appear to perform a phenomenal version of Spirit's eventual dominion over nature, but here the dominion is, as it is not in Hegel's famous account of the becoming that transforms the relation between lord and bondsman, all negation (Hegel, *PS*, 117; *Werke*, 3:153). The next lemma puts the Earth's negativity abruptly aside, rather suddenly abandoning "organic nature" to "simple determinateness" of the sort that inevitably brings taxonomy to its knees. That is to say, the Earth's Demogorgon-like well of negativity exists apart from organic nature, which seems caught in a positivist web of unknowing: it "has no history"; it is fallen from the universal as surely as one of Blake's Zoas; it has only "the singleness of existence, and the moments of simple determinateness, and the single organic life united in this actuality." These elements together "produce the process of Becoming," but "merely as a contingent movement" (*PS*, 178–79; *Werke*, 3: 224). Contingency, that enemy of the universal Idea, and ever the companion of phenomenal nature, is here made to constrict the degree of vitality, movement, and becoming available to organic nature. If Hegel has traveled far from Schlegel's *Naturphilosophie*, where nature and mind share a world soul, he is careful to avoid a version of idealism that is cut off from world and consciousness. If there is clearly a price to be paid (by nature) for this outcome, it is also the case that Hegel's idealism is robust, involved as it is (or as it purports to be) with the world as an object of consciousness and the necessary ground for the emergence of spirit.[19]

In the *Logic* of 1830, part 1 of the *Encyclopaedie*, Hegel argues that, without our reflection, nature would only be known as that site where things go to die: "Nature shows us a countless number of individual forms and phenomena. Into this variety we feel a need of introducing unity: we compare, consequently, and try to find the universal of each single case. Individuals are born and perish: the species abides and recurs in them all: and its existence is only visible to reflection" (*Logic*, lemma 21, p. 34; *Werke*, 8: 77). Precisely because Hegel wishes here to foreground the work of logic as the rational engine that will insure the development of Spirit, he insists that nature cannot do taxonomic work, whereas logic can: "the syllogistic form is a universal form of all things. Everything that exists is a particular, which couples together the universal and the particular. But Nature is weak and fails to exhibit the logical forms in their purity" (*Logic*, lemma 24, pp. 39–40; *Werke*, 8: 84).

Having put Spirit in clear dominion over nature, Hegel can afford to recognize the special, because phenomenally grounded, character of his idealism as distinct from the empty abstractions he elsewhere mocks:

> Ideality only has a meaning when it is the ideality of something: but this something is not a mere indefinite this or that, but existence characterized as reality, which, if retained in isolation, possesses no truth. The distinction between Nature and Mind is not improperly conceived, when the former is traced back to reality, and the latter to ideality as a fundamental category. Nature however is far from being so fixed and complete, as to subsist even without Mind: in Mind it first, as it were, attains its goal and its truth. And similarly, Mind on its part is not merely a world beyond Nature and nothing more: it is really, and with full proof, seen to be mind, only when it involves Nature as absorbed in itself. Apropos of this, we should note the double meaning of the German word *aufheben*. We mean by it (1) to clear away, or annul; thus, we say, a law or a regulation is set aside; (2) to keep, or preserve: in which sense we use it when we say: something is well put by. This double usage of language, which gives to the same word a positive and negative meaning, is not an accident, and gives no ground for reproaching language as a cause of confusion. We should rather recognize in it the speculative spirit of our language rising above the mere "either—or" of understanding. (*Logic*, lemma 96, 141–42; *Werke*, 8: 204–5)

Here, Nature—rather than taxonomy or positivist empiricism—becomes the good example of the philosophical work summarized by the verb *aufheben*. The analogy is precise: as nature is lifted up by the work of spirit, so does philosophical or speculative thought rise above the either/or of understanding, which for Hegel repeats the fixed categories of taxonomic differentiation. Presented as an example, not a rival, nature is taken up, rather than put down, in a politic rewriting of the rhetoric of struggle elsewhere in Hegel's writing on spirit and nature.

In the Jena lectures of 1805–6, Hegel had been primarily concerned, Leo Rauch argues, with the "absolute materiality" of nature because it "does not merely penetrate everything, it itself *is* everything." Noting Hegel's claim that materiality can only be so "by virtue of its mirroring of spirit," Rauch asks a bit sharply: "How real is all this?"[20] From Hegel's perspective in 1830, the answer would be that this is as real as the world gets because the reflective and speculative Spirit recognizes Nature as it cannot recognize itself. But the backstory of this story still commands interest. The absolute materiality of nature can only be met by a counterimage in

Spirit, whose more profound understanding of the absolute will displace its material shadow.

One barrier to this outcome is Goethe's view of botanical nature, which insists that the mechanical processes of plants maintain a congress between matter and a spiritualized and animate nature. By the time Hegel takes up the project of writing the *Logic* and *Philosophy of Nature* that became parts 1 and 2 of his *Encyclopedia of the Philosophical Sciences*, his mature account of nature and spirit is already in place. What eddies beneath that argument is materiality, which had a near magnetic attraction among his contemporaries who tried to join hidden forces to material processes. Hegel has all this in the rearview mirror as he continues to write on nature and Spirit. When he began lecturing on the philosophy of nature, among the figures he saw in that mirror was Goethe.

ANALOGY AND FORM IN GOETHE'S METAMORPHOSIS

Writing at the end of the nineteenth century, the plant physiologist Julius von Sachs charged Goethe with taking up "the position of the so-called nature philosophy."[21] Although it is rarely invoked now, echoes of this charge reverberated for long enough in writing on the history of science to obscure the differences between Goethe and *Naturphilosophie* writers that, up to a point, make him a more amenable figure to Hegel. I return to this older charge because it addresses what troubled Hegel about Goethe's writing on plants, and it lays bare a confusion about the role of figurality in scientific thinking that is more silent now than it once was, but that is still very much alive. Sachs insisted that Goethe uses the terms *metamorphosis* and *Urpflanze*, or ur-plant, as analogies, figures, abstractions, but he also imagines them to be true, or as Sachs puts it, facts. Although more is at stake than the question of whether or not Goethe uses such figures (he does), I begin with what I take to be the lesser charge, that Goethe is as loose and easy with analogies between nature and everything else as nature philosophy had been. I ask first whether he used analogy with *Naturphilosophie* abandon. I then consider what we might yet learn about the role of figure and analogy in his writing on plants.

When Goethe begin writing himself brief notes on his botanical experiments in the 1780s, Jocelyn Holland has explained, he at first relied on analogies between plants and human anatomy that recalled those in Johann Gottfried Herder's multivolume *Ideen zur Philosophie der Geschichte der Menschheit (Ideas on a Philosophy of the History of Mankind*, 1784–91), which Goethe helped to assemble. "The seed's first structure develops plant-like in the womb," says Herder. In his notes for a planned essay "On the Cotyledons," Goethe writes about "the trace of the placenta which is so visible on the palms."[22] Yet in the early botanical notes he soon changed his mind, arguing instead that naming the "first leaves" of the plant after the placenta or any other part of the human female anatomy encourages "illusory" comparisons between "the various hulls of the seed and the little skin of animal births" that prevent observers from "getting to know the nature and quality of such parts."[23] Even in Linnaeus, Goethe adds, such comparisons are really no better than "the cover of the undiscovered," an unkind but witty allusion to the class of plants that Linnaeus named the Cryptogamia because their reproductive parts were hidden.[24] In the *Metamorphosis* essay of 1790, Goethe lists several names for this part of the plant, but all are presented as lesser synonyms for "*Cotyledons*": "they have also been called seed-valves, kernel-pieces, seed-lobes, and seed-leaves; attempting thus to designate the various forms in which we perceive them." The English translation preserves the possibilities suggested by the Goethe's German compound words (*Kernstücke* for seed-valves, etc). *Placenta* does not appear even as a synonym.[25] Goethe suggests that those parts were not hidden, just undiscovered.

Goethe's sense that some comparisons are dangerous turns on their figurative power: when such figures divert attention from the physical characteristics under scrutiny, they cannot be used. When Goethe published the *Metamorphosis* essay in 1790, his caution about the use of analogy was in some measure strategic. As a poet who presented his experimental conclusions about plant development to a reading public, he needed to distinguish his scientific writing from the cosmic analogies of early *Naturphilosophie* writing. Yet his attention in the 1790 essay to minute differences in form that he had studied for years seems to register something that goes beyond a market strategy. In Goethe's scientific writing over many decades and many more volumes, minute differences together with slow, even reluctant, generalization drive and structure inquiry.

Even so, Goethe in 1790 risked the term *metamorphosis*. Sachs charged him with using the term *Urpflanze* as well, but this term is not in the 1790 essay, although Goethe had certainly begun to think about a logic of formal development in plants whereby different articulations of a leaf structure might carry or articulate the principle or intuition he had noted in his 1787 Italian journal: "Alles ist Blatt" ("all is leaf").[26] The *Metamorphosis* essay also introduces the figure and idea of polarity to which Goethe returned in later in poems, *Faust*, and *Die Wahlverwandtschaften*, or *Elective Affinities*, the 1809 novel whose characters and plot dramatize contemporary botanical but particularly chemical polarities.[27]

In the 1790 essay, the idea of metamorphosis depends on Goethe's initial parsing of formal rhythms or polarities of development that he would later characterize explicitly in term of polarity. Goethe argues that leaf-like structures mark processes and changes that are consistent from the seed to the flowering plant and further that these processes, not the reproductive parts of plants, merit notice. Although he assents to the terms of Linnaeus's understanding of the sexual system, noting that "nature steadfastly does its eternal work of propagating vegetation by two genders" in the way the female pistil develops, Goethe also insists that "we might likewise say of the stem that it is an expanded flower and fruit, just as we assumed that the flower and fruit are a contracted stem."[28] Male and female Linnaeus (and others) created them. But for Goethe plant reproductions display the same rhythm of contraction and expansion that operate at all stages of the plant's development, including its culminating blossom and fruit.

Convinced, as Elaine Miller notes, that form, or *morphe*, cannot be understood except in terms of its constant metamorphosis, Goethe presents the growth of the plant as a series of remarkable, even inventive, adaptations and new directions directed by "inner force as well as outer form" (*MP*, 89).[29] In words that Sachs heard perfectly well, Goethe says that the notion of metamorphosis makes "a hidden relationship among various external parts of the plant" visible (*MP*, 76). Yet he makes this point provisionally, as though trying out local versions of this hypothesis to see if they work for distinct moments in the development of the plant before he advances a more capacious hypothesis. The first of these local versions is the term *anastomosis* (*OED*, "to furnish a mouth or outlet"), presented to name how leaves develop as the "vessels . . . which start from the ribs, find one another with their ends, and form the leaf skin" (*MP*, in Goethe, *Scientific Studies*, 79). Fully anastomosed, the leaf gathers nourishment that it returns to the stem of the plant. An incomplete or "half-anastomosed" plant would be one that has the threadlike formations, but lacks the full leaf, as do some aquatic plants (Goethe's example is *Ranunculus aquaticus*).

For Goethe, *anastomosis* is not so much a digestive or assimilative process as it is an opening to formal and organic differences that continue to appear as the leaf form that will continue to metamorphose into other, related forms as the plant grows.[30] Goethe here understands plant *anastomosis* in a sense that is echoed by the usage of the term in Greek comedy to characterize the function of the chorus when it speaks during the intermission to remind the audience of the desired social equilibrium that will return by the end of the play. The *Metamorphosis* essay is equally committed to lawful and directed outcomes, eschewing irregular and monstrous variations in plant development. In modern biology, *anastomosis* refers to the way that distinct species may recombine in later evolutionary moments. Goethe was no evolutionist in the modern sense of this term, but his attention to how leaves develop also understands this local process as a rhythm or pattern of development that he finds elsewhere in the plant's development.

In a prefatory essay published with the reprinted *Metamorphosis* in 1817, he argues that although *Gestalt* is typically understood to refer to a completed and fixed form, "if we look at these *Gestalten*, especially the organic ones, we will discover nothing in them is permanent, nothing is at rest or defined, everything is in a flux of continual motion." Better, suggests Goethe, to use the word *Bildung* because it can refer to both the end product and what is in process of being produced. Better, that is, to use a word that refuses to distinguish a visible, apparently completed form from the process of its development.[31] In the essay itself, *Bildung* summarizes two processes, contraction and expansion, that make up anastomosis. Explaining that "Nature forms a composite calyx [outer set of leaves that enclose the base of the flower] out of many leaves compacted around a single axis," Goethe says that this process is "driven by the same strong growth impulse" that has propelled growth all along. Now, the plant "suddenly develops an endless stem, so to speak" (*MP*, in Goethe, *Scientific Studies*, 92). This expansion, about as sublime as plants get to be, is achieved by an "enormous contraction" of the flowers and the seed vessels they fertilize. Here Goethe advises his observer to recognize in these events the work of several principles, to be used as one would work with "algebraic formulas." The presentation of the principles looks like analogy by chiasmus: "expansion and contraction, compaction and anastomosis" (*MP* in Goethe, *Scientific Studies*, 92). Goethe's conclusion lists all these once more as processes that "fulfill . . . Nature's laws throughout."

Ever the unfulfilled law of Goethe's essay, the authority of the Linnaean system is at such moments put aside and left unnamed. Nature is instead the grand if benevolent enforcer, even when the plant arrives at reproduction: "Nature does not create a new organ in the calyx; it merely gathers and modifies the organs we are already familiar with, and thereby comes a step closer to its goal" (*MP*, in Goethe, *Scientific Studies*, 82). Goethe emphasizes what follows from this claim, calling attention to the shifting Bildung of the plant as it adapts the leaf form to different purposes. Goethe's resistance to Linnaeus goes still further as he considers whether anastomosis may involve a union that is higher, more refined, than the various sexual unions Linnaeus had described; it may not even involve the male stamen-female pistil template for plant "marriage," a point made more inviting by Goethe's repeated use of hermaphroditic plants as examples and made explicit when he argues that the "union of two genders is anastomosis on a spiritual level" (*MP*, 86).[32]

When Goethe suggests that anastomosis might also be at work in the surprising resemblance between the female (style) and male (stamen) reproductive parts of some species, he questions the grounding principles of Linnaean taxonomy: first, that male organs of plants are the leading indicators for taxonomic location; and second, that specification is determined by reproductive relationships between parts whose sexual distinctiveness is the rule. Linnaeus had admitted exceptions reluctantly, viewing them as brief deformations of the rule. Goethe observes:

> In many instances the style looks almost like a filament without anthers; the two resemble one another in external form more than any of the other parts. Since both are produced by spiral vessels, we can see plainly that the female part is no more a separate organ than the male part. When our observation has given us a clearer picture of the precise relationship between the female and male parts, we will find that the idea of calling their union an anastomosis becomes even more appropriate and instructive. (*MP*, in Goethe, *Scientific Studies*, 87)

As he continues to offer examples of how the male stamen of plants develop, the standard-bearer of the Linnaean system begins to look a little less, well, masculine. The stamen, Goethe begins, arises "when the organs, which earlier expanded as petals, reappear in a highly contracted and refined state." So far, so good for the priority of the strong, male stamen. But its importance for Goethe is rather that it illustrates "the alternating effects of contraction and expansion by which nature finally attains its goal" (*MP*, in *Scientific Studies*, 84). He adds that since both the pistils and stamens are produced by spiral vessels that carry nourishment up the plant, "we will find the idea of calling their union an anastomosis . . . even more appropriate and instructive" (*MP*, in *Scientific Studies*, 87). In the 1820s, Goethe finally proposed *polarity* as the term, or hypothesis, that summarizes the succession of intensifying contractions and expansions in plant development.[33] But in the *Metamorphosis* essay, he nudges hypothesizing forward by degrees, entertaining at each step a degree of terminological and material particularity that he then folds into a loose bundle of terms that are congruent although not clearly synonymous, noting at each stage how mechanical processes as well as less material ones are present in plant development.

Goethe's discussion of spiral vessels is a case in point. Like many botanists before and after him, he was fascinated by the work performed by spiral vessels, which exchange nourishment among the parts of the plant, in ways that many, far too many from Hegel's perspective, compared with the human circulatory system. In the *Metamorphosis* essay, Goethe avoids this comparison, remarking instead on the role these vessels play in the rhythm of expanding and contracting forms of plant metamorphosis. Yet in a 1796 essay, he calls the spiral tendency the "vital" principle of the plant that allows it to "live out its life."[34] It does so, he argues, by two complementary processes: the first is a vertical tendency, made possible by longitudinal fibers that hold the plant up like "a spiritual staff" and become the woody stem; the second is a spiral motion that "develops, expands, nourishes" the growing plant ("Spiral Tendency," in *Scientific Studies*, 106). As the essay ends, he declares again that the spiral vessels "bring life to the plant organism," invoking contemporary authority to supplement his own: "In our time, researchers have insisted that these vessels themselves should be recognized as alive, and described as such" (*Scientific Studies*, 107). With its refinement of the contracting and expanding rhythms of plant development, the spiral tendency captures the doubled rhythm of Goethe's thinking about plant nature—at once minute in its attention and insistent, particularly in this essay, on the vitality of plants as such, not by analogy to human life. The one human analogy Goethe allows and indeed produces is himself. Much as he matches his rhythm of hypothesis with that of the plant's development in the *Metamorphosis* essay, here his analysis of the spiral tendency echoes Nature's standard operating procedure as he understands it, in which Nature's "large-scale intentions are realized in the smallest detail" (*Scientific Studies*, 106).

By insisting that Goethe can only call a leaf a leaf, and further, that the notion of a leaf form is inappropriate because it is at some level abstract, Sachs fundamentally misunderstood what Goethe's figure of metamorphosis makes available. Much as string theory uses the figure of mathematical strings of relationships to hypothesize a set of claims about space, so does metamorphosis imagine and thus hypothesize processes that are not otherwise mappable or fully visible. Not to hypothesize, to be instead content with counting sepals, classifying species, is a failure of intellectual imagination.

Goethe's poem "The Metamorphosis of Plants," republished in 1799 to solicit a better reception for the 1790 essay, is, unlike later poems in which some version of nature is the subject, protective of the earlier essay's characteristic mix of figural restraint and botanical detail. The poem, which stages a lover's one-sided "conversation" with his beloved, whose presence and relation are conveyed by the lover's conversational gestures, begins with his notice of his beloved's confusion about the variety of flowers and their names. Unlike so many other pedagogical botanical poems and discourses addressed to women, Goethe's poem, published in a women's journal, revisits this subgenre without staging the usual Linnaean or pre-Linnaean explication of classificatory procedures and the logic of plant names.[35] Instead, its argument concerns parallels between the metamorphosis of the plant and the stages of their courtship. It would be hard to imagine a better framework for making analogies between plant and human life and love than this.

Yet Goethe's speaker is unwilling to enroll figures that might distract his auditor from seeing the process of plant metamorphosis clearly, much as Goethe had earlier insisted that such figures be excluded. Where one modern translation introduces the figure of the "Womb of Earth" that nourishes the seed, Goethe relies on a figure that is less insistent about human analogy: "der Erde / Stille befruchtender

Schoss" ("the quietly fertilizing lap of Earth"). By way of synecdoche, the phrase avoids giving Earth a womb, keeping the spatial analogy loose enough to allow the observer, here the beloved, to imagine the space in which the seed develops. A degree of restraint marks even the moment in the poem when the speaker does compare the first appearance of the seed to the birth of a child.

Aber einfach bleibt die Gestalt der ersten Erscheinung,
Und so bezeichnet sich auch unter den Pflanzen das Kind.
(Single, simple, however, remains the first visible structure;
[literally, "the form in its first appearance"]
So that what first appears, even in plants, is the child.)[36]

The syntax is striking: "auch unter den Pflanzen" (literally, "among plants, too") indicates the judiciousness of this comparison across different kinds, perhaps to model a restraint that the speaker will sustain in a poem that is both a botany lesson and a love poem. The likeness could have been more boldly declared and extended. The speaker might have said that the plant's development is like that of a child who becomes an adult. As the poem's account of the plant's metamorphosis reaches its flowering and fruit stages, Goethe's language also intensifies, becoming more mythological and more disposed toward human/plant analogy. Hymen, amid marvelous odors, officiates in a union that will produce a swelling of new seeds and finally fruit in a space Goethe now names the "womb," or literally, "Mother's lap" (*Mutterschoss*), a more expressly figurative term than *Schoss* ("MP," in Goethe, *Selected Poems*, 157). Goethe's wavering figural restraint is easy to miss as the rhetoric of the poem becomes increasingly caught up in a remarkable mix of botanical and sexual specificity and an explicitly spiritual diction:

Blattlos aber und schnell erhebt sich der zätere Stengel,
Und ein Wundergebild zieht den Betrachenden an.
Rings im Kreise stellet sich nun, gezählet und ohne
Zahl, das keinere Blatt neben dem ähnlichen hin.
Um die Achse gedrängt, entscheidet der bergende Kelch sich,
Der zur höchsten Gestalt farbirge Kronen entlässt.
Also prangt die Nature in hoher, voller Erscheinung,
Und sie zeiget, gereiht, Glieder an Glieder gestuft.
Immer staunst du aufs neue, sobald sich am Stengel die Blume
Über dem schlanken Gerüst wechselnder Blätter bewegt.
Aber die Herrlichkeit wird des neuen Schaffens Verkündung;
Ja, das farbige Blatt fühlet die göttliche Hand.
(Leafless, though, and swift the more delicate stem rises up now,

And, a miracle wrought [*Wundergebild*], catches the onlooker's eye.
In a circular cluster, all counted and yet without number,
Smaller leaves take their place, next to a similar leaf.
Pushed close up to the hub now, the harbouring calyx develops
Which to the highest of forms rises in colourful crowns.
Thus Nature stamps with higher, fuller being,
Then takes and orders each gradation, each stage.
Time after time you wonder as soon as the stalk-crowning blossom
Sways on its slender support, gamut of mutable leaves.
Yet each splendour is obscured by another creation.
Yes, the colorful leaf feels the godly Hand.)

("MP," ll. 39–50, in Goethe, *Selected Poems*, 157)[37]

From this moment on, the poem more closely resembles others in which Goethe records his view of an ever-changing nature that reproduces the doctrine of the many and the one in its own romantic image. In this nature "Nothing's outside that's not within," for "one and every creature," and "No thing is single, if it lives / But multiple in its being."[38]

In the *Metamorphosis* essay, Goethe only glancingly considered *Zufall*, the role of contingency or accident in nature, and the problem of system. In his post-*Metamorphosis* writing on nature, Goethe aligned contingency with nature in terms that Hegel would eventually refute in his *Philosophy of Nature*. He also discusses "only the annual plant," no trees, no perennials that might complicate the story of single and determined bounds that the essay offers. Excluded, too, from this story are examples of what he calls "irregular" or "regressive" metamorphosis in which plants miss a stage or go back a stage into an "indeterminate, malleable state" that unfits them for achieving those "works of love" for which nature has equipped flowers. The prohibition against something like queer plant sex seems here to involve the possibility that the flowering plant might fail to produce/reproduce the ascending metamorphosis of stages that he emphasizes.

In brief, no irregular or regressive metamorphosis need apply, unless it shows what is otherwise hidden in regular metamorphosis. "Accidental" metamorphosis, like that caused by insects that pollinate plants, gets even briefer mention. Yet Goethe's excursion to versions of plant metamorphosis that are in different ways "not regular" suggests what the rest of the essay does not address: the role of chance or accident in plant development and nature at large. Put beyond the boundary of the *Metamorphosis* essay, then, is what might happen whenever each point of anastomosis in the plant's development becomes an opening or outlet for an unanticipated, irregular, or accidental metamorphosis. This possibility resembles the endpoint that Émile Benveniste's definition of *rhythmos* imagines: "what is moving, mobile, fluid, the form of that which does not have organic consistency; it fits the pattern of a fluid element, of a letter arbitrarily shaped, of a robe which one arranges at one's will."[39] In the *Metamorphosis* essay, Nature is ever at the ready to redirect the plant precisely to avoid the possibilities Benveniste describes: metamorphoses that could intervene and thereby short circuit, supervene or contingently redirect the plant, death among them. Goethe's decision to emphasize the annual plant oddly sidelines death by making it part of the story, well before he concludes that story by noting that after fruiting "nature usually stops the growth process at the flower and closes the account there, so to speak: nature precludes the possibility of growth in endless stages, for it wants to hasten toward its goal by forming seeds" (*MP*, in *Scientific Studies*, 94). Death, yes, but death to get on with life once again. So, just off the edges of Goethe's essay is the recognition that nature and plants survive and even flourish, exotically, mysteriously, with *Zufall* at their heels and nested inside their operations—like the squash blossom bore that insures a zucchini plant will never set fruit.

But there is more. As Goethe remarked to the botanist Ernst Meyer in 1823, beyond the project of the *Metamorphosis* essay he recognized, that the desire to order botanical Nature, to find a systematic adequate to mapping its species into higher orders, beginning with genera, must fail because

> Nature has no system; she has—she is—life and development from an unknown center toward an unknowable periphery. Thus observation of nature is limitless, whether we make distinctions about the least particles or pursue the whole by following the trait far and wide. ("Problems," in Goethe, *Scientific Studies*, 43)

The problem begins with the logic of genera. It is "impossible," Goethe argues, "to treat one genus like another" because genera simply do not behave like interchangeable logical sets. He adds: "I would say there are genera with a character which is expressed throughout all their species; we can approach them in a rational way. They rarely dissolve into varieties, and thus they deserve to be treated with respect. I will mention the gentians, but the observant botanist may add several more" ("Problems," in *Scientific Studies*, 43). Among other genera that are less manageable or, as I think Goethe would say (with Hegel) not rational, are

> characterless genera in which species may become hard to distinguish as they dissolve into endless varieties. If we make a serious attempt to apply the scientific approach to these, we will never reach an end; instead, we will only meet with confusion, for they elude any definition, any law. I have occasionally ventured to call these the wanton genera, and have even applied this epithet to the rose,

> although this in no way detracts from its graceful quality. ("Problems," in *Scientific Studies*, 44)

Genera deserving of respect vs. wanton and graceful ones: Goethe's diction is, taxonomically speaking, pursed, ready to judge what will and will not do from the vantage point of rational classification. It is a moment in Goethe worth remarking, for its tone, which echoes Hegel's disapproval of flowering plants, and for the implicit gendering that would have genera in which defining characters do not appear be wanton yet graceful, like a beautiful being who flaunts our will.

The problem, he then suggests to Meyer, is not nature but mind. The fault he had earlier charged against Linnaeus is now presented as a general human tendency: "Wherever the human being plays a significant role, he acts as a lawgiver. . . . In the sciences we find an indication of this in the innumerable attempts to systemize, to schematize. But our full attention must be focused on the task of listening to nature to overhear the secret of her process, so that we neither frighten her off with coercive imperatives, nor allow her whims to divert us from our goal." Finally, in an essay found among his personal papers and published posthumously in 1831, he argues that the "progress of natural philosophy has been obstructed for many centuries by the conception that a living being is created for certain external purposes and that its form is so determined by an intentional primal force" (*Scientific Studies*, 53). Contending that this way of imagining nature is at once "trivial," "comforting," "satisfying," and wrong, Goethe blames human inquiry for the error:

> Man is in the habit of valuing things according to how well they serve his purposes. It lies in the nature of the human condition that man must think of himself as the last stage of creation. Why, then, should he not also believe that he is its ultimate purpose? Why should his vanity not be allowed this small deception? Given his need for objects and his use for them, he draws the conclusion that they have been created to serve him. Why should he not resolve the inner contradictions here with a fiction rather than abandon the claims he holds so dear? Why should he not ignore a plant which is useless to him and dismiss it as a weed, since it really does not exist for him? (*Scientific Studies*, 53)

He inverts this scenario to show how humans imagine themselves as the object of nature's arrangements: the "sportsman" who has had a hunting rifle made for him will "praise the forethought of Mother Nature in preparing the dog to fetch his prey" (*Scientific Studies*, 54). The irony is biting: man, the tool-making animal, supposes that Nature will provide whatever tool man needs to dominate or kill creatures. Goethe's most philosophical analysis of why, in his view, this way of thinking about nature is wrong turns precisely on what happens when man assumes that things have a purpose he has assigned them:

> In relating all things to himself man is forced to lend these things an inner purpose which is manifested externally, and all the more so because nothing alive can be imagined as existing without a complete structure. Since this complete structure develops inwardly in a fully specialized and specific way, it needs an external environment which is just as specialized. It can only exist in the outer world under certain conditions and in certain contexts. (*Scientific Studies*, 54)

Goethe argues that we stand to gain more insight into what he calls the "mysterious architecture of the formative process" if we grant that this process is "built on a single pattern" that is both "determined" by other elements and itself "a determinant" (*Scientific Studies*, 55). This declaration joins a distinctly global and Humboldtian view of morphology with a call to notice particularity: look at how this plant, this animal responds to external, or what we would call environmental, differences. It also reiterates, with extraordinary concision, his repeated notice in the *Metamorphosis* essay that the plant develops via processes that appear to be mechanical, hence "determined," but also "determinant," in some sense self-directing or, as Goethe puts it in the *Metamorphosis* essay, nature-directed. Hegel readdresses this claim by finding in the same essay a plant nature more amenable to Hegelian *Aufhebung*.

HEGEL READS PLANT NATURE

As Hegel lectured on the philosophy of nature and revised those lectures between 1817 and 1830, he returned repeatedly to the relation between spirit and nature. Hans-Christian Lucas shows that this return was marked by a continued sense of struggle as Hegel sought to demarcate both the point of spirit's emergence from and ultimate domination of nature. In his *Lectures on the Philosophy of Spirit* (1827–28), Hegel describes a process of degrading nature by degrees, marked by the spirit's ingratitude as it performs this reduction:

> The impression that spirit is mediated by something else is removed by spirit itself, since the latter has, so to speak, the sovereign ingratitude [*Undankbarkeit*] to remove, to mediate the very thing it appears to be mediated by, to lower it to the level of something that exists only through it, and, in this wake, to make itself completely independent.[40]

Hegel's reading of Goethe may seek, at some level, to moderate this ingratitude, as if by doing so Hegel might emulate the older poet's reverence for nature and plants. Quite apart from this speculation, Hegel had several reasons for approaching "vegetable or plant nature" by way of Goethe. Both were wary of the early *Naturphilosophie* for, as Hegel put it, the "humbug" of its "philosophizing without knowledge of fact and by mere force of imagination, and treating mere fancies, even imbecile fancies, as Ideas."[41] As a poet who nonetheless put the experimental study of nature first, Goethe was a desirable model for how one could be attracted to *Naturphilosophie* and yet recover from it, in particular from its effort to homogenize spirit with nature. Given Hegel's own critique of empirical thinking, he may also have recognized in Goethe's empiricism a model easily vanquished. And finally, both Goethe and Hegel supposed that matter and nature were in some sense implicated with and in spirit, although Goethe pulls toward this argument while Hegel pulls away from it. For all these reasons, Goethe becomes something like an internal combatant for Hegel, a disturbing marker for what continues to fester in Hegel's system.

Goethe first turns up in the *Philosophy of Nature* in two poetic excerpts. The first, in which Hegel slightly rearranges lines from *Faust*, proclaims that one cannot find spirit by describing life as a chemist does; the second, taken from Goethe's monograph *On Morphology* (1820), claims that Nature supplies both inner and outer (core and rind). Hegel provides an instructive application of the second passage to the work at hand: "In grasping this inner side, the one-sidedness of the theoretical and practical approaches is transcended, and at the same time receives its due" (Hegel, *PN*, trans. Miller, 12; *Werke*, 9: 21–22). When Hegel returns to Goethe later in the *Philosophy of Nature*, he reads the *Metamorphosis* essay more critically.

Hegel begins his discussion of plant nature in this work with an overview of plant metamorphosis that he takes from Goethe, without attribution, then subtly recasts. In Hegel's redaction, the metamorphosis from seed to flower and fruit looks more particulate and disassociative than it does in Goethe's account. Hegel writes: "The process whereby the plant differentiates itself into distinct parts and sustains itself, is one in which it comes forth from itself and falls apart into a number of individuals, the whole plant being rather the basis [*Boden*] for these individuals than a subjective unity of members" (*PN*, trans. Miller, 303; *Werke*, 9: 371). This metamorphosis is only, Hegel charges, "superficial," a turn of phrase that slyly repurposes Goethe's view that *metamorphosis* names a hidden process made glancingly visible with each anastomosis, contraction or expansion.

When Hegel again takes up Goethe's essay, now praising it and its author, Goethe's insistence on the great work of these processes disappears under Hegel's rhetoric: "The parts exist as intrinsically the same and Goethe grasps the difference between them merely as an expansion or contraction."[42] Just before this, Hegel suggests that Goethe's argument authorizes this reduction because he "aims to show . . . [that] all the forms remain only outer transformations of one and the same identical fundamental nature, not only in the Idea but also in their existence, so that each member can quite easily pass into the other; a spiritual, fleeting, whispered breath of forms (Ein geistiger flüchtiger Hauch der Formen), which does not attain to qualitative fundamental difference, but is only an ideal metamorphosis in the material aspect of the plant."[43] As readings go, this one is remarkably untuned to its source, even perverse. Hegel tellingly dismisses what Goethe argues is most recognizably spiritual in the plant's metamorphosis—that its changing shape rises up, intensifies, until it reaches the end toward which it has all along been driven by its own processes (at once determined and determinant) and by nature. What troubles Hegel, that spiritual but fleeting, whispered breath of forms, is what convinces Goethe that nature has spirit.

As Hegel works through the stages of the plant, he treats its development as wholly material even as he quotes Goethe's essay extensively, amplifying it with other scientific materials and apparently as interested as Goethe in questioning Linnaeus's insistence that plants are fertilized by sexual congress between male and female parts. But whereas Goethe suggests that when this union occurs, it culminates a process of intensification that had begun with the germinated seed, Hegel rather unsportingly asks if fertilization is even necessary, calling it "a play, a luxury, a superfluity for propagation" (*PN*, trans. Miller, 345; *Werke*, 9: 422). Hegel supposes that because the buds of the plant are individuals "and need only to come into contact with the earth in order to be themselves fertile as distinct individuals," sex is not necessary. Instead, he suggests that precisely because "the preservation of the plant is itself onto a multiplication of itself," this process rather haplessly requires the work of insects and the wind to make fertilization occur. Hegel concludes that in the plant (unlike the animal), digestion and generation amount to the same thing. The basis for this reading, which Goethe would have called "monstrous" had he read it, is Goethe's remark that just prior to the budding of the flower the plant halts its development, a moment of contraction that precedes and makes possible the expansion into fruit. Inevitably, Hegel says of this last moment that the plant "falls asunder into many seeds which are capable of living independently" and thus fails to sustain the unity that animals have after they reproduce (*PN*, trans. Miller 347; *Werke*, 9: 425).

The target of Hegel's interpretation is Goethe's reiterated and exuberant argument for the unity and spiritual vitality of the plant, more intensely rendered as it develops. Hegel puts it this way: "Goethe has ingeniously represented the unity of the plant as a spiritual conductor (*geistige Leiter*). But metamorphosis is only one side which does not exhaust the whole; we must also pay attention to the difference of the organs." Hegel proceeds to do so in spades, specifying each in order to claim its fundamental differentiation from the rest of the plant (*PN*, trans. Miller 320; *Werke*, 9: 392). Temporarily absent from his account is the role of the spiral vessels that carry nourishment throughout and thus act as a material version of that spiritual conductor. When he does take up the those vessels, he presents them by introducing an array of recent scientific authorities who say that some plants lack spiral vessels altogether, and, further, that in those plants that have spiral vessels a diversity of judgments exist about the mechanism whereby they transmit liquids or sap (*PN*, trans. Miller, 326–36; *Werke*, 9: 400–11). Hegel's lengthy excursus on this point submerges Goethe's argument by invoking current academic opinions gathered from Hegel's scientific colleagues and a wide assortment of comparable experts.

The differences that separate Goethe from Hegel specify precisely those aspects of botany that are, depending on one's perspective, the most troublesome or the most liberating. Whereas Goethe emphasizes the individuality, particularity, and metamorphosis of plant form, and the contingent, unsystematic energy of nature in general, Hegel argues that individuality (notably that of plants and small dogs) must sacrifice itself to Spirit, thereby inaugurating one version of the self-estrangement that assists the work and individuation of Spirit and, further, that it is the progress of Spirit that insures self-conscious order and development, over against the vagaries of an unsystematic, contingent nature. These positions are oddly, hauntingly inverse: where Goethe sees the metamorphosis of individual parts of plants as constitutive of its inner-and outer-directed being and conducive to a higher union of matter and spirit, Hegel sees a random set of individuals bizarrely presented as a composite outcome called the flowering plant. Metamorphosis is for Hegel a monstrously disaggregating set of events and parts. For Goethe such a reading transforms his intuitive figure for plant development and unity into something monstrous.

That Hegel chooses to shadow Goethe as long as he does says something commanding about the tug of this set of differences on Hegel's desire to perform for himself the operation of Spirit over nature, or the nature that Goethe presents, where spirit and matter and plants are aligned in ways that Hegel must reject. The differences between them convey what is as stake in romantic theories of nature as alive or dead, mechanical or not, not because Goethe takes one pole and Hegel the other, but because Goethe's understanding of the vitality of nature as both mechanical and sturdy enough to withstand or adapt to contingency robs Hegel's Spirit of its job: to see nature not in its contingent reality but its ideality. It is not that Hegel cannot, as it were, "take" contingency, but rather that it is useful and malleable to him only as an occasion for Spirit to overcome. He tells us that nature is weak; that its reason is not reason enough; that its spirit is not spirit enough; that it lacks the collected force of will and ego. Yet how weak is Nature if it can withstand the rack of contingency that universal reason can only manage by subjugating it? Goethe imagines instead a nature that is less pliable, less likely to sacrifice itself to Spirit and more able to live with what the world throws in its way. So construed, nature reels away from the driving engine of Hegelian spirit, blocking or deflecting its world historical progress.

It is hardly accidental that plants are the allegorical, if belated, focus of Hegel's philosophic critique of *Naturphilosophie*. Because they inhabit the kingdom of nature between minerals and animals, plants carry attributes of both such that they trouble Hegel's effort to locate them as a kind of halfway point for the development of spirit from and then away from nature. Indeed, in the tripartite division of nature ratified by Linnaeus and others, in which plants were expected to hold a middle ground between inanimate minerals, on the one hand, and the animate animal kingdom, on the other, claims about the vitality or life (let alone loves) of plants could and did cause trouble. Seemingly unconstrained by their purported difference from animal and human vitality, romantic plants operate like free radicals, unsettling claims about the ascendancy of spirit and mind over nature. Hegel's claim that "many an innocent flower" will be crushed by the world historical spirit may, in short, ward off another, nearly as likely, scene of sacrifice in which spirit finds itself enthralled by the material life of plants. The apparent inadvertence with which the Hegelian world spirit crushes flowers on its way to spiritual greatness suggests something more aggressive and defensive, a casting off that seems strangely unnecessary, unless plants in fact matter.

That they do and that they were relatively easy targets for Hegel has in part to do with their increasingly troubled role in systems of knowledge. Precisely because botany was still, in the early nineteenth century, the most accessible branch of natural history, its taxonomic troubles made it a ready target for popular as well as learned critique. A wedge, as it were, that Hegel may have hoped to use to discredit the romantic inclination toward a quasi-Orientalist flower religion.

The near-Deleuzian dance of repetition and difference in Hegel's restaging of Goethe's essay on plant metamorphosis begins by echoing Goethe then veers away to insist that the plant develops in a wholly mechanical and disconnected series of steps or jerks. This antagonism to Goethe and vegetable nature marks the moment when the preference for synthesis over particulars recognizes nature itself as the contingent and stubborn substrate of that particularity. As Theodor Adorno would later remark, although particulars and the phenomenal world specify the nonidentity that Hegelian dialectics purports to keep in view, Hegel himself fails to sustain that dialectic in his presentation of Spirit and nature. Adorno's analysis of this failure and his call for a negative suggests why nature has been for romantics and moderns the carrier of a spirit of nonidentity that philosophical thought about systems and individuals has yet to fully grant.

ADORNO'S RABBITS

In *Negative Dialectics*, Adorno charges Hegel with abandoning the nonidentity of objects and particulars in nature that is necessary to sustaining a genuinely negative dialectics, a project that Adorno understands to have been Hegel's and to have become his own. Adorno thus argues for a restless dialectics that would, or ought to, remain ever on the watch for moments of otherness and negativity that keep philosophy from becoming all idea, all abstraction. In significant measure, Adorno's critique concerns the degree to which Hegel's brief for the World Spirit veers away from this project.

I consider here essays in which Adorno revisits Hegel, much as Blake, in his poem *Milton*, revisits Milton, to restart or recall the dialectics that Hegel lets slip. In these essays Adorno proceeds obliquely, elliptically, even at times poetically. In the last of them, a very brief piece in *Minima Moralia*, he echoes Hegel's phrase from the *Philosophy of History*, "*die List der Vernunft*," usually translated "the ruse [or cunning] of reason." Hegel explains:

> This may be called the *cunning of reason*–that it sets the passions to work for itself, while that which develops its existence through such impulsion pays the penalty, and suffers loss. For it is *phenomenal* being that is so treated, and of this, part is of no value, part is positive and real. The particular is for the most part of too trifling value as compared with the general: individuals are sacrificed and abandoned. The Idea pays the penalty of determinate existence and of corruptibility, not from itself, but from the passions of individuals. (*Philosophy of History*, 33; *Werke*, 12: 49)[44]

Reinhart Koselleck argued that Hegel here excludes contingency even as its role in history becomes difficult to ignore.[45] In Hegel's text this claim specifies the lesson to be learned from the lines that precede it, in which he declares that the "World-historical individual" is "so mighty a form" and so "devoted to the One Aim," that it "must trample down many an innocent flower." As economic systems go, this one is stunning in its ability to make others pay for their losses, while reason takes the jackpot, cunningly disguised as Spirit.

Adorno returns more than once to Hegel's ability to put the passions of individuals and the work of particulars to his own uses. Yet in "Aspects of Hegel's Philosophy," delivered on the occasion of the 125th anniversary of Hegel's death, he lets Hegelian cunning off the hook: "Hegel introduced the cunning of reason into the philosophy of history in order to provide a plausible demonstration of the way objective reason, the realization of freedom, succeeds by means of the blind, irrational passions of historical individuals." How to grasp why Adorno here admires what outrages him at other moments? He goes on to suggest that in overpowering what is blind and irrational in individuals—including, I surmise, particulars in nature—Hegel creates the engine of a dialectic that the "slave" can finally use to overthrow the "master" (Adorno, "Aspects," *Hegel*, 42; *Zur Metakritik*, 286–87). Here the spirit does to an (apparently) dominant, powerful nature what the dialectic allows the slave to do, eventually, to his master, at least in the realm of absolute, ideal freedom of spirit. Absent from this analysis is the heavy materiality of being a slave or servant to one's master or lord, or of being just a particular of no consequence to the rule of abstraction or spirit that masters all. Adorno's wariness of what he calls in "Skoteinos" the bourgeois tendency to hypostasize the "blinders of the particular, the belief that its contingency ("Zufälligkeit," or chance) is its law" suggests another view of Hegel's brief for Spirit over particulars ("Skoteinos," *Hegel*, 112; *Zur Metakritik*, 345).

When Adorno returns in *Minima Moralia* to the matter of philosophical ruses, or cunning, he performs a way of reading from Hegel forward to late philosophical modernity and after Auschwitz that is hauntingly about, once again, saving particulars. My argument begins by joining Hegel's phrase "*Die List der Vernunft*" to the elliptical parsing of it in Adorno's short piece about two rabbits and a song, quoted below:

> As long as I have been able to think, I have derived happiness from the song: 'Between the mountain and the deep, deep vale': about the two rabbits who, regaling themselves on the grass, were shot down by the hunter, and, on realizing that they were still alive, made off in haste. But only later did I understand the moral of this: sense [*Vernunft*, that is reason] can only endure in despair and extremity; it needs absurdity, in order not to fall victim to objective madness. One ought to follow the example of the two rabbits; [who] when the shot comes, fall down giddily, half-dead with fright, collect one's wits and then, if one still has breath, show a clean pair of heels. The capacity for fear and for happiness are the same, the unrestricted openness to experience amounting to self-abandonment in which the vanquished discovers himself. What would happiness be that was not measured by the immeasurable grief at what is? For the world is deeply ailing. He who cautiously adapts to it by this very act shares in its madness, while the eccentric alone would stand his ground and bid it rave no more. He alone could pause to think on the illusoriness of disaster, the "unreality of despair," and realize not merely that he is still alive but that there is still life. The ruse of the

dazed rabbits [“*Die List der ohnmächtigen Hasen*”] redeems, with them, even the hunter, whose guilt they purloin.[46]

Although the standard English translation, reproduced here, gives “sense” for *Vernunft*, as in “good sense,” or “smarts,” meaning perhaps rabbit cunning, the whole passage urges its application to the work of reason (*Vernunft*), here signaled by the rabbits’ quickness to profit from their unexpected survival, and also to the eccentric’s decision to accept the world’s madness and grief and then go on living. Both recognize in the contingent or unexpected an opportunity for swerving away from those ready to take life from the world. The rabbits’ ruse, or cunning, is their eccentric, dazed, yet effective flight, which teaches the lesson reason needs to learn from contingency. For the fact that the rabbits happen to still be alive is an occasion they should and do profit from by running away. In the case of the dialectic as Adorno understands its capacities, one can profit from what is contingent, in nature, in the eventness of the world.

The weight on this anecdote is at once event-ridden and philosophical. For here the guilt Adorno feels for having survived the Holocaust, for having gotten away, can be addressed, or perhaps borne, only by finding in this story a dialectical swerve or style that repairs the failure of history and of philosophy. I have suggested elsewhere that the other essay to be profitably read with this brief one is Adorno’s “Parataxis,” which shows how Hölderlin’s use of poetic parataxis holds in suspension what dialectics might overrun as it forgets its own negativity and takes charge.[47] This lesson, Adorno’s writing on Hegel makes clear, is drawn from Hegel’s failure to sustain his own project. Better to shift focus, away from the hunter to the hunted. Better to save those particulars, be they rabbits or vegetable nature, than make them innocent sacrifices, duped into giving up life so that Spirit might triumph.

Like those rabbits, Adorno steals the hunter’s guilt and assumes it, much as particulars in Hegel’s scheme for world Spirit pay the price for their own loss. As Adorno reinhabits this strange economy, he suggests that rabbits, like other particulars, need to have some place in philosophy as in life. The story takes him this far, at any rate, after Auschwitz and many years after Hegel. Here, in a way that is perhaps more subtle than his rescue of Hegel in “Aspects,” Adorno assents to the structure of the work of Hegelian spirit, but understands from inside it how incriminating, how damaging that structure can be to particular beings when, in this case, it lets the hunter off the hook on the grounds that his intention misfired.

In the randomness of those who lived and those who did not in and after Auschwitz, Adorno’s assumption of guilt may also register his resistance to “working through the past,” a program that he rejects in the essay with that title because it appears to sanction forgetting the Nazi era.[48] Yet the story he tells in *Minima*

Moralia also urges the passional understanding of what it means to be a particular, alive or dead, that takes on the suffering, the loss without recompense, which is the consequence of holding the position given to nature and its contingent particulars in Hegel's dialectic on behalf of the world Spirit. It may be absurd for the rabbits to have escaped with their hides and lives intact, but in that absurdity lies a logic that might at least recognize the cunning of Hegelian reason not as a cost to be borne but as one to try to outrun, elliptically. Here the elliptical gesture is to side with those rabbits and to understand, in his time and philosophy, the cost as well as the unexpected outcome entailed by doing so.

In sum, I read Adorno's rabbits as figures for the resilience of plant nature in the way of change and chance, not because plants can move the way rabbits do or think as humans or the Spirit thinks (distinctions Hegel would no doubt insist on) but because plants and the figured names they invite convey life forms that exist, as best they can, on their own, with an inner directedness that is not necessarily well aligned to a universal purpose or spirit of nature. Particular, various, and as such uneasily assimilated to the species, genera, and systems to which they are said to belong, plants register the philosophical and material pressure of remainders, points of excess, in romantic system-making.

CHAPTER 8

Conclusion

Wild Orchids

> We'd sit with our field-guides and our ragged bundles of leaves and flowers spread out on the table, and talk them through, taking every opportunity to be side-tracked. Orchids (though we never picked these) had special fascination. Many had the look of elaborate conceits in porcelain, or colonies of hatching insects. The botanists who named them saw likenesses too; to lizards, bees, bugs, butterflies, spiders, even pyramids. But in the tribe (*Orchis*) with which we were most familiar, they saw chiefly little homuncules, formed from the heads, or helmets, and arms and trailing legs of individual flowers. The size of the hoods, the sinuousness of the limbs, the waisting, the elegant shape of the lobes, made them variously men, soldiers, ladies, and in one particularly gangly armed type, monkeys. But we could scarcely tell them apart. The hills were alive with unresolved manikins and hermaphrodites. Orchids are one of the most recently evolved groups of plants, still hazy about their identity, and they hybridise quite promiscuously.

Richard Mabey's reminiscence in *Nature Cure* of plant hunting with friends in the Cévennes in southwestern France begins here and then returns, in echoing folds of seductive prose, to the impulse to identify and name orchids—species of the *Orchis* genus—as though in naming them one might for a moment fix identities that may shift suddenly, wildly, in another direction, hybridizing with abandon in despite of taxonomy and nomenclature, inviting as well as taking every opportunity to be sidetracked and inviting sharp botanical eyes to do likewise. This mimetic disposition works, with no small degree of pleasure, to bind, however lightly, a shifting array of orchid species to human or, at least, animal names and body parts, among them "the heads, or helmets, and arms and trailing legs of the individual flowers" and the "sinuousness of the limbs, the waisting, the elegant drape of the lobes." Eating other wild plants in a restaurant, Mabey and his companions look over two albums of orchid photos left there by a German photographer, and again find themselves driven by an "atavistic urge" to name these orchids, to pin them down, plant by plant, these "*orchidées du pays*," so that their indigenous savor might be anchored with a global name and taxonomic place. This urge becomes irresistible when they look at "one variety in particular, the rare and almost a mythical cross between the man and monkey orchid. . . . We nicknamed it the missing link orchid, but doubt that we were the first."[1]

Mabey's narrative echoes several strands in the nineteenth-century's fascination with orchids: their variousness; their resemblances to insects, birds, animals, humans—the arresting mimetic crosses among these kingdoms of nature; and the impulse to name and organize them into species. One degree of difference separates Mabey from nineteenth-century predecessors who also, and as eagerly, pursued the wild orchids of Europe. Despite their keen attention to *Orchis* varieties, none appears to have recognized that they were natural hybrids. Charlotte Smith's poetic notice of these "mimic flowers" registers the flicker of agency as well as resemblance that contemporary botanists sought mostly to repel. John Clare's quizzical account of differences between the orchises he saw in the field and in published botanical descriptions points toward the later recognition that orchises do not, at least in the field, maintain the distinctions given them in books. What Mabey's story of orchid hunting in the Cévennes shares unequivocally with earlier botanical inquiry is a complex understanding of why it is still, in this era of orchid cultivars, appropriate to talk about "wild orchids." Long before their hybridic tendencies prompted horticulturalists to create their own hybrids, crosses, and new cultivars, to the delight of orchid growers, thieves, and buyers ever since, orchids were considered wild, whether found in the fields of England and Continental Europe or in more exotic places, from the Americas to India and Asia.

The phrase *wild orchids* conceals a pungent irony. Because they hybridize in the wild they are good candidates for creating new cultivated varieties, or cultivars, those invented varieties that signal domestication by horticultural and taxonomic means. This outcome is familiar to readers of colonial natural histories. It is at issue in every story one might tell about botany in the romantic era and beyond. Then, too, and with a figurative inversion that is also endemic to what orchids do or invite human observers to do, wild orchids prompt wild thoughts that escape more deep seclusion when they enter poetry and prose, as names and images and as ideas about agency and plants for which orchids seem to be poster children or poster men and women, and much else. As such, wild orchids summarize, in a heightened key, the inquiries of this book—among them classification, plant versus human capacities, nomenclature, exotic and local floras, imperial botany, local versus global botanical purposes, and, running through all these, the figurative invention that botany prompted in the early nineteenth century, despite every effort to write about plants without flying off in the direction of figures.

I reprise this book's argument via James Bateman's *The Orchidaceae of Mexico and Guatemala* (1838–43), which consolidated an "orchid-mania" that persists into our horticultural present. Long promised to a hefty list of subscribers and finally published in 1843, the *Orchidaceae* was, as Bateman boasted to its readers, the biggest ever published, each page measuring 70 × 53 cm (28.5 × 18 in.).[2] With the visual wit that he exercised repeatedly in the *Orchidaceae*, Bateman advertised its size in the frontispiece, a caricature by George Cruikshank, which depicts a large group of tiny human beings trying to haul the book upright, using ropes looped over a tall, forked pole. Despite their efforts, the book is still very much on its side. Individual figures try to prop the book up by pushing on its spine, relying on a pulley system that looks unreliable. One reading of this visual boast might note that book and topic exceed the human capacity to get them right as well as upright. The size of Bateman's book was, like that of Robert Thornton's *Temple of Flora* nearly a half-century before, material evidence of the ambitiousness of the publishing project.

Bateman's list of subscribers suggests that prestige was as much the bottom line as profit: titled subscribers lead the pack, beginning with Queen Dowager Adelaide, to whom the work is also dedicated, the King of the Belgians, and the Grand Duke of Tuscany, followed by ten English dukes and earls, among them the Duke of Devonshire, whose family had long been involved in botanical matters and print culture, Lord Radstock, Earl Fitzwilliam, who had been John Clare's patron and friend, Earl Powis, whose gardens and East Indies wealth and artifacts remain famous (or infamous, depending on your point of view).[3] Lady Grey, of Groby, subscribed. The naturalist and inventor of plant geography Baron Alexander von Humboldt, subscribed, as did the French Baron Benjamin Delessert, a descendant of the Madeleine-Catherine Delessert to whom Rousseau wrote his botanical letters.

A healthy list of untitled subscribers followed, among them nurserymen, libraries, and booksellers. Acquiring the book for a library meant having a shelf big enough to hold it and, presumably, someone to guard it. The aristocratic pitch of Bateman's book, apparent in the order of subscribers and sustained throughout by his repeated notice of aristocratic gardens and gardeners that cultivate orchids constitutes, admittedly, a curious form of advertisement. Most readers, with the exception of those specifically invited, would not have visited orchids grown in private estate gardens. The fact that the artist for more than half of the colored plates was Mrs. Augusta Withers, flower painter to the queen, would have pointed up the aristocratic tinge (or twinge) associated with growing orchids.

Considered in these terms, Bateman's *Orchidaceae* is a rarefied picture book that invited its 120 or so buyers (and many more readers who couldn't buy it, then or now) to participate in a genteel or elite encounter with orchids in nurseries and books. In his opening remarks, Bateman makes explicit (little is left implicit in this work) the affinity between aristocrats and orchids by noting that another botanist,

Plumier by name (yes, really), "paid his court to this tribe" by presenting "figures of some of the West Indian species" (*Orchidaceae*, 1).

Priced at twenty guineas, *The Orchidaceae of Mexico and Guatemala* was hardly a steal, but it gave value for gold: forty colored plates engraved after original paintings, most of them by Mrs. Withers and Miss Sarah Drake, both with established reputations as botanical painters; related vignettes and caricatures and a text that links plates and caricatures and botanical information in an allusive, emblematic texture that folds in poetry, stories, and traditions indicated by engravings of Mexican and Guatemalan scenes and customs into its presentation of orchids found during several botanical excursions to Central America that Bateman had financed. If the *Orchidaceae* was evidently aimed at a high-paying audience, their book borrowing friends, or those with lending library subscriptions, it was also attentive to scholarly debates about orchids at the close of the romantic era, just as the popular craze for orchid and orchid cultivation had begun to take off. Unabashedly gleeful about "orchid-mania," Bateman used the occasion of his very big, very expensive book to remind readers of the "cool" cultivation method he had developed so that English growers could grow (instead of killing them after a season) the Central American orchids his book introduced to English audiences.

Like so many illustrated books and magazines on botany during the period and like Bateman's study of the genus *Odontoglossia*, the *Orchidaceae* addresses two audiences.[4] For those readers who wished to grow orchids, Bateman lists the hundred best orchids available or potentially available for cultivation in British collections, tantalizing ambitious gardeners by identifying three in the list "that have not yet flowered in England," a gambit that would almost certainly have reminded readers of the late eighteenth-century race to be the first to grow a pineapple in an English greenhouse (Thomas Pennant won that one), and probably many since then. Bateman followed this list with a detailed account of his own method for cultivating tropical orchids. For professional and highly skilled amateur botanists (like himself), he includes near the front of the work a synopsis of all described species of Orchidaceae thus far discovered in Mexico and Guatemala, organized by genus, although he refers to orchids in introductory remarks as a family, an order, and a tribe, a wandering nomenclature that effectively covers all the bases in the unsettled nineteenth-century debate about taxonomic hierarchies. The large format and plates evidently mark this book as a commercial venture. Bateman was working within a long tradition of creating botanical works for a crossover audience, although the size and expense of *The Orchidaceae of Mexico and Guatemala* insured that its buying audience would be few.

The text accompanying each plate of the *Orchidaceae* begins as professional botanists and learned amateurs would have expected, with the Latin binomial name, its tribe, and the name of the botanist whose classification it is, followed by a Latin description of the principal parts of the orchid in question and its tuber, the latter specified so that growers would know which orchid they were buying by looking at the tuber alone. The English description that follows, a free rather than literal translation of the Latin, introduces a longer account of the orchid's origin, rarity, and how to grow it. Many, but not all, of these accompanying texts end with a vignette that may present a Central American scene or native. To each vignette Bateman attached a poetic tag or brief phrase. In formal terms the plate/text relationship in the *Orchidaceae* looks like those in Renaissance emblem books that rely on images and assorted texts to explain a single allegorical idea or virtue. This emblematic structure is driven by an associative principle: traits of individual orchids or some aspect of their collection triggers a visual echo or a series of verbal/visual puns.

This looping, mimetic play Bateman blames on orchids, explaining that their distinctive structure, grotesque forms, and "the imitative character of their flowers" allow them to take "liberties"(6). The language of sexual excess and putative agency marks another transgression: "accustomed as we are to look upon the animal and vegetable kingdoms as altogether distinct, our astonishment may well be awakened, when we see the various forms of one appropriated by the flowers of the other" (6). As taking liberties tends to do, this transgression across the kingdoms of nature leads to others: "for, as if it were too simple a matter to imitate the works of *Nature* only, they mimic, absolutely mimic, the productions of *art*!" (6). Elaborating this motif, Bateman himself takes liberties:

> But not content to rest even here, they display a restless faculty of invention, fully equal to their powers of imitation, and after having, like Shakespeare, "exhausted worlds," like him too, they seem to have "imagined new;" and thus we find their flowers exhibiting a variety of strange and unearthly objects, such as bear no resemblance to created things, nor yet to any of the works of man. (6)

The examples presented to support this extraordinary claim begin with the insect theme and variations of the English and European orchis genus, including species named for their resemblance to bees, birds, and spiders, then continue to more exotic variations, among them one from Trinidad, "the gorgeous vegetable butterfly [identified as *Oncidium papilio* in a note], whose blossoms," writes Bateman, are "poised at the extremity of their long elastic scapes, wanton gaily in the wind," and "seem impatient of that fixture by which they are differenced in kind from the flower-shaped Psyche that flutters with free wing above them."

The embedded phrase, lifted from Coleridge's *Aids to Reflection* (1839), takes off in an unsanctioned direction from its quoted source. For where Coleridge argues that the flower's apparent longing to become a butterfly belongs to a series of differentiations in the chain of being that insists on the human as the highest form, Bateman has already claimed that orchids resemble anything they wish to resemble, even things not yet created. The difference between Coleridge's conservatively romantic view of the ladder of nature and Bateman's wild suggestion that flowers assume shapes that resemble members of all created kingdoms and then those not yet created turns in part on the degree of agency each allots to plants. Bateman suggests that they are as inventive as the greatest of English dramatists, and the very type for creative imagination. His yet more extravagant claim is that orchids create nature in free form not bound by mimetic protocols. No wonder Bateman, Mabey, and many writers between them take degrees of license in naming orchid structures.

This relay between orchids and humans as well as animal and insect forms amplifies and deliberately exaggerates an impulse that carries through the modern history of the European, but especially British, response to exotic plants. Rarely, it would seem, did plant hunters leave them alone to grow in peace. Explorers tramped all over the Americas, India, Ceylon, Sumatra, Australasia to find them, document them, name them, dry them, bring them home, and, in the case of orchids, domesticate them as the very epitome of cultivar excellence, replaying yet again James Wallace's description of the monstrous, confounding plants he tried to take back from the failed Scottish colony in Darien.

In Bateman's *Orchidaceae* and other works on botany published between 1699 and 1843, this rhythm of repetition and difference turns and returns: plants are both recognizable, like us, and monstrous. Here is Bateman, introducing exotic Central American orchids to an eager public by way of the seventeenth-century Dutch botanist Rumphius:

> After noticing, in terms of due commendation, the dignified habits of *most* of the tribe, he proceeds with a sigh, to remark: —"that among these vegetable nobles, just as among the nobles of mankind, some degenerate individuals are ever to be found, who are on the ground always, and seem to constitute a class of their own." But, it is not merely in their "habits" that the terrestrial species are placed *below* the Epiphytes [orchids that grow on trees]: they are also greatly inferior to them in singularity and beauty. (2)

Everything is possible here: the nobility of class, the degeneracy of the lower orders, a fantasy of high and low that riffs off the distinction between tropical epiphytic orchids and their groundling relatives. Moreover, orchids mean sex, but also, Bateman's rehearsal of orchid names in Mexico and elsewhere insists, love and sympathy, and, inevitably, death, likewise "highly romantic" in the tumbled logic of a single paragraph, which opens with a deadly poison found in Demerara (no sweetness here) that is combined with the juice of the Catasetum orchids and ends with notice that Catasetums and Cyrtopodiums exude a viscid matter used both for making the soles of shoes and for poisoned arrows. In between, Bateman describes a love elixir that is prepared from the minute seeds of the *Grammatophyllum speciosum*, no doubt with a degree of syntactic correctness; and, near the end of the paragraph, he reminds readers of a more familiar claim, long circulated in ancient and European herbals, that the roots of the orchis species native to Europe and England produce salep, whose "nutritive substance," as Bateman puts it, had for centuries been said to ensure male sexual vigor.

In the middle of all this, Bateman choreographs an exchange of sympathy between orchids and the natives who seek them "with an avidity," which would seem to say that there was "no sympathy like *theirs*." The flicker of resemblance between Amerindian greed for orchids and the British rage for the same orchids is slipped into and under the fantasy of a shared sympathy between orchids and those "sentimental natives" who want them as much as Bateman's readers and growers do, though they are reserved for "princesses or ladies of high degree" in the East Indies. Back in Mexico, local orchid names, which offer "nearly the entire alphabet of the language of flowers," make sympathy a watchword: Flor de los Santos, Flor de Corpus, Flor de los Muertos, Flor de Maio, and No me olvides (do not forget me, or forget-me-not).

Bateman's habit of careening across the globe to add instances, resemblances, exotic details to the subject at hand is echoed in the notes, which on the bottom of a single page reel from recent botanical discoveries in Africa to Rumphius's original Latin to notice of an elixir-producing plant having made its way to England to a coy reference to the use of rare orchids in the headdresses of some English ladies of "high degree." Like Thornton's earlier and equally emblematic structure in the *Temple of Flora*, Bateman's *Orchidaceae* uses similar additions (poems, commentary). But whereas Thornton is insistent on imperial adventures in other regions, Bateman identifies plant explorers and the artists whose vignettes of local scenery dot the work without making much of what we would characterize as his imperial gaze. Bateman's narrative attention is instead imitative of the errant, hybridizing behavior of orchids. Even the apparently rigid class sensibility of this book is put in ironic relief by those orchids that Bateman suggests do or do not live up to their inherent nobility. The outcome is a work that is extravagantly alive to the most volatile figurative possibilities that accrue to botany in romantic writing.

Listing the genera of Central American orchids that can be grown, under the right conditions, in England, Bateman concludes,

> They bear carriage remarkably well, and while in blow [bloom] they may be removed with safety (and sometimes even with advantage to themselves), into

> apartments of ordinary temperature, where their duration is much greater than in a stove. Indeed, it is easy to foresee the arrival, and that too at no distant period, of the time when their flowers will appear as much "at home" in the British drawing-room as in the Mexican temple, and when they will be prized as highly by the English as by the Indian belle. (4)

Having a good carriage or bearing carriage—the difference is muted in the surround of Bateman's rhetoric. Will they stay longer if they are given apartments of ordinary temperature? How adaptable these natives are! And soon they will be at home among the English. If this were not the book it is, one might almost hear here an echo of the alien/familiar tone of the narrator of Ishiguro's *When We Were Orphans*, so effective by now is Bateman's personification of exotic orchids. Recent postcolonialist writing has rightly suggested that similar moves at other levels of British colonial purpose lead to allied contradictions as what seems native is made to appear natural and English, vice versa, or some odd mix of this too-neat opposition. What interests me in this wry spectacle of the English making exotic plants feel at home (as well as bought and paid for) is the role of poetic lineage and forward trajectory in that most romantic of natural histories, botanical nature.

Consider, for example, Bateman's use of lines from Percy Bysshe Shelley's *Alastor* on the title page of the *Orchidaceae*. Identified without elaboration as Shelley's, the lines suggest the figure that Bateman visualizes as the undulation of stylized serpents around an elaborated, pseudo Aztec/Mayan frieze, with prickly pear cactus, *Cactus Grandiflora*, erupting from the volcano at the top (why not?), and volcanoes at the bottom in careful array, with a band of quasi-floral, quasi-sculptural figures twining around or supporting figures of stone gods that Bateman explains on his first page are Olmec, Toltec, and Aztec:

> like restless serpents clothed
> In Rainbow and in fire, the parasites,
> Starred with ten thousand blossoms, flow around
> The grey trunks.
>
> (Shelley, *Alastor*, 438–42, *SPP*, p. 84)

Bateman's use of these lines is suggestively parasitic, at once locally attentive to the poem just here, yet uninterested in what surrounds these lines in Shelley's poem. On the title page of the *Orchidaceae*, where the lines appear, a visual border of twining serpents echoes Shelley, making its text the parasitic host for the title-page design. Bateman appears to suppose that Shelley describes epiphytic orchids as "parasites" whose "ten thousand blossoms" cover the trunks of trees. The mistaken belief that such orchids were parasites, common enough in the early decades of the nineteenth century, assumes that because orchids, like parasites, grow on trees, they, too, must be parasitic. We could settle all this, excuse Shelley of imputed error, and suppose that these "parasites" were mistletoe, but for all those brightly colored blossoms. On the next page, Bateman says that these orchids are not parasitic at all but unique "air-plants," so described because they take nourishment entirely from the air.

Two interpretive events or itineraries emerge from this brief interplay between Bateman and Shelley. First, Bateman's expansive recirculation of other texts and ideas is itself parasitic on Shelley's verse. Bateman takes only what interests him, with little attention to the botanical and figurative possibilities that Shelley's language offers. Second, Bateman's text, here and everywhere else, becomes an engine of affiliate, linking segments and ideas that can go anywhere, metamorphose into whatever way is available, and dazzle the reader. Is this a literary embodiment of rhizomatic development, a figure quite literally rooted in botany and alive to those thousand plateaus of meaning and affinity that Guattari and Deleuze described for readers weary of genetic models based on single trees?[5] Yes, with precisely the degree of inorganic modeling that their rhizomatic protocols encourage. What Bateman produces on the title page and throughout the *Orchidaceae* is a knowing, far from organic, and often fantastic chain reaction among orchids, orchid tales, and a series of caricatures that are visual echoes of orchids already displayed, but in a lowered tonality, with a nudge toward play and replay that feeds, as it were, on air. If epiphytic orchids and all the other kinds of Central American orchids seem to have been the occasion for this invention, they do not anchor or ground it. This is a big book about orchids, but it is categorically not the book of nature. Like some other botanical enterprises of the era, Bateman's *Orchidaceace* exceeds its natural plot of ground. So much the better.

The second itinerary that opens from this brief convergence between Bateman and Shelley is the startling mix of poetic strategy and botanical ground in Shelley's poem. In the passage from which Bateman extracts the lines quoted above, the poet Alastor is drawn toward a death in nature that is, as Shelley's mythic revisioning of the poet Wordsworth as Alastor insists, at once an achievement and a failure. Shelley's poet arrives at the forest that will draw him in and where he will die with his Wordsworthian poetic identity trailing behind him as he longs, then refuses, to deck himself with bright strands of "falling spear-grass," the same grass that decked Margaret's decaying cottage walls in Wordsworth's *Ruined Cottage*:

The meeting boughs and implicated leaves
Wove twilight o'er the Poet's path, as led
By love, or dream, or god, or mightier Death,

He sought in Nature's dearest haunt, some bank,
Her cradle, and his sepulcher. More dark
And dark the shades accumulate. The oak,
Expanding its immense and knotty arms,
Embraces the light beech. The pyramids
Of the tall cedar overarching, frame
Most solemn domes within, and far below,
Like clouds suspended in an emerald sky,
The ash and the acacia floating hang
Tremulous and pale. Like restless serpents, clothed
In rainbow and in fire, the parasites,
Starred with ten thousand blossoms, flow around
The grey trunks, and, as gamesome infants' eyes,
With gentle meanings, and most innocent wiles,
Fold their beams round the hearts of those that love,
These twine their tendrils with the wedded boughs
Uniting their close union; the woven leaves
Make net-work of the dark blue light of day,
And the night's noontide clearness, mutable
As shapes in the weird clouds.

(Shelley, *Alastor*, 426–48, *SPP*, p. 84)[6]

The "net-work" of Shelley's language folds similes into botanical figures (clouds, the ash, and the acacia) and these into each other. In the implicated mix of figure across these lines, "parasites" is a moving target, gesturing in one direction toward the way the ash and acacia hang, doubling those clouds suspended in the simile that introduces them; and in the other toward those blossoms that twine around, their amorous enfolding that echoes that of the oak's arms around the light beech above. A poet could get lost here, and does. What this figuration also traps here is a layered botanical mimesis that constitutes the nature that first attracts and in the end embalms the poet. This is not to say that Shelley is, in some evident or instructive way, against nature or plants. The impact of this figured, botanical array is nothing so programmatic, even in a poem that purports to chastise the older Wordsworth for sinking into nature instead of politics or love or social problems and political solutions. Rather, what Shelley works out is a highly figural wandering that is mimetic several times over as it folds botanical figures to each other and to still other figures that are outliers, but only just, to the figurative texture of the scene. If all this gestures toward Wordsworth's inward turn, abetted or perhaps enabled by nature, it also marks out a dense relay of botanical figure that creates the very depth of field in this place that is Shelley's argument about where the nature poet goes to die. Few romantic poems convey as much in so small an enclosure of the poetic reach of botanical names.

Bateman's project is clearly different, informed by a spectatorial visual culture and book history, together with a botanical story and material culture that seemed to call out to and for extravagance. From the 1780s through the 1820s, elaborate engines for visual display, at times serial and often phantasmatic, such as the magic lantern, the eidofusikon, the diorama, and the panorama, fueled a public taste for spectacle and display that elite and less elite viewers shared, even as the display of art in exhibition halls open to a (paying) public transformed the experience of art for romantic-era viewers.[7] The botanical books of this period, big and small, from magazines to large books like Bateman's, convey an allied mix of extravagance and technical innovation. As Bateman's contribution to this extraordinary publication and visual history, *The Orchidaceae of Mexico and Guatemala* is distinguished by its vignettes and caricatures, which punch up the mirror relation between the engraved, colored plates and their accompanying texts in ways that give those wild orchids of Central America an extended run for the money. This from an amateur, moneyed horticulturalist who during the 1840s created at Biddulph Grange in Staffordshire a series of gardens, now more or less restored, that "contains whole continents, including China and Ancient Egypt—not to mention Italian terraces and a Scottish glen."[8]

Bateman's geographical fervor is at times less precise in the *Orchidaceae*. To conclude his account of *Oncidium leucochilum*, Bateman offers a vignette of Istapa [Ixtapa], on the Pacific coast of Mexico, with volcanic mountains in the background, after a sketch supplied by his orchid hunter George Skinner. Appended lines from John Campbell's poem *Pleasures of Hope* address the Andes as "giant of the Western Star," presumably not visible from Istapa, several thousand miles to the north of the Peruvian Andes. Bateman is, however, vigilant in his management of visual jokes or puns in other vignettes, beginning with a pair of cockroaches as the scourge of orchid shippers, linked to the *Catasetum maculatum* orchid under discussion because it is the only species that tends to survive being shipped with cockroaches. The first vignette of the pair, both attributed to Isaac Cruikshank, depicts the shell of a cockroach, to which Bateman appends the Virgilian phrase "Monstrum horrendum." The next vignette is also a cockroach, flipped onto its back. The thematic that links the two comes to us from Wordsworth's phrase "strange contrast doth afford." The contrast is between the first cockroach, which has eaten most of the orchids packed in a nearly empty box—"a portly, well-conditioned insect, happy, to all appearance, in the resources of his well-stored stomach"—and the second creature, who looks cadaverous, all legs and shallow-ribbed. Looks,

Bateman goes on to explain by examples, are deceiving. The contrast is between the top and bottom of the creature, which appears to have a "face" on both sides. Yet Bateman goes on, relishing the reversal of theme and variation:

> Look at him as he lies before you, and you pity his cadaverous countenance and admire his self-denial; turn him over, and you have the very *εἴδωλον* of plumpness and sensuality; on one side all is "roses," while all is "thorns" on the other: reverse him once more, and he who but a moment since "looked every inch an alderman," is now the picture of an insect anchorite. (plate 3, vignette, n.p.)

The Greek word Bateman offers to learned readers, *εἴδωλον* [eidolon], literally means "image," yet its complex history of usage from Aristotelian image to Platonic *phantasia* [*φαντασία*] joins sight to illusion, to chimera, and to the art of making things seem to appear before the eyes of listeners.[9] No wonder Bateman moved from the figure of the cockroach personified as an alderman or, by repeating and transforming the initial vowel, an anchorite. Played out across two vignettes, the joke or pun captures the labile movement between visual shape and figurative implication for which orchids are standard-bearers, as Bateman suggests when he reads "the rich and redundant vegetation" of the tropical forest depicted in another vignette as an "endless profusion of climbers and twiners, epiphytes and parasites, *et id genus omne*!! And, then, what a strange variety of animated beings"—a serpent, a monkey, parroquets, and "the pendent purse-like nests of the orioles, or cornbirds" (plate 6 vignette, n.p.). Purse-like nests, orioles that are cornbirds—the metonymic engine of Bateman's prose rarely stops. Another tropical forest vignette that bookends *Schomburgkia tibicinis* echoes and enlarges on its species name, *tibicinis*, which in botanical Latin (a species all its own) means the "piper's or flute-player's" orchid. The name honors Robert Schomburghk, who discovered the plant source for the poison strychnine while working as a colonial explorer in the West Indies and Central America.[10] In the vignette, a naked boy blows a trumpet as two other boys, made adult size or bigger by the skewed perspective of the vignette, climb up to reach this orchid, and one falls. Bateman echoes the exaggerated size of the boys climbing and falling, calling the musical instrument created from the long, hollow stem of this orchid a trumpet, not a flute or simple pipe, following the Honduran practice: these "vegetable trumpets" are so coveted by "wild urchins" that they call this orchid the "trumpet-plant." Bateman is not, I think, so much careful of local plant names as he is willingly suggestible that local practices invite an expansiveness that is linguistic as well as scalar. He follows the lead of this orchid's long spike and "conspicuous" flowers.

Three other vignettes (in some but not all copies), including one that is on the last page of the *Orchidaceae*, are engravings identified as the work of "Lady Grey of Groby," based on Katherine Grey's pressed-orchid album of the early 1840s (see chapter 4). All of them provide Bateman with ample opportunity for offering a punning, figurative relay between text and image and plant. The first of them in some copies depicts *Cycnoches ventricosum* (plate 5), a species belonging to the genus of "swan" orchids. Bateman's discussion of this one compares it to another in the genus, calling them "rival swans." Remarking that a combination of the best traits of each would create a "vegetable swan, as perfect in its parts as were the flies and bees with which other Orchises of English meadows present us," he reiterates the special attraction of orchids that look like insects or birds or people. For those who insist on exact resemblance, he adds a note: "To catch the resemblance of the two *Cycnoches* species to swans, it is necessary to *reverse* their flowers: this, however, merely restores them to their natural position, which they have lost by the circumstance of the raceme growing *downwards* instead of *upwards*." As the plant raceme joins the main stalk, it is bent down and the "swan" is overturned, with its neck and head below its body. Although this position is "natural" for this species, Bateman's instance to the contrary supposes that what is natural is the way real swans look. So flagrant a substitution of swan for orchid registers the considerable enticement or slippage of names that mimic flowers and birds.

The vignette that closes this description in some copies is Katherine Grey's drawing of another swan orchid, which she identifies as *Cycnoches loddigesii*, here assigned a more generic name tag, "rara avis," which evidently gestures toward common bird names for flowers that reappear in Latin binomials. Bateman asserts that this vignette "is taken from a specimen of Cycnoches Ventricosum, preserved in the rich herbarium of Lady Grey of Groby, under whose 'plastic hands' a suitable curve has been given to the neck of the vegetable mummy, which has thus rendered its resemblance to a swan quite obvious." In the same herbarium, "other specimens of this rare bird are to be seen, which have every appearance of being on the wing, an effect is produced by merely throwing open the sepals and the petals." Calling a pressed-flower album an "herbarium," which it both is and is not, insists on the botanical rather than artful aspect of the unspecified original, even as Bateman specifies how it is that those "plastic hands" transform the swan-orchid flower so that its resemblance is evident—even, as he puts it, to those of "limited capacity." The fact that this vignette does not appear in all copies may indicate a decision to add this plate for some buyers. The other two vignettes, which do appear in other copies, are equally careful not to specify precisely who the artist is and the nature of her "herbarium." The confusion about what kind of orchid plate 5 depicts may not be Grey's. The colored plate displays an orchid with male and female flowers that could be instead *Cycnoches edgertonianum*, the declared subject of the last plate in the book.[11]

For plate 19, *Caleandra baueri*, an orchid named after Lindley's other great illustrator, Francis Bauer, who drew it for Lindley, Bateman offers another "Lady Grey of Groby" vignette, here identified as a "tableau" in her "*Herbarium* or, to speak more correctly, the *Museum siccum*." The synonymic gesture more nearly recognizes the format of Katherine Grey's orchid album, where pressed-flower tableaux create a landscape for each of the orchids depicted. As Bateman describes it, the scene "is evidently laid on the shores of some Indian stream, whither part of the numerous progeny of *Cycnoches* (a genus exclusively American) are supposed—by poetic license—to have migrated: attracted, perhaps, by the well-known dainties, which are so plentifully provided in the pitchers of the *Nepenthes distillatoria*." The poetic license at work in this account is as much Bateman's as Grey's: The word *pitchers*, used for the container from which the swan drinks, signals the common name of the American pitcher plant, long a favorite subject of botanical illustration. In this verbal tableau, the idea of a pitcher precedes its identification as being in fact part of a plant and its common name.

Plate 40, *Cycnoches Egertonianum*, which, Bateman allows, is often mistaken for the *Cycnoches Ventricosum* of plate 5 (which it may in fact be), concludes with the third and final Grey vignette, a tableau in which a near-riot of orchids cavort, all looking like something else. Bateman calls it "a most ingenious device, compounded of divers Orchidaceous flowers, which, with very gentle violence, had been induced to assume to attitudes in which they appear below." The scene includes large bird-like insects, two more swans, crawling insects, two small kangaroolike animals, other insects, and a few blossoms that actually look like orchids, all supervised by a hooded female figure. As if to moderate the wit and glee of the tableau, Bateman appends this Miltonic tag:

> Nature breeds
> Perverse, all monstrous, all prodigious things:
> Abominable, unutterable, and worse
> Than fables yet have feigned, or fear conceived,
> Gorgons and hydras and chimæras dire.
> —Milton[12]

For Bateman and for many others who wrote about "monstrous" plants, the Miltonic evocation of such forms in Pandemonium, the capital of Hell, belies the image and Bateman's obvious delight in the shape-shifting of this final vignette. For although he calls the female figure who oversees it all a "hag," nothing in the image suggests that she is. Bateman's note carefully names all the species depicted, from the "hag," who is a flower of Cypripedium (a well-cared for Venus's slipper orchid, and no hag), to *Masdevallias, Cycnoches, Epipendra*, and other orchids identified by their genus and species names. A circular vignette with the two *Cycnoches* (whatever their precise species) in the center and mostly orchid pairs clustered around them, the image confirms Bateman's sustained delight in orchids that look like x or y or occasionally x and y.

The attention to botanical monsters from early to late, or more precisely, from 1699 to the middle of the nineteenth century, pursues two, sometimes adjacent, tracks. For botanists like Robert Brown and his contemporaries, a monstrosity is a plant or species with aspects or tendencies that are, in the current systematic scheme of things, unclassifiable, such as the enormous *Rafflesia arnoldii*, the cryptogamic plant that gave Brown the evidence he needed to overturn English support for the Linnaean systematic. But for many, plant "monsters" invited a pale, dreamy version of Miltonic horror more akin to frissons of delight than terror. Bateman here has his eye on these less serious readers (and their pocketbooks). Yet Milton's lines also predict the strange new botanical world of plants that Wallace found in 1699, and the effort then and long afterward to create taxonomic homes and names for plants whose unfamiliarity seemed at once fair and monstrous.

Whether or not orchids were monstrous was then and long thereafter a less compelling question than whether their putative resemblances to insects registered some degree of plant agency in their method of fertilization. In the nineteenth century, the orchids that prompted this question most persistently were not exotic species but those native to England and Europe, the orchis genus that Bateman lingers over on his way to more exotic orchids from Mexico and Guatemala. This question was entertained (or dismissed) from the first volume of the *Botanical Magazine* in 1787 to Charles Darwin's 1862 monograph *On the Various Contrivances by which British and foreign orchids are fertilised by insects* and a later companion volume, *The Effects of Cross- and Self Fertilisation in the Vegetable Kingdom* (1877). In the early 1960s, the botanist C. H. Dodson published the results of his study of how certain bees pollinate tropical orchids.[13]

Bateman takes the question of orchid agency as far as anyone could, well beyond Erasmus Darwin's poetic analogies between plants and humans, which hover as conjecture and figure. I want to step briefly beyond the frame of my argument to take this analysis of the language of plant agency to its nineteenth-century end in Charles Darwin's two monographs and his earlier experiments and letters on this topic. The younger Darwin's resistance to more than a century of earnest and playful thinking about plants as sentient and deliberative must negotiate between such claims and the generative evolutionary reality suggested by adaptive plant mechanisms involved in fertilization. What is most compelling is where Darwin draws the line in presenting those "various contrivances."

By the late 1850s, Charles Darwin was experimenting with orchids, asking for new and old specimens, exchanging speculation and information with a number of correspondents. Much of that attention concerned orchises, the English genus that Clare and so many others found intriguing. Of the two hypotheses that were typically made about these species—that they looked like the insects that fertilized them and in some way cooperated in or perhaps engineered their own fertilization—Darwin was far more driven to argue against the latter, though he was not, like J. E. Smith long before, willing to insist on orchid-insect resemblances as a key to cross-fertilization among the *Orchis* species. In a preliminary report published in the *Gardeners' Chronicle* in 1860, he summarized a series of experiments that proved that the bee orchis did indeed self-fertilize. In one sense, this was good news. Here, at least, the fact that this orchis looked like a bee did not mean that it was fertilized by bees. Yet Darwin's reservations about what he had found were oddly interested in being proved wrong, at least suppositionally:

> We must admit that the natural falling out of the pollen-masses of this Orchis is a special contrivance for its self-fertilisation; and as far as my experience goes, a perfectly successful contrivance; nevertheless a long course of observation has made me greatly doubt whether the flowers of any kind of plant are for a perpetuity of generations fertilised by their own pollen.[14]

Darwin goes on to imagine that the bee orchis's "sticky glands" could be used, as they are used in other orchises, to assist insect fertilization. In part this supposition is impelled by something like evolutionary logic: if there are sticky glands, they might be good for something. Darwin's logic may also have another concern in view. In the long history of its usage among botanists and amateurs, the word *contrivance* implies that plants might contrive. True, such mechanisms are for many evidence of divine or evolutionary design, but the tendency to use the word *contrivance* to talk about remarkable plant mechanisms tends to muffle this conceptualization. Romantic fascination with what plants do, as individuals, as botanical particulars, may on these occasions trump claims about higher agency. This, I surmise, is why Charles Darwin wished to find evidence that the bee orchis does not always accomplish its own fertilization.

But what of fertilization of orchises by insects? Darwin dismisses Robert Brown's suggestion that the bee orchis looks like a bee to deter other insects from visiting it, remarking "I cannot think this probable."[15] Perhaps not, but what is probable and is by now widely recognized is evidence that would have been no less distasteful: that some orchids look like female insects of an identifiable species so that the males of that species will try to mate with them, thereby triggering the orchid's fertilization.[16] For Darwin, the remarkable character of such adaptations is in some ways a close call. In one sense, what could better demonstrate evolutionary adaptability?

Yet natural mimicry as deft as this inevitably raises questions, or invites myths, about the cleverness of plants. Perhaps individual plants, rather than a Creator or an evolutionary mechanism, can make adjustments to circumstances that cannot be attributed to species adaptation. If this were so, then it could be said that plants have agency, not in the aggregate but in particular instances. Questions about agency lead in turn to questions about conceptual power. Is it the plant, a name, or a concept such as evolutionary adaptation that drives the analysis of how plants do things?

Put concisely, romantic era inquiries and fictions about the lives and loves of plants trouble the effort to separate material things and concepts, from Hegel's Spirit to Darwin's evolutionary model. Romantic botany provides instead a sustained and varied demonstration of how matter, cultural material, poetic figuration, and questions about the status of beings and categories work in productive friction. The specificity of that friction, as well as its turns and halts, constitutes a diversely tuned argument for the nature of romantic nature.

Notes

CHAPTER 1: INTRODUCTION

Epigraph: Wallace, "Part of a Journal" (1700), 539, quoted by William T. Stearn, "Linnaeus's Sexual System of Classification," in Linnaeus, *Species Plantarum*, 1: 25.

1. Bill Brown, "The Secret Life of Things," 3. My reading of botany as plant matter recalls, in a different time and place, Brown's analysis of labor in *The Material Unconscious*, 11.

2. Brown, "The Secret Life of Things," 10, quoting Adorno, *Negative Dialectics*, 191.

3. Oerlemans uses the phrase *material sublime* in *Romanticism* to characterize the romantic response to an alienated material world (202–210). I use it rather to emphasize the pull of the material and the singular against romanticism's most elevated conceptual ambitions as Schwartz describes it in *Wormwork*. Gigante examines ugly or monstrous aesthetics in *Life*, 223–224. In *Ugly Feelings*, Ngai assesses the conjunction of aesthetic feeling and ugliness, 6 and 279–297.

4. Wallace, "Journal," 539–540. Wallace was the first European naturalist to work in Panama. The next botanical exploration there occurred in 1790. See Dwyer, "A Note on Plant Collectors," 107–108. Wallace's rather cheerful view of Darien contrasts sharply with that of earlier visitors. In 1700 he asserted, "the Soil is rich, the Air good and temperate, the Water is sweet, and every thing contributes to make it healthful and convenient." In 1514 Peter Martyn warned, "*Darien* is pernicious, unwholesome and outrageous," and in 1715 Rev. Francis Borland, a survivor of the ill-fated Scottish colony, bitterly quoted Martyn, then addressed Darien itself as "thou Land devourest men and eatest up thy inhabitants." See Borland, *Memoirs of Darien*, 18. Wallace's enthusiasm for the place may convey his delight with its tropical flora and fauna. A half-century later, after exotic plants had found their way into many English gardens, Horace Walpole wryly noted the reverse confusion: "My present and sole occupation is planting, in which I have made great progress, and talk very learnedly with the nurserymen, except that now and then a lettuce run to seed overturns all my botany, as I have more than once taken it for a curious West Indian flowering shrub." Walpole is here writing to Henry Seymour Conway, 29 Aug. 1748, quoted by Laird from the epigraph to *The Flowering of the Landscape Garden*.

5. Brown, "The Secret Life of Things," 5.

6. Daston and Galison, *Objectivity*, 374–375.

7. Drayton, *Nature's Government*; Grove, *Green Imperialism*; Bewell, "Erasmus Darwin's Cosmopolitan Nature," 19–48; Kumar, *Science and Empire*; Miller and Reill, *Visions of Empire*; Casid, *Sowing Empire*; and Tobin, *Colonizing Nature*.

8. Foucault, *Les Mots et les Choses*, 129–130, 135–156.

9. Daston and Galison, among many others, argue against Foucault for a more gradualist account of the rupture that Foucault describes: *Objectivity*, 375.

10. Foucault, *Order of Things*, xv, 135, 155.

11. François, *Open Secrets*, 8–9 and 37. The argument made by François about plants is muted, half-glimpsed in her trenchant account of how openness and a wise passivity (of the kind that Wordsworth's nature invites and the redwoods of Northern California offer) oppose a more aggressive and violent wrenching of meaning and putative secrets.

12. Mitchell reflects on speculative consequences of romantic fascination with plants in "Cryptogamia," 631–651.

13. Cesalpino, "Dedication," quoted by Pavord, *Naming of Names*, 239.

14. Larson, *Reason and Experience*, 36.

15. William T. Stearn, introduction, Linnaeus, *Species Plantarum*, 1:24–35.

16. Jardine discusses amateur and professional engagement with natural history in the eighteenth century in *Scenes of Inquiry*, 16–17.

17. Gigante echoes the antimony between mechanism and vitalism that has long shaped critical discussion of romantic nature(s) in *Life: Organic Form and Romanticism*.

18. Cesalpino, "Dedication," *De plantis libri XVI*; quoted by Pavord, *Naming of Names*, 232.

19. Stevens notes that for A.-L. de Jussieu, who worked out the first one hundred families of the Natural System in 1789, *differentiae* refers to the external features of plants. The term *characters* is sometimes used as a synonym: Stevens, *The Development of Biological Systematics*, 30; Morton, *History of Botanical Science*, 126, refers to internal features that "indicated the organization of plants and hence their true nature."

20. Ray, preface to *Methodus plantarum nova*, n.p. I have slightly altered the English translation from that presented in Pavord, *Naming of Names*, 282–283. Early in their long correspondence, Collinson told Bartram, "It Is impossible for any One Author to give a General History of plants": Collinson, letter to Bartram, 24 Jan. 1734, in *'Forget not Mee & My Garden*, 11.

21. Wilkins, *Essay towards a Real Character*, 13 and 67.

22. Stevens, *Development of Biological Systematics*, xx–xxi.

23. Quoted by William T. Stearn, "Alexander von Humboldt and Plant Geography," 116.

24. Alexander von Humboldt published the French edition, *Essai sur la géographie des plantes*, in 1806, the German edition, *Ideen zu einer Geographie der Pflanzen*, in 1807. He dedicated the latter to Goethe: William T. Stearn, "Humboldt's *Essai*," 121–127.

25. Stearn, "Humboldt's *Essai*," in *Humboldt, Bonpland, Kunth*, 126–127.

26. Derrida often considers the problem of singularity's necessary (and impossible) relation to universal claims and concepts. I cite here his most resonant formulation in "Force of Law," 3–67; Adorno, "Aspects," 42. See my chapter 7.

27. Wordsworth, *The Prelude* (1805 text), book 7, ll. 228–238, p. 238; hereafter cited parenthetically as *Prelude*.

28. Several scholars have written about botanic figures in romantic poetry, among them Bewell, "Keats's 'Realm of Flora,'" 71–98; Simpson, *Wordsworth, Commodification, and Social Concerns*, 91–105; Gidal, *Poetic Exhibitions*, 163–207; Porter, "Scientific Analogy and Literary Taxonomy in Darwin's *Loves of the Plants*"; Pascoe, "'Unsex'd Females'"; and Mahood's ecocritical study of three romantic poets, *The Poet as Botanist*.

29. Pollan, *The Botany of Desire*, 58, 69, 78; Deleuze and Guattari, *A Thousand Plateaus*, 9–10.

30. De Zegher, "Ocean Flowers and Their Drawings."

31. The full title of Abbé Guillaume-Thomas Raynal's influential study is *L'Histoire philosophique et politique des établissements et du commerce des Européeans dans les deux Indes*, first published anonymously in 1700. I cite the expanded fifth edition of 1777.

CHAPTER 2: BOTANICAL MATTERS

1. Baillon, *Étude générale du groupe des euphorbiacées*, 280, quoted by Stevens, *The Development of Biological Systematics*, n.p.

2. Mabberley assesses the taxonomic significance of Robert Brown's *Prodromus Florae Hovae Hollandiae* in *Jupiter Botanicus*, 164–170; Morton characterizes Brown's achievements in *History of Botanical Science*, 373; Julius von Sachs notes Brown's reluctance to comment on the theoretical implications of his work, then offers an extended discussion of Brown's taxonomic contributions in *History of Botany, 1530–1860*, 140–142.

3. Aristotle, *Topics*, in *Complete Works of Aristotle*, book 1, sec. 16, 1: 179; Morton, *History of Botanical Science*, 126; Stevens, *The Development of Biological Systematics*, 30; Larson, *Reason and Experience*, 46; Atran, *The Cognitive Foundations of Natural History*, 142; and Pavord, *Naming of Names*, 232.

4. Linnaeus, *Philosophia Botanica*, 219. See also Stafleu, *Linnaeus and the Linnaeans*, 84.

5. In Aristotle, the Greek term for species is *eidos*. See *Topics*, in *Complete Works of Aristotle*, book 4. sec. 2ff, 1: 204, and the translator's index, 2: 2485. Pavord reiterates this key point in *Naming of Names*, p. 24.

6. Larson, *Interpreting Nature*, 32; Secord notes that artisans and amateur naturalists used the Linnaean system until nearly the middle of the nineteenth century: "Science in the Pub," 269–315. Linnaean categories persist in botanical notebooks as late as 1837. King tracks the long nineteenth-century career of Linnaean sexual distinctions in *Bloom*.

7. Among the natural groups or families that have survived in modern taxonomy, Larson, in *Reason and Experience*, 4, lists Umbelliferae, Coniferae, Cruciferae, and Graminaceae. Pavord notes that the Umbelliferae, so identified by Robert Morison in the seventeenth century, are described by Aristotle's student Theophrastus: Pavord, *Naming of Names*, 41 and 387. Some natural groups reflect what botanists now recognize as an evolutionary inheritance; that is to say, members of some of the natural groups identified prior to Charles Darwin's evolutionary hypothesis share certain characters because they share a common ancestor. Similarity is, however, no guarantee of common ancestry. Species or individuals that have or appear to have many characters in common may not constitute a "natural" group at all; rather, they may have, by a process of convergent evolution, ended up merely looking similar because they have responded (i.e., adapted) to a common set of factors (Henry J. Noltie, correspondence, 2006). For a brief account of these evolutionary questions, see Gould, *Dinosaur in a Haystack*, 116–118.

8. Stafleu, *Linnaeus and the Linneans*, 26.

9. Larson makes this observation in *Interpreting Nature*, 34.

10. Koerner, *Linnaeus*, 44; Larson, *Reason and Experience*, 98–99. The list of those who believed or did not believe in the constancy of species is difficult to construe with accuracy. Prior to Linnaeus and beginning with Aristotle, most agree, everyone did. It appears that Linnaeus did, then didn't; that de Candolle did; and so on up to about 1850. Julius von Sachs summarized this view of the evidence in *History of Botany, 1530–1860*, 108–54. Atran argues a rather different view of Aristotle's views on species or natural kinds in *Cognitive Foundations*, 87.

11. Ereshefsky presents both versions of the Linnaean hierarchy in *The Poverty of the Linnaean Hierarchy*, 213. The current taxonomic hierarchy from Kingdoms down is reported on a website for "Science 101" as kingdom, phylum, class, subclass, order, suborder, family, genus, and species. See www.101science.com.

12. Bicheno, "On Systems and Methods in Natural History," 479–496, cited by Gordon McQuat, "Species, Rules, and Meaning," 483n.

13. Dupré, *The Disorder of Things*, 7.

14. Bicheno, "On Systems and Methods," 490–491.

15. Atran, *Cognitive Foundations*, 318n1.

16. J. E. Smith identifies the "Ophrys Loefelli or Dwarf Ophrys as a distinct species from that of Linnaeus's 'Ophrys lilifolia' " in Sowerby and Smith, *English Botany*, 1: 47; Smith distinguishes two other Ophrys species that one of Linnaeus's English editors had confused: ibid., 1:383. Smith further disagrees with Linnaeus's tendency to lump all the insect resembling species of Ophrys (fly, bee, spider) into one species, ibid., 1: 64.

17. Figuier, *The Vegetable World*, 203.

18. Gledhill describes arguments about the difficulty of ascertaining a taxonomic species, similar concerns about identifying Natural Orders and genera, and current taxonomic questions and uncertainties in *The Names of Plants*, 14–15 and 20–21.

19. Foucault, *Les Mots et les Choses*; in English, *The Order of Things*, 71–77.

20. Collinson to Linnaeus, May 13, 1739, in *"Forget not Mee & My Garden,"* 72.

21. Collinson also uses the language of debt and gratitude in a letter to John Custis, thanking him for plants. See no. 18, 15 Dec. 1735, in *"Forget not Mee & My Garden,"* 36–38.

22. Collinson to Linnaeus, 18 Jan. 1744, in *"Forget not Mee & My Garden,"* 111.

23. Ritvo surveys arguments against this practice in the 1830s and 1840s in *The Platypus and the Mermaid*, 64. Modern nomenclatural procedures insist only that the use of persons' names be restricted to the explorer who found the species or subspecies being named, the taxonomist/botanist who identified it as such, or, in the case of plants propagated by horticulturists, the person who did so. See Gledhill, *The Names of Plants*, 25.

24. Aristotle, *Analytica Posteriora*, book 1, sec. 19, 100a3–9, and *de Anima*, book 3, sec. 8, 432a10–14, in *Complete Works*, 1: 165 and 1: 687; Atran, *Cognitive Foundations*, 143.

25. Samuel Johnson, *Dictionary*; s.v. "Species."

26. Hull, *Science and Selection*, 205–206; Mayr, "Another Look at the Species Problem," 174.

27. Atran, *Cognitive Foundations*, 87.

28. Atran, *Cognitive Foundations*, 92–118; Larson, *Interpreting Nature*, 49–52.

29. Locke, *Essay concerning human understanding*, cited parenthetically in the text by book, chapter, section, and page. Locke's definition appears in the *Oxford English Dictionary*, s.v. "Species," under this definition: "a class composed of individuals, hav-

ing some common qualities or characteristics." For an astute reading of Locke on the problem of species, see Festa, *Sentimental Figures of Empire*, 40–42.

30. Kant, *Critique of the Power of Judgment*, 276.

31. Coleridge, *Statesman's Manual*, in *Lay Sermons*, 30.

32. *Oxford English Dictionary* (1971), s.v. "Species."

33. Schiller, *Naïve and Sentimental Poetry*, 158.

34. Hegel, *Philosophy of History*, 32–33; in German, *Werke*, 12: 49.

35. Stafford, *Voyage into Substance*.

36. Derrida, "Force of Law, 955. See also Gasché, *Inventions of Difference*, 227–250.

37. Hull, *Metaphysics of Evolution*, 183.

38. Hull, "Biological Species, 64.

39. Gledhill, *The Names of Plants*, 27.

40. Gledhill, *The Names of Plants*, 20.

41. Ruse reviews the evolutionary usefulness of such plants in his introduction to *What the Philosophy of Biology Is*, 8–9; Hull considers evolution and the edge of species distinctions in "Biological Species," 64.

42. Mayr, "The Species Category," 319, and "Another Look at the Species Problem," 177.

43. Mayr, "Species Concepts and Definitions," 11–12; Hull, *Science and Selection*, 205–207.

44. Hull, *Science and Selection*, 112.

45. Ereshefsky, *The Poverty of the Linnaean Hierarchy*, 131, 132, and 143.

46. McQuat reviews these debates in "Species, Rules, and Meaning," 473–519.

47. Darwin, *Charles Darwin's Natural Selection*, ed. Stauffer, 98; cited by McQuat, "Species, Rules, and Meaning," 515.

48. Gledhill, *The Names of Plants*, 14.

49. Ereshevsky, *The Poverty of the Linnaean Hierarchy*, 208–237.

50. Other botanists had noted this point, among them Nehemiah Grew, who identified stamens as male sexual organs, John Ray, who developed a fuller account of plant sexuality, and Sébastien Vaillant, whose early eighteenth-century experiments became the basis for Linnaeus's presentation of plant sexuality as a taxonomic principle. See Gledhill, *The Names of Plants*, 17; Pavord, *Naming of Names*, 372 and 389.

51. Linnaeus, *Species Plantarum*, 2: 1061–1186.

52. Graham, Graham, and Wilcox, *Plant Biology* does not mention cryptogams; McLean and Cook, *Textbook of Theoretical Botany*, 1: 487 presents the term as "an old name which covers all the non-flowering plants." The term *cryptogams* reappears in two twentieth-century botanical dictionaries: Jackson's *A Glossary of Botanic Terms*, s.v. *cryptogamia*, and Allaby's *A Dictionary of the Plant Sciences*, s.v. *cryptogam*, defined there as "a plant that reproduces by spores or gametes rather than seeds," without notice of the meaning of the term or its class in the Linnaean system. In current biological research, the cryptogamic plants include all nonflowering plants, such as ferns, horsetails, and clubmosses (vascular) as well as mosses and liverworts (nonvascular), and Protista (less elegantly, algae and slime molds). Biologists emphasize the diversity, complexity, and evolutionary importance of these plants, as well as their continued importance to life on Earth. Now excluded from the cryptogamic category, the Fungi (which includes fungi and lichens) are not considered to be plants at all. I am indebted to H. J. Noltie for guidance through the tangled taxonomy of biological kingdoms. Cryptogamics are dis-

cussed on the science Web page for the Royal Botanic Gardens, Edinburgh, www.rbge.org.uk/rbge/web/science/research/crypto/index.jsp.

53. Soltis, Soltis, Endress, and Chase, *Phylogeny and Evolution of Angiosperms*, 305.

54. Schiebinger, *Nature's Body*, 12–39.

55. Erasmus Darwin, "The Loves of the Plants," part 2 of *The Botanic Garden*, 62–65, p. 6.

56. For the full Linnaean systematic in Latin and English, see William T. Stearn, ed., "An Introduction to Linnaeus, *Species Plantarum*, 1: 24–35. Stearn notes the changed Latin status of unmarried females in subsequent editions of *Species Plantarum* (1: 29).

57. Stearn translates these remarks by Johann Amman and the St. Petersburg scholar Johann G. Siegesbeck, noting that by the early nineteenth century this prudishness was more marked than it had been in the urbane eighteenth (Stearn, introduction to Linnaeus, *Species Plantarum*, 1: 25). In 1821, the botanist Samuel F. Gray began his account of the natural system by taking issue with "the prurient mind of Linnaeus" (*A Natural Arrangement of British Plants*, 1: vii). Yet Stearn's comparison may miss a key point about the climate of public opinion at the end of the eighteenth century, when the chorus of prudish horror was especially loud. By the 1790s, those who did botany or wrote about it, especially women, were greeted by so much sexual prurience and innuendo that they were loathe to use the terms *male* and *female*. Whether the details of the sexual system would ever have been so widely discussed had the Lichfield Society English translation not been published is a moot point now. See my chapter 3 for an account of this controversy.

58. At times, John Bartram asked correspondents to help him puzzle through another naturalist's Latin descriptions. See Bartram, *The Correspondence of John Bartram*, 271, 380.

59. Miller to Bartram, 12 Jan. and 28 Aug. 1758, in Bartram, *The Correspondence of John Bartram*, 434 and 438. Miller's *Gardeners Dictionary* was frequently reissued in new editions and abridgments from 1739 until at least 1775.

60. Bartram to Miller, Feb. 1859, in Bartram, *The Correspondence of John Bartram*, 456.

61. Koerner, *Linnaeus*, 117–139.

62. Stevens argues that any of these systematic possibilities could be invoked to assert the continuity of nature: *The Development of Biological Systematics*, 4.

63. Buffon, *Histoire naturelle*, 13, 40, and 19. William Smellie's frequently reissued and partial translation, *Natural History: General and Particular*, omits Buffon's "Premier Discours," in which these criticisms appear. Sloan summarizes Buffon's critique in "The Buffon-Linnaeus Controversy," 356–375.

64. *Encyclopédie; ou Dictionnaire raisonné des sciences, des arts et des métiers* (1751), s.v. "Botanique," 2: 340–345; Jardine quotes and translates Buffon's comments in this article and another on "Histoire naturelle" in *Scenes of Inquiry*, 12–13.

65. William T. Stearn, introduction to *Species Plantarum*, 1:34.

66. Noltie notes that Ondaatje's use of the Linnaean system is surprising, since A. P. de Candolle's presentation of the natural system was the acknowledged authority by 1847. Robert Wight's and George Walker Arnott's (never completed) *Prodromus Florae Peninsulae India Orientalis*, 1834, which had covered about half of the Indian flora by this time, had used the de Candolle system.

67. Mabberley discusses Robert Brown's preface to *Prodromus Florae Novae Hollan-*

diae in *Jupiter Botanicus*, 161. For an English translation of the Latin text, see William T. Stearn, *Historiae naturalis classica*.

68. Bicheno, "On Systems and Methods," 490–491.

69. Stafleu, introduction to Jussieu, *Genera Plantarum*, xxiii; Atran, *Cognitive Foundations*, 196.

70. Stafleu, introduction to Jussieu, *Genera Plantarum*, v–xiv.

71. Stafleu provides this count in introduction to Jussieu, *Genera Plantarum*, xxiv.

72. Morton, *History of Botanical Science*, 257–258.

73. Stafleu, introduction to Jussieu, *Genera Plantarum*, x.

74. William T. Stearn, introduction to Linnaeus, *Species Plantarum*, 1: 34.

75. Sachs, *History of Botany*, 137.

76. Jussieu, *Genera Plantarum*, 16.

77. A. P. de Candolle, *Théorie*, 76: "Je tirerai donc, de ces considerations élémentaires, ce premier théorême, que dans la classification des êtres organisés, le degré d'importance de chaque organe ne peut être calculé exactement, que relativement aux organes qui se rapportent à la même classe de fonctions."

78. A. P. de Candolle, *Théorie*, 78.

79. A. P. de Candolle puts it this way: "It is certainly remarkable that in comparing Linnaeus's writings with those of Jussieu on this point that it is Linnaeus who should be reproached for having exaggerated the advantages of the natural method. Indeed, he repeated after a thousand others that 'Nature does not make leaps,' when in fact the most zealous partisans of the natural orders today acknowledge that there are leaps or gaps in the series of beings" (my translation) ("It est même assez remarquable qu'en comparant, à cet égard, les écrits de Linné avec ceux de Jussieu, ce soit à Linné qu'on soit obligé de reprocher trop d'exagération en faveur des principes de la méthode naturelle. Ainsi, il a répété après mille autres, la nature ne fait pas de sauts: tandis que les plus zélés partisans des ordres naturels reconnaissent aujourd'hui qu'il y a des sauts ou des interruptions dans la série des êtres."); A. P. de Candolle, *Théorie*, 60.

80. A. P. de Candolle, *Théorie*, 60; quoted by Atran, *Cognitive Foundations*, 210.

81. Stafleu, introduction to Jussieu, *Genera Plantarum*, xx–xxii.

82. Jardine, *Scenes of Inquiry*, 2–3 and 286.

83. Jardine, *Scenes of Inquiry*, 15–20.

84. Stevens, *The Development of Biological Systematics*, 81.

85. Alphonse de Candolle, *Introduction*, 1: 469.

86. Alphonse de Candolle, *Introduction*, 1: v: "La botanique présente ces difficultés comme toutes les sciences. Elle en a d'autres qui lui sont propres. Au lieu de se subdiviser chaque année davantage en sciences distinctes, comme la physique, par exemple, qui se partage aujourd'hui en optique, électro-magnétisme, etc., on sent plus que jamais la nécessité de lier en un faisceau [bundle, as in a bundle of sticks—with the implication that these constitute a group] compact les branches, autrefois séparées, de l'étude des végétaux."

87. Derrida in part understood *différance* to register inherent resistance of singularity to categories. See his "*Différance*," in *Margins of Philosophy*, 1–28.

88. Bradbury, *The Microscope Past and Present*, 134–151, and *The Evolution of the Microscope*, 92–199; Crary, *Techniques of the Observer*, 199–129 and 131n50.

89. Stafleu, introduction to Jussieu, *Genera Plantarum*, xxii.

90. Ford, "Brownian Movement in Clarkia Pollen," 235–241.

91. Smith, review of *Flora Britannica*, 362.

92. Mabberley, *Jupiter Botanicus*, 136.

93. "Review of Lindley's *Ladies' Botany* and Edwards's *Botanical Register*,"140.

94. Henry J. Noltie, correspondence with author, 15 Dec. 2005, and 17 Jan. 2006. See also Raven, Evert, and Eichhorn, *Biology of Plants*, v–vi, and *Tiscali Encyclopaedia*, s.v. "Protist," www.tiscali.co.uk/reference/encyclopaedia/hutchinson.

95. *Encyclopaedia Britannica*, 7th ed. (1842), 3: 30–31, s.v. "Botany."

96. Mabberley, *Jupiter Botanicus*, 376; Sachs, *History of Botany*, 155–159; Morton, *History of Botanical Science*, 399–403.

97. Mabberley, *Jupiter Botanicus*, 177–178.

98. Morton, *History of Botanical Science*, 374.

99. Mabberley, *Jupiter Botanicus*, 400 and 162–166.

100. Mabberley, *Jupiter Botanicus*, 289 and 404.

101 Mabberley, *Jupiter Botanicus*, 285 and 404. Although he followed developments in the microscope, Brown made many of his discoveries with a dissecting microscope of relatively low power (268).

102. Mabberley, *Jupiter Botanicus*, 252; Morton, *History of Botanical Science*, 375.

103. Collinson to Cadwallader Colden, 9 Mar. 1744, Collinson, "*Forget not Mee & my Garden*," 114.

104. Stewart, *On Longing*. Gidal notes that a letter to the *Weekly Advertiser* in December 1755 discusses terror of novelty and the sublime and the "extreme of littleness" See Gidal, *Poetic Exhibitions*, 55.

105. Mabberley, *Jupiter Botanicus*, 81 and 225, 315–116, 362, 263.

106. Mabberley, *Jupiter Botanicus*, 252.

107. The study of plant "monstrosities" remains an important aspect of taxonomy, now informed by DNA analysis: Noltie, correspondence, Dec. 2005.

108. Edgeworth, *Belinda*, 319.

109. Kant, *Critique of the Power of Judgment*, 136.

110. Aristotle, *De Generation Animalium*, 769b11–b21, in *Complete Works*, 2: 1191; quoted in Atran, *Cognitive Foundations*, 143.

111. Ritvo, *Platypus and the Mermaid*, 132–177.

112. Adam Smith, *Essays on Philosophical Subjects*, 7; quoted by Gidal, *Poetic Exhibitions*, 58.

113. In "Objections to the appointing of Public days for admitting all Persons to the Museum without distinction," John Ward insisted to the Trustees of the new British Museum in 1755 that the "Eye of the Spectator" would experience "Terror" in confronting an array of specimens, BL Add MS 6179, f.61, British Library; quoted by Gidal, *Poetic Exhibitions*, 59–60.

114. Collinson to Charles Cadwallader Colden, 4 Sept. 1743, Collinson, "*Forget not Mee & my Garden*," 110.

115. Newton notes the anecdote in "Review of *Echolalias*," 25.

116. In the late eighteenth century, the botanist Carl Ludwig Willdenow identified an Indian herb that occurs in the subtropics of the old and new worlds and is part of the pharmacopia of the Ghats of Southern India by its local names, "ruktroora, sheral." Noltie notes that the first of these names was also used for other, unrelated plants with similar medicinal properties: Noltie, *The Dapuri Drawings*, 142.

117. Koerner, *Linnaeus*, 24–25.

118. Pavord, *The Naming of Names*, 396.

119. Gledhill, *The Names of Plants*, 18–19.

120. McQuat tracks this dispute in "Species, Rules, and Meaning," 494–500. For one anti-Latin salvo, see Wood, "On the Study of Latin," 46–49, and "On Making the English Generic Names of Birds Correspond to the Latin Ones," 238–239; cited by McQuat, 493n.

121. Koerner, *Linnaeus*, 52–53.

122. Collinson to Linnaeus, 20 Apr. 1754, and Bute to Collinson, n.d., in *"Forget Not Mee & My Garden,"* 177 and 187n4; cited by Koerner, *Linnaeus*, 54.

123. Ereshefsky, *The Poverty of the Linnaean Hierarchy*, 212–247.

124. De Candolle, *Théorie*, 280–282. A. P. de Candolle's systematic achievement is widely recognized, see Sachs, *History of Botany*, 126–239; Morton, *History of Botanical Science*, 1–373; and Gledhill, *The Names of Plants*, 25.

125. A. P. de Candolle, *Théorie*, 262–63; Gledhill, *The Names of Plants*, 25.

126. Gledhill, *The Names of Plants*, 18 and 25.

127. Pavord, *The Naming of Names*, 392.

128. Alcock, *Botanical Names for English Readers*, 69–70.

129. Noltie, *Dapuri Drawings*, no. 150, 206.

130. Gledhill, *The Names of Plants*, 1.

131. Linnaeus, *Philosophia Botanica*, 169.

132. Gledhill, *The Names of Plants*, 2.

133. Gledhill, *The Names of Plants*, 2, and Pavord, *The Naming of Names*, 265.

134. Gledhill, *The Names of Plants*, 4.

135. Figuier, *The Vegetable World*, 398.

136. Drayton, *Nature's Government*, 47.

137. *Zander Handwörterbuch dem Pflanzennamen*, 884.

138. Smith, Correspondence, unpublished letter to Sowerby, 14 Aug. 1801, Library of the Linnean Society, London.

139. William S. Stearn, *Botanical Latin*.

140. Gledhill, *The Names of Plants*, 52.

141. Pavord, *Naming of Names*, 263.

142. Gledhill, *The Names of Plants*, 44.

143. "Solitarius," "Remarks upon Zoological Nomenclature," 523; quoted by Ritvo, *Platypus and Mermaid*, 59 and 234n.

144. Ritvo notes the survival of "trivial" zoological names in the British Museum into the 1860s in *Platypus and Mermaid*, 68.

145. J. E. Smith, "Fly Orchis," in Sowerby and Smith, *English Botany*, 1: 64.

146. Curtis, *Against Autonomy*, 14. Curtis discusses Lyotard's use of this phrase to imagine what idea of justice might preserve dissent; see Lyotard, *Peregrinations*.

CHAPTER 3: BOTANY'S PUBLICS AND PRIVATES

Epigraph. Quoted by Pratt, *Imperial Eyes*, 109.

1. "Botanicus," Folder 1561, Osborn MSS Files, Beinecke Library, Yale University. The *Anti-Gallican* appeared in twelve monthly installments in 1804. The "Botanicus" letter refers to another *Anti-Gallican* at the time of Napoleon's final defeat that appears to have circulated in manuscript among Tory readers.

2. Coats, "The Empress Joséphine," 40–46.

3. Here and elsewhere, the marchioness does not record the Latin botanical name—in this case, aka *Fritillaria imperialis*, of the Lily family, and one of the plants that Erasmus Darwin and Maria Jackson introduce to young botanists.

4. Devil in a Bush, aka *Nigella damascena*.

5. Bill Brown, "Thing Theory," 1–16.

6. Noltie, *Robert Wight*, 1: 46.

7. Noltie, *Robert Wight*, 1: 30.

8. Walter Benjamin, "G," in *The Arcades Project*, 181.

9. Levine, "The Criticism of Purpose."

10. Carr, *Sydney Parkinson*, 79.

11. Miller and Reill, *Visions of Empire*.

12. Blunt and Stearn, *The Art of Botanical Illustration*, 230.

13. Blunt and Stearn, *The Art of Botanical Illustration*, 220.

14. Secord, "Corresponding Interests," 383–403; Spary, *Utopia's Garden*; Noltie, *Robert Wight*, vol. 1; the exchanges between Collinson and Bartram appear in Bartram, *The Correspondence of John Bartram*, and Collinson, *"Forget not Mee and My Garden"*; Tilkin, *Le Groupe de Coppet*.

15. Uglow, *The Lunar Men*, 268–270; Mackay, "Agents of Empire: The Banksian Collectors," 38–57.

16. Gascoigne, *Joseph Banks*, 77–78, 88; Raistrick, *Quakers in Science and Industry*, 243–275; Watts, *The Dissenters*, 2: 27–29, 344.

17. A copy of the 1731 edition of Miller's *Gardeners Dictionary* that once belonged to Thomas Pennant includes a 1766 letter from James Lee to Pennant listing fruit trees available for purchase at the Lee and Kennedy Hammersmith nursery near London. This copy belongs to the Harry Ransom Humanities Research Center at the University of Texas at Austin.

18. Shteir, *Cultivating Women, Cultivating Science*, 52.

19. Fothergill, *Some anecdotes of the late Peter Collinson*.

20. Sowerby and Smith, *English Botany*.

21. Quaker circumspection apparently extended to other matters. When Banks returned from the voyage of the *Endeavour*, he jilted a lady who was under James Lee's protection to marry an heiress. Lee thereafter declined to be named among Banks's friends. The rupture and the reasons for it were reported not by Lee himself but by Dr. Robert Thornton, in a self-serving "Memoirs of the late James Lee" that Thornton printed at the beginning of a posthumous edition of Lee's *An Introduction to Botany*.

22. Fothergill, *An Account of life and travels*, 218–219.

23. Collinson, "Collinson's Commonplace Book," ed. Massey, Item 122–25, 6 Mar. 1752, MSS 323b, Linnean Society, London.

24. Sowerby, "Notes and Correspondence about Fungi," MSS SOW, BMNH. The taxonomy speculation is noted on the back of a drawing dated 1805; Sowerby's dedication to Davies is dated 1809.

25. Shelley, *Frankenstein*, 86.

26. Daston and Park, *Wonder and the Order of Nature*, 9–10, 303–316; Benedict, *Curiosity*, 3, 117, 201, 244.

27. Bartram, *The Travels of William Bartram*, 216.

28. Slaughter, *The Natures of John and William Bartram*, 172–174.

29. *Encyclopaedia Britannica*, 3 vols. (1771), 1: 627–653; 10 vols. (1778), 2: 1283–1315, plus an extended table of generic plant names, 18 vols. (1797), 3: 417–73, "s.v. "Botany"; *Annual Register* 2 (1759): 372–376.

30. Roper, *Reviewing before the Edinburgh*, 27.

31. Keats, "La Belle Dame sans Merci," 246–247; for a transcription of this version and analysis of its "Caviare" signature, see Kelley, "Poetics and the Politics of Reception," 333–362.

32. Gilmartin, *Writing against Revolution*, 98, and Roper, *Reviewing before the Edinburgh*, 12, 20, 21, and 175.

33. *Gentleman's Magazine* 40 (1770): 612–613; 45 (1775): 564–565, "Progress of Botany in England"; 53, pt. 1 (1783): 131–133; 54, pt. 1 (1784): 20–21, 256–257; 54, pt. 2 (1784): 486, 603, 657, 659, 660–664, 729–730, 851–852, 970–972; 57. Subsequent highlights cited below.

34. *Gentleman's Magazine* 54, pt. 2 (1784): 851–852: "Religious Uses of Botany."

35. *Gentleman's Magazine* 62, pt. 1 (1792): 305–306 and 438.

36. *Gentleman's Magazine*, supplement (1784): 970–972; 57, pt. 1 (1787): 14–15, 44–45, 110–115, 204–205, 297, 471–472; 57, pt. 2 (1787): 666–670, 880, 885–886, 996–999; 58, pt. 1 (1788): 19–21, 495–496, 760–772; 58, pt. 2 (1788): 975; 59, pt. 1 (1789): 14; 127–128; 59, pt. 2 (1789): 685; 59, pt.2 (1789): 827, 872; 60, no. 2 (1790): 791; 61, pt. 1 (1791): 25–26, 202–204; 61, pt. 2 (1791), 113, 305–306, 438; 62, pt. 2 (1791): 912–913, 1102–1103; 63 (1793), pt.1 304, 399, 493; 64, pt. 1 (1794), 180, 326–327; 65, pt. 2 (1795), 1051.

37. *Gentleman's Magazine*, 66, pt. 1 (1796), 9–10; 66, pt. 2 (1796), 561, 730, 739, 832, 918, 996, 998, 999, 1000, 1001, 1017; 67, pt. 1 (1797), 18, 94, 95, 102, 215, 304, 306–307, 400, 488; 67, pt. 2 (1797), 542; 68, pt. 1 (1798), 234; 68, pt. 2 (1798), 597, 1010, 1060; 69, pt. 1 (1799), 96; 69, pt. 2 (1799), 573, 600, 635, 656, 832, 964.

38. *Monthly Review* 2–4 (1790), 256; 5–7 (1791), 210, 361; 8–9 (1792), 165–166, 325; 10–11 (1793), 182, 290–291, 289; 12–16 (1794), 448–450; 17 (1795), 226–227; 21 (1796), 348; 26 (1797), 450; 27 (1798), 468–469; 33 (1800), 220, 380–382; 36 (1801), 201–202, 351–354, 533–535; 38 (1801), 310; 39 (1802), 211–212, 269–276; 41 (1803), 30–32, 208–211; 42 (1803), 374–378, 437, 511–514; 43 (1804), 113–127 (review of Erasmus Darwin's *Temple of Nature*), 324, 481–491, 44 (1804), 439, 464–468; 45 (1804), 282–288, 490–500.

39. "Account of the New Botanic Institution in Ireland," 208–210.

40. Saunders, *Picturing Plants*; Blunt and Stearn, *The Art of Botanical Illustration*, 194–260.

41. Thornton, *Temple of Flora*.

42. Bermingham, *Learning to Draw*, 208–224.

43. Noltie, *Indian Botanical Drawings*, 13–14.

44. Linnaeus, *Philosophia Botanica*, appendix; Nickelson, *Draughstmen, Botanists, and Nature*, 32–34.

45. Secord, "Botany on a Plate," 28–57.

46. Hill, preface to *The Vegetable System*, n,p., quoted by Nickelson, *Draughtsmen, Botanists and Nature*, 205.

47. Kopytoff, "The Cultural Biography of Things," 64–69.

48. Nickelsen, *Draughtsmen, Botanists and Nature*, 188–215.

49. Atran, *Cognitive Foundations*, 87–98; Daston and Galison, *Objectivity*, 69–82; Nickelsen, *Draughtmen, Botanists and Nature*, 90–93; Sivarajan and Robson, *Introduction to the Principles of Plant Taxonomy*, 205; Noltie, correspondence, 2009.

50. Nickelsen, *Draughtsmen, Botanists, and Nature*, 4–10, 273–276.

51. Smith to Sowerby letters, in Sowerby, Correspondence, Boxes 19 and 20, A56 through A59, written between 1799 and 1810, BMNH, London.

52. Trattinnick, *Archiv der Gewächskunde*, xv, quoted in Nickelsen, *Draughtsmen, Botanists, and Nature*, 176–177.

53. Nickelsen, *Draughtsmen, Botanists, and Nature*, 162–167.

54. Harley, *Artists' Pigments*, 42.

55. Kuehni and Schwartz, *Color Ordered*, 59–84.

56. Robert Brown, "An account of a new Genus of Plants," 1: 367–98 and 399–431—reprints of its first publication in *Transactions of the Linnean Society*: "An Account of a new Genus," 13 (Apr. 1821): 201–34, and "On the Female Flower," 19 (1844): 221–247.

57. Mjoberg's remark is reported on a *Rafflesia* website: http://homepages.wmich.edu/~tbarkman/rafflesia/Rafflesia.html.

58. For Merian's career and art, see Blunt and Stearn, *The Art of Botanical Illustration*, 142–146; Davis suggests the ecological understanding of Merian's paintings in *Women on the Margins*, 140–202.

59. Thornton, *Temple of Flora*, "Dragon Arum," plate 16; other plates discussed here are "The American Cowslip," plate 22, and "Curious American Bog Plants," plate 21.

60. Blunt and Stearn, *The Art of Botanical Illustration*, 236–43; Handasyde Buchanan, bibliographic notes in Thornton, *The Temple of Flora*, 13–20; Klonk, *Science and the Perception of Nature*, 38–65; Doherty, "Robert Thornton's A *New Illustration*, 49–82.

61. All of Bartram's drawings are presented for the first time in Magee, *The Art and Science of William Bartram*.

62. Charlotte Smith to Sir J. E. Smith, 15 Mar. 1798, in Charlotte Smith, *Collected Letters*, 314.

63. Erasmus Darwin, part 2 of *The Botanic Garden*, 54: "The Loves of the Plants," This part, "The Loves of the Plants," was published separately in 1789. It appears as part 2 of the third edition, published in 1791 as *The Botanic Garden; a Poem, in Two Parts*. Abbreviated in subsequent citations as *BG* 2, followed by page number.

64. Porter, "Scientific Analogy and Literary Taxonomy," 213–221; Page, "The Darwin before Darwin," 146–169; Packham, "The Science and Poetry of Animation," 191–208; Fulford, "Coleridge, Darwin, Linnaeus," 124–130.

65. Bewell, "Erasmus Darwin's Cosmopolitan Nature," 3 and 33.

66. Darwin, *BG* 2, 89.

67. Bewell, "'Jacobin Plants,'"132–139.

68. King-Hele, *Erasmus Darwin*, 124; Uglow, *Lunar Men*, 381–383.

69. Erasmus Darwin, *Zoonomia*, 2: 408, quoted by Uglow, *Lunar Men*, 482.

70. Darwin, *BG* 1, vii.

71. Uglow, *Lunar Men*, 270.

72. King-Hele (*Erasmus Darwin*, 316–318) argues that Darwin is the implied target in Wordsworth's 1798 advertisement to *Lyrical Ballads* that singles out "gaudy and inane phraseology" among the contemporary ailments of poetry. In Darwin's poem, the descriptive phrase "gaudy band" turns up twice. The famous verse parody of Darwin's poem is [George Canning and J. H. Frere] "Loves of the Triangles," 108–129, 134–141; the poem was also caricatured in Gillray's "New Morality"; for another critique, see Thomas Brown, *Observations on the Zoonomia of Erasmus Darwin*.

73. Gould, "Four Metaphors in Three Generations," in Gould, *Dinosaur in a Haystack*, 445–452.

74. Uglow, *Lunar Men*, 272–275.

75. Lindley, *An Introductory Lecture*, 4, quoted by Stevens, *The Development of Biological Systematics*, 107.

76. Darwin, *BG*, Additional Notes, 94–112; Omitted Notes, 124–128.

77. I wonder whether the fickle, unctuously pious Collins in *Pride and Prejudice* might be Jane Austen's pointed reply to Darwin's complaint about adulterous females in the plant *Collinsonia*.

78. Erasmus Darwin, *Phytologia*, 564.

79. Uglow, *Lunar Men*, 152.

80. Gould, "Ordering Nature by Budding and Full Breasted Sexuality," in Gould, *Dinosaur in a Haystack*, 427–431. Janet Browne offers a taxonomy of Darwin's women figures in "Botany for Gentlemen," 593–621.

81. *Ficus Indica* (called "Fica" in Darwin's poem to mark a shift in gender) also develops via an extensive root system, although one created when branches root where they touch the ground. Identified by Darwin as belonging to the class polygamy, Fica apparently puts in suspension the amorous inclinations of the female and male flowers in plants and trees of the class, refusing to listen to the "amorous sighs" of her Suitor-train (*BG* 2, 99).

82. Erasmus Darwin, *System of Vegetables*, 1: 11.

83. Erasmus Darwin, *System of Vegetables*, 1: 9; Goethe, *Versuch die Metamorphose der Pflanzen zu erklären*.

84. Stevens reproduces nonlinear diagrams that attempt to graph systematic relationships in *The Development of Biological Systematics*, 185–198.

CHAPTER 4: BOTANIZING WOMEN

1. David Richards discusses Le Moyne's work and its dissemination as an engraving in *Masks of Difference*, 80–82; Hulton's catalogue raisonné, *Jacques Le Moyne de Morgues*, reproduces this watercolor as plate 7 as well as other drawings of flowers and fruits. The Victorian language of flowers consolidates a somewhat more scattered sentimental tradition that includes, among others, Génlis, *La Botanique historique et littéraire*, and Phillips, *Flora Historica*. The copy of Phillips's book owned by the University of Wisconsin–Madison is extra-illustrated with 145 watercolor drawings from nature by Mrs. W. Hall, probably an amateur artist.

2. With thanks to Deidre Lynch for noting what Pamela risks by staying to embroider flowers on Mr. B.'s waistcoat.

3. Hulton, *Jacques Le Moyne*, 1: 164.

4. Chaudhuri, *The Trading World of Asia*, 547-548, and Marshall, *The Oxford History of the British Empire*, 487.

5. Richards, *Masks of Difference*, 79–80.

6. Grandville and Geoffroy, *Les Fleurs animées*, 2: 55–56. My translation. The French text:

> Je ne puis traverser un marché aux fleurs sans me sentie saisi d'une amère tristesse. Il me semble que je suis dans un bazar d'esclaves, à Constantinople ou au Caire. Les esclaves sont les fleurs.

Voilà les riches qui viennent les marchander; ils les regardent, ils les touchent, ils examinent si elles sont dans des conditions suffisantes de jeunesse, de santé et de beauté. Le marché est conclu. Suis ton maître, pauvre fleur, sers à ses plaisirs, orne son sérail, tu auras une belle robe de porcelaine, un joli manteau de mousse, tu habiteras un appartement somptueux; mais, adieu le soleil, la brise et la liberté: tu es esclave!

7. Thornton, *Temple of Flora*, plate 12, "Mimosa Grandiflora, Large-Flowering Sensitive Plant," with an extract from Erasmus Darwin's *The Botanic Garden* that begins, "Chaste Mimosa."

8. My translation of Grandville and Geoffroy, *Les Fleurs animées*, 1: 11–12, quoted in Bermingham, *Learning to Draw*, 209. The French text:

[Narrateur] La pauvre Sensitive n'était pas faite pour le monde; je m'en suis trop tôt aperçu.

[Sensitive] A peine eus-je revêtu le costume de femme, que ma sensibilité me causa des tourments affreux. J'en parle pas de l'amour, ma pudeur devait me défendre.

Je souffrais par bien d'autres motifs! Au théâtre, la musique me faisait tomber en pamoison [sic]; les emotions du drame me jetaient en des évanouissements prolongés; le moindre changement de température agissait sur mes nerfs.

9. Saint Aubin's *Livre de caricatures* circulated anonymously. As an artist and botanist, his drawings of plants not surprisingly include some he invented. See Jones and Richardson, "How Not to Laugh in the French Enlightenment."

10. Moore, *Lalla Rookh*, 116.

11. Bourdieu, *The Logic of Practice*, 52–66.

12. Bermingham, *Learning to Draw*, 209 and 192–194.

13. Shteir, *Cultivating Women, Cultivating Science*, and Secord, "Science in the Pub," 269–315.

14. Grosz, *The Nick of Time*, 17, citing Deleuze, *Difference and Repetition*, 248.

15. English translations of Rousseau's writings on botany, including the *Lettres élémentaires*, are included in *The Collected Writings of Rousseau*, 8: 91–257. The editors' introduction describes Rousseau's network of botanical correspondents in England and on the Continent. See Kuhn's "'A Chain of Marvels,'" 1–20, and Scott, "Rousseau and Flowers," 73–86; Cheyron, "L'Amour de la botanique," 53–95, and idem, "Ray et Sauvages annotés par Jean-Jacques Rousseau," 82–99.

16. Christopher Kelly and Alexandra Cook discuss the reception of Rousseau's *Lettres élémentaires* in their introduction to *Collected Writings of Rousseau*, 8: xiii and 273.

17. In *Cultivating Women, Cultivating Science*, Shteir discusses women who studied or wrote about botany or were botanical artists and writers who used the epistolary format, among them Wakefield's *An Introduction to Botany*, 79–89. Guest considers Wakefield among other women writers in *Small Change*, 319. In volume 8 of *Collected Writings of Rousseau* (273n.), Kelly and Cook add Almira Phelps's *Familiar Letters on Botany* to the list.

18. Bewell, "'Jacobin Plants,'"132–139; Shteir, *Cultivating Women, Cultivating Science*, 35–57; Browne, "Botany for Gentlemen," 593–621; Schiebinger and Gronim, "What Jane Knew," 33–59; King, *Bloom*; Ruwe, "Charlotte Smith's Sublime," 117–132; Pascoe, "'Unsex'd Females,'" 211–226, and "Female Botanists," 193–209; Cook, "'Perfect' Flowers," 259–279.

19. Bewell, "'On the Banks of the South Sea,'"173–196.

20. Shteir, *Cultivating Women, Cultivating Science*, 120–135. Gina Douglas, former librarian of the Linnean Society, explained in conversation that Ibbetson's microscopes appear to have introduced distortions that carry over into her drawings and notes.

21. Shteir, "Gender and 'Modern' Botany in Victorian England," 37.

22. Carlson, *England's First Family of Writers*, 50–51.

23. Godwin, *Fleetwood*, 198–200.

24. Godwin, *Fleetwood*, 200–201.

25. Uglow, *Lunar Men*, 271.

26. Uglow, *Lunar Men*, 70–75 and 353, quoting Lindsey's introduction to *Autobiography of Joseph Priestley*, 86–87.

27. Quoted by Uglow, *Lunar Men*, 354, whose manuscript source is H. Moyes to WW, MS 534.79 Addition, 30 Apr. 1783, Royal Society of Medicine Manuscripts, London. A note in *Letters of Anna Seward* characterizes Mary Knowles as "the celebrated Quaker lady who worked the King's picture so admirably in worsted." Seward's frequent letters to Knowles say nothing of her needlework but a great deal about her wit and judgment. Seward, *Letters*, 1: 47.

28. Guest considers Barbauld's political reputation and her argument with Wollstonecraft in *Small Change*, 222–227; Janowitz, *Women Romantic Poets*, notes Mary Priestley's role in the Warrington circle (19, 48–49); Keach, "A Regency Prophecy," 569–577.

29. Barbauld, "To a Lady, with some painted flowers," in *The Poems of Anna Letitia Barbauld*, 77; hereafter abbreviated as Barbauld, *Poems*; quoted by Wollstonecraft, *A Vindication of the Rights of Woman*, 168; hereafter cited as *VRW*.

30. Wollstonecraft, *VRW*, 324.

31. Barbauld, *Poems*, 267n.

32. Rousseau, *Lettres élementaires*, 133.

33. Armstrong reads Barbauld's "Inscription for an Ice-House" as a riddle about women and power: "The Gush of the Feminine," 13–32.

34. Herbert McLachlan, *Warrington Academy*, 22–31.

35. Barbauld, *Poems*, 121–122.

36. Lynch writes about hothouse cultivation and the horticultural overtone series it produces among romantic writers in "'Young ladies are delicate plants,'" 689–729.

37. Paine, *Rights of Man*, 2: x.

38. Shteir traces some of women's botanical careers during this era, including that of Wakefield and, more briefly, Jackson, in *Cultivating Women, Cultivating Science*, 81–145.

39. Wakefield, *A Brief Memoir of the Life of William Penn*, 3–8.

40. Withering, *An Arrangement of British Plants*.

41. The second edition (1792) of Withering's *Arrangement* is anonymously reviewed in *Monthly Review* 11 (May–Aug., 1793): 284–290; Uglow, *Lunar Men*, 381.

42. Maria Jackson published anonymously, but the title pages of her books identify their shared authorshop by same "Lady": *Botanical Dialogues, Botanical Lectures*, in which Jackson repurposed material from *Botanical Dialogues, Physiology of Vegetable Life*, and *Florist's Manual*.

43. Shteir, *Cultivating Women, Cultivating Science*, 110.

44. White, *English Romanticism and Religious Dissent*, 69–73 and 114–116. Wollstonecraft was, and Jackson would soon become, one of the writers published by the radical publisher Joseph Johnson.

45. Sachs, *History of Botany*, 253.

46. Jackson, *Physiology of Vegetable Life*, 84, referring to Sir J. E. Smith's *Introduction to Physiological and Systematical Botany*.

47. Richardson, *Literature, Education, Romanticism*, 113.

48. Charlotte Brontë, *Jane Eyre*, 8–9.

49. Birmingham, *Learning to Draw*, 204–215.

50. De Certeau, *The Practice of Everyday Life*, 41.

51. A summary of the Waller household that Frances Beaufort managed appears in an online archive maintained by Trent University, http://trentu.ca/admin/library/archives/94–006.htm.

52. Quoted by Uglow, *Lunar Men*, 481.

53. Reeder, "The 'Paper Mosaick' Practice of Mrs. Delany," 228.

54. Moore, *Sister Arts*, 23–43.

55. Hayden, *Mrs Delany*, 188. Edmundson, who notes that the invention of paper mosaics was unprecedented even in an era of rampant botanical illustration, identifies the sources of Delany's plant names and discusses her botanical interests and contacts in "Novelty in Nomenclature," 188–203.

56. Edmundson, "Novelty in Nomenclature," 195.

57. Reeder, "The 'Paper Mosaick,'" 228.

58. Katherine Charteris Grey, who married the heir of Earl Grey of Groby, may have seen Delany's mosaics, inasmuch as earlier family members among the Greys of Groby were related by marriage to Delany's patron, the Duchess of Portland, through her late husband. Edmundson notes this familial connection in "Novelty in Nomenclature," 201.

59. See, for example, PlantNet: http://plantnet.rbgsyd.nsw.gov.au/cgi-bin/NSWfl.pl?page=nswfl&lvl=sp&name=Pterostylis~obtusa.

60. *Erica pubescens* at the Linnean Society: www.linnean-online.org/5506.

61. Charlotte Smith, *The Young Philosopher*, 114, 119, 128, 174–175.

62. Smith, *Poems*, 198. Sowerby's *Index* in Sowerby and Smith, *English Botany*, includes an extensive list of *Erica* species.

63. Smith, *Poems*, 208n. See also entries in the 1817 cumulative index for early volumes of Curtis, *Botanical Magazine*.

64. Smith refers principally to Withering's *Arrangement*, William Mason's *English Garden*, William Woodville's *Medical Botany*, Sowerby and Smith's *English Botany*, Gilbert White's *Natural History of Selborne*, Thomas Martyn's adaptation of Philip Miller's *The Gardeners Dictionary*, and Erasmus Darwin's *The Economy of Vegetation*; Smith, *Poems*, 52, 54, 66, 107, 194, 210, 232, 237, 276, and 295–298.

65. See also Crawford, "Lyrical Strategies," 199–222. Relationships among botanists, dissenters, and women during the romantic period were probably deeper and broader than even Polwhele was prepared to acknowledge. Women artists and a few explorers belonged to Quaker networks of nurserymen and shipboard naturalist-artists; working-class male botanists in Lancashire met in pubs; and there existed at least one network for the amiable exchange of botanical information between dissenters and nondissenters at Warrington Academy, where the Barbaulds, Priestleys, and others lived or congregated in the 1760s. Among the early tutors at the academy was Jean-Paul Marat. See Herbert McLachan, *Warrington Academy*, 72–84; Secord, "Science in the Pub."

66. See the dated examples in the *OED*; Atran, *Cognitive Foundations*, 205.

67. Charlotte Smith, *Minor Morals*, 1: 20.

68. Clare, *The Natural History Prose Writings*, 20; hereafter cited as *NHPW*.

69. William Cowper, *The Task* 6: 165–167, quoted by Smith, *Poems*, 283n.

70. Linnaeus, *Species Plantarum*, Pl. 1370, and Smith's commentary in Sowerby and Smith, *English Botany*, 20: 1298.

71. Smith, *Poems*, 236n.

72. Aikin, *An Essay on the Application of Natural History to Poetry*, 33.

73. Rei Terada uses this phrase to describe the uneasy response of philosophers and other writers to Kant. See Terada, *Looking Away*, 15–34.

CHAPTER 5: CLARE'S COMMONABLE PLANTS

1. *John Clare by Himself*, 38; hereafter *JC*. All quoted extracts preserve Clare's non-standard orthography, which reflects his dialect speech and variable orthography then in use.

2. Richard Mabey, "Pastoral Suite," *Guardian Unlimited*, Saturday, July 29, 2006. www.guardian.co.uk/books/2006/jul/29/featuresreviews.guardianreview.

3. Clare, *The Shepherd's Calendar*, 40; hereafter *SC*.

4. Heaney, "John Clare's 'Prog,'" 63–82. Clare varies this dialect word, too: *proggles* as well as *prod*, meaning something between a nudge, a prodding, and a poke.

5. Williams, *Keywords*, 70–72. Sorenson assesses shifting eighteenth-century notice of vulgar, cant, or dialect language in "Vulgar Tongues," 435–454.

6. Ray, *Dictionariolum Trilinguae*. ed. Stearn, 6–14.

7. Clare, *Poems of the Middle Period*, vol. 1: 254–75; hereafter *MP* 1. In an 1825 journal entry, Clare calls this the "break day when the Fen commons used to be broke . . . by turning in the stock": *NHPW*, 236.

8. Mabey, *Flora Britannica*, 174–175.

9. Mabey, *Nature Cure*, 75–83, 93–103.

10. Clare uses the standard spelling in commenting on Elizabeth Kent's description of this flower in *Flora Domestica*: *NHPW*, 17. Except for poems published in *The Midsummer Cushion*, where some errant spellings of this flower name are regularized to "cowslip," Clare mostly uses other spellings: the variant "cows lips" is suppressed in the published version of The Wild Bull," in *Midsummer Cushion Poems of the Middle Period*, vol. 3:523n; hereafter, this volume cited as *MP* 3. In Clare, MS A37 is "cowslap," and in Pforzheimer MS 196, "cows lap", in "The Eternity of Nature." Also in *Middle Cushion*, "Cows laps": Clare *MP* 3:528, l. 27; and see Clare's poem "To the Cows Lip (in ms. "To the Cows lip"), in Clare, *MP* 1:323ff.

11. Miller, "Enclosure and Taxonomy in John Clare," 635–657.

12. Edgerton MSS 2247, F 303/4, 25 June 1827; Edgerton MSS 2248, F 245/6, 9 July 1830, British Library, London. Clare's friendships with others interested in natural history began early and locally with Ned Simpson, Tom Porter, John Billings, the bakers, and the village doctress. George and Anne Baker consulted Clare about plant names they included in their *Glossary of Northamptonshire Words and Phrases*, (see 1: xv). Early in the 1820s, Clare met Edmund Tyrell Artis and Joseph Henderson, employees of the Fitzwilliams. Both men were formally educated and later published scientific papers—Artis on archeological subjects that he began to explore in the vicinity of Helpston, sometimes with Clare's help; Henderson on ferns, a topic that he investi-

gated on botanical excursions with Clare. Henderson read and revised Clare's poetry and showed him new acquisitions in the library at Milton Hall, which had an extensive collection in natural history, and Lord Radstock continued to add important works. Clare reciprocated, loaning Henderson magazines and giving him plants. For detailed histories of Clare's friendship with Artis and Henderson, see Vardy, *John Clare: Politics and Poetry*, 130, 146, 150–158, Heyes, "'Looking to Futurity,'" and manuscript letters between Clare and these friends, now in the British Library and identified in these notes.

13. Kain, Chapman, and Oliver, *The Enclosure Maps of England and Wales*, 102.

14. Wordsworth, "The Thorn," in *Lyrical Ballads*, 350–53; Wordsworth, *The Fenwick Notes*, 19.

15. Deleuze, *Difference and Repetition*, 208.

16. In *John Clare: Politics and Poetry* (119–120, 59–76, and 93–102), Vardy discusses Clare's penchant for literary forgery, which riffs on a widespread romantic interest in impersonation, as one aspect of Clare's highly articulated understanding of sociability and friendship.

17. Clare, *JC*, 257–258.

18. Morton, *Ecology without Nature*, 38. My use of the phrase *ambient poetics* argues against Morton that Clare's sense of "being there" is more mobile and more insistently figurative than readers have often supposed.

19. Simpson, *Situatedness*. Simpson considers the philosophical tradition that concerns biography and private life, from Descartes forward (146–91), without noting Clare's focused poetics of site and situation.

20. Goodman, "Romantic Poetry and the Science of Nostalgia," 195–216.

21. Clare, *JC*, 41–42.

22. Jameson, *Postmodernism*, 8, 34, and 119; and idem, *A Singular Modernity*, 40–57; Latour, *Reassembling the Social*, 74–75; and de Certeau, *The Practice of Everyday Life*, 39–42.

23. Vardy, *John Clare: Politics and Poetry*, 168, 172–181.

24. Ngai, *Ugly Feelings*, 43 and 226–227; Heidegger, *Being and Time*, 390, 221, and German text, *Sein und Zeit*, 340; Guignon, "Moods in Heidegger's *Being and Time*," 235; Morton, *Ecology without Nature*, 180.

25. Barrell, *The Idea of Landscape*, 98–109. Barrell provides sketches of Helpstone lands before and after enclosure, but does not reprints several enclosure and pre-enclosure maps, though not the 1819 enclosure map listed in Kain, Chapman, and Oliver, *Enclosure Maps*.

26. Mabey, *Nature Cure*, 75; quoting Gary Snyder, *The Practice of the Wild*, 40.

27. Bate gives what economic details are known about Clare's cottage life in "Night Light on the Life of Clare," 41–48; Clare, "A Sunday with Shepherds and Herdboys," *Poems of the Middle Period*, vol. 2:20; hereafter *MP* 2.

28. Vardy, *Clare's Poetry and Politics*, 167–182.

29. Clare, *Poems of the Middle Period*, vol. 4:112; hereafter *MP* 4.

30. Neeson, *Commoners*, 64. Neeson's analysis relies on evidence drawn from Northamptonshire tax rolls, enclosure maps, and other archival evidence; hereafter, *Commoners*.

31. Robinson, "John Clare: 'Searching for Blackberries,'" 208–212.

32. Chambers and Rees, *Cyclopaedia*, s.v. "*pig-tail*"; *OED*, s.v. "*pightle*"and "*assarts*"; Neeson, *Commoners*, 65.

33. Neeson, *Commoners*, 29, 67, 95, 106–7, 115, 142.
34. Loewenstein, "Digger Writing," 74–88.
35. Neeson, *Commoners*, 69, 188–194, 201. Scholars have suggested that the pro-enclosure rhetoric of the late eighteenth and early nineteenth centuries was tailored retrospectively so that enclosure came to look like a successful event in the Whig narrative of the English agricultural revolution of the mid-eighteenth century. Historians, including some who might be expected to question claims about the advantages that would accrue from enclosure (better agricultural yields over larger sections of land, higher rents, a larger wage-labor force) have suggested that common rights or commoners themselves were long gone or much diminished in number before enclosure. Others who acknowledge that enclosure displaced commoners and effaced common rights have minimized the economic as well as social value of commonability. Neeson reviews this debate in *Commoners* (see 6, 301). My summary of commonable right also draws on Turner, *English Parliamentary Enclosure*, 33–37, 72–75, 94–109; Mingay, *Parliamentary Enclosure in England*, 18–21; and Yelling, *Common Field and Enclosure in England*, 86–87, 152–153, 163, 190–191, and 214–215.
36. Clare, "The Mores," in *MP* 2:347–50.
37. Barrell, *The Idea of Landscape*, 106, 210–224n.
38. Barrell, *The Idea of Landscape*, 103.
39. Neeson, *Commoners*, 30, 9–15.
40. Clare, *The Early Poems of John Clare*, vol 1:161, ll. 122–123; vol. 1 hereafter *EP* 1.
41. Neeson, *Commoners*, 2–3.
42. Clare, "The Mores," in *MP* 2:348 and 22–27.
43. Neeson, *Commoners*, 32.
44. Neeson, *Commoners*, 184.
45. Nancy, *The Inoperable Community*, 27–37 and 73–78.
46. Mayall, *English Gypsies and State Policies*, 25; Yahav-Brown, "Gypsies, Nomadism, and the Limits of Realism," 1126; citing appendix 1 of Mayall, *Gypsy Travellers in Nineteenth-Century Society*. See also Simpson, *Wordsworth's Historical Imagination*, 44–46; Martin, "John Clare's Gypsies," 48–59; Janowitz, "The Transit of Gypsies in Romantic Period Poetry," 218. Langan explores the idealization of mobility in *Romantic Vagrancy*, 18–19.
47. "The Gipsey's Song," *Midsummer Cushion*, *MP* 4:55, ll. 85–88.
48. Goodridge and Thornton, "John Clare: The Trespasser," 106.
49. *John Clare*, *LP* 1: 29 and *LP* 2: 11–14.
50. Garrison, "'Disdaining Bounds of Place and Time,'" 376–87. Garrison quotes from the first of Alice Becker-Ho's recent books on the role of borrowings from Romani in modern European languages. See Becker-Ho, *Les Princes du jargon* and *L'Essence du jargon*.
51. Clare, *JC*, 334n. Clare's editors suggest that he consulted Pennant's *Genera of Birds*.
52. Gledhill, *The Names of Plants*, 133.
53. In *Politics and Poetry*, 132–134, Vardy rightly insists that Henderson was close to Clare as a botanist and as a reader of Clare's poems who offered him detailed suggestions and admiration. Heyes makes a similar point in "Writing Clare's Poems," 33–45. In a letter, Henderson advises Clare: "I wish you would study a little more of the systematic arrangement of plants, very little trouble, if you could bend your mind to it,

would enable you to become acquainted with the subject, do not be frightened at a few technical terms a knowledge of one will lead you insensibly to others & as you proceed your prejudices will vanish, & you will only have time to attend to and admire the beautiful chain which links the whole into system and order, it is not from the mere circumstance of being able to call a plant by a particular name that you will be enabled to derive the greatest pleasure, a knowledge of its habits, the wonderful provisions of nature for its protection and propagation, the beautiful construction of the various parts which compose it, all bearing a just proportion in the individuals, [?] endless variety in the system, but at the same time to enable you to trace each individual plant to its proper place, I have not time to say more on the subject for John is waiting at the door. . . . I will send you an Introduction to Botany which will put you in the way" (Edgerton MSS 2249, Dec. 25, 1825, F 255/6, BL, London).

Henderson's letters to Clare document the character of their long friendship and intellectual exchange. After Clare's triumphant visit to London in 1824, Henderson, addressing "My Dear Clare," playfully congratulated him: "If this note is fortunate enough to find you returned from your London literary Tour, & you should find yourself sufficiently recovered to participate in a more humble excursion, the favour of your company would add great satisfaction to the following party, viz. Mr Artis, Mr Richardson—Peterboro', Mr Almond & your humble servant; on a sort of Plant-hunting, Butterfly-catching expedition to Whittlesea-mere, we will expect you ("if it please God & the weather") about nine o'clock tomorrow morning."

In the late 1820s and 1830s, as Clare's health and spirits fluctuate, Henderson wrote frequently, asking Clare how he is, why he hasn't written, sending him plants. In May 1835, Henderson wrote, "I am truly rejoiced to find that there is a little improvement in your health and that you are able to use the pen again, And I do hope that the favourable and well deserved reception which your poems have met with will cheer your spirits and make John Clare himself again" (Edgerton MSS 2249, F 297/8, BL, London).

54. Clare, *NHPW*, 272–273. Margaret Grainger, the *NHPW* editor, notes that Clare makes a similar comment in an 1824 journal entry. Charles Darwin's monograph *The Movements and Habits of Climbing Plants* was first published in 1865.

55. Although Elizabeth Kent and Clare differed on the value of Linnaean names, her letters to him emphasize their common interest in differentiation among species of birds and plants not sufficiently recognized in published works. I am indebted to Alan Vardy for his transcriptions of Kent's letters in Edgerton MSS 2247, BL, London.

56. Clare lists the other columbines in *NHPW*, 18.

57. Vardy, *Poetry and Politics*, 154.

58. Joseph Henderson to Clare, 10 Apr. 1827, BL Edgerton MSS 2247, F 276–277.

59. Clare's and Henderson's handlist of English orchises is MS 45 of those owned by the Peterborough Museum and Gallery. Although Grainger's transcription of this handlist is otherwise accurate, it does not identify Clare's hand in the addition of this Latin name in the left margin.

60. Summerhayes, *Wild Orchids of Britain*, 112; following other lexicons and her knowledge of local plants, Grainger identifies Clare's "cuckoo bud" as O. mascula in her editorial notes for *NHPW*, 18n. Although Anne Baker identifies Clare as one of her informants, her coauthored *Glossary of Northamptonshire Words and Phrases* declares (following Elizabeth Kent instead of Clare) that his cuckoo-buds is a crowfoot (also a plant), and uses lines from one of Clare's poems to make this case. Druce's *The Flora*

of Northamptonshire lists (by Latin name) the plants Clare mentions, illustrating that list with copious quotations from Clare's poetry. Druce identifies three orchises by their Latin names, although the passages from Clare call them all cuckoos, further specifying cuckoo-bud for one, which Druce identifies as *O. mascula* (cxii).

61. "Natural Mimicry," Henry E. Huntington Botanical Garden Exhibition. Kitty Connolly on the Huntington botanical staff confirmed these plant names and traits.

62. Grainger identifies "lords and ladies" in her edition of Clare, *NHPW* as the plant that Clare calls aron; so does the consolidated glossary provided in *MP* at 5: 646; and so does Mabey in *Flora Britannica*, 385–386. Mabey also notes the etymologies and common synonyms for this arum.

63. Mabey, *Nature Cure*, 188.

64. Clare, "To The Snipe," *MP* 4:574, ll. 10–16.

65. Heaney, "John Clare's 'Prog,'" 63–82.

66. Simpson, "Speaking Place," 131–134.

67. William Carlos Williams, *Paterson*, 9.

INTERLUDE 1: MALA'S GARDEN

1. Mootoo, *Cereus Blooms at Night*, 127–128.

2. Raynal, *L'Histoire des deux Indes*.

3. Casid, *Sowing Empire*, xvii–xxi. Casid argues that the itinerary of this exotic plant through the lives of the characters signals the presence of hybridizing botanical activity in Caribbean history and sites where one might otherwise see only the colonial imposition of plants and governance.

4. Hong, "Colonialism, Transnationalism, and Sexuality," 73–103.

5. Mootoo, *Cereus Blooms at Night*, 126.

6. Recent scholarship on botany and empire includes Kriz, "Curiosities, Commodities, and Transplanted Bodies," 85–105; Broadway, *Science and Colonial Expansion*; Grove, *Green Imperialism*; Drayton, *Nature's Government*; essays collected in Miller and Reill, *Visions of Empire*; Fara, *Sex, Botany, and Empire*; and Tobin, *Colonizing Nature*.

7. Casid, *Sowing Empire*, 190–236. For further discussion of the role of local informants and plant names, see Ruiz, *The Journals of Hipólito Ruiz*, 18–19; Chabran and Varey, "Entr'Acte," 106–107; Engstrand, "Mexico's Pioneer Naturalist," 17–32; Schiebinger and Swan, *Colonial Botany*; Schiebinger, *Plants and Empire*, 73–103; and Kriz, "Curiosities, Commodities, and Transplanted Bodies," 85–105.

8. Browne, *The Civil and Natural History of Jamaica*, 113; Descourtilz, *Voyages d'un Naturaliste*, 1: 40, and plate on opposite page, 2: 19. I am indebted to Deborah Jenson for directing me to Descourtilz's work and the Taino engravings. The debate about Alexander von Humboldt's coloniality—was he anti or pro?—remains unresolved. Pratt argues in *Imperial Eyes* that Humboldt looked at South America with a colonialist eye for profit and profited from the work of local naturalists: 117, 126–130. Aaron Sachs defends Humboldt in "The Ultimate 'Other'" 111–135.

CHAPTER 6: READING MATTER AND PAINT

1. Sir William Jones, "Design of a Treatise on the Plants of India," in *Asiatic Researches*, hereafter *AR*, 2:270.

2. In *Science and the Raj*, Kumar bluntly observes that "dominance by one meant subordination of another, and this non-binary equation existed at both the centre and circumference of colonialism" (36); Arnold, in *The Tropics and the Traveling Gaze*, adds that whatever might be called cooperation was orchestrated by the British and thus not genuinely cooperative (176–184). Chatterjee, in *Nationalist Thought and the Colonial World*, disagrees, suggesting that such readings of the colonial relationship are so locked into a binary of difference and subordination that they cannot see anything else (5–81). In a remark that Kumar quotes (*Science and the Raj*, 17), Kapil Raj suggests that "imperium and subaltern materials are not like grain and chaff to be winnowed." Others who argue for a more blended relation between the imperial British and Indians include Harris, "Long-Distance Corporations," 269–304; Agrawal, "Dismantling the Divide"; Raina, "Betwixt Jesuit Enlightenment Historiography"; Baber, "Colonizing Nature"; Hensen, "The Medical Skills of the Malabar Doctors"; and Chakrabarti, "Medical Marketplace beyond the West."

3. Hegel, *Aesthetics*, 1:343–344; idem, *Werke*, 13: 443–445.

4. Metcalf, *Ideologies of the Raj*, ix–x.

5. Grove, *Green Imperialism*, 14. In *Castes of Mind*, Dirks discusses the Western invention of castes in India as they are today understood, noting the European and then British effort to define caste narrowly, relying on Brahmanic accounts that also tended to present caste as transnational and metahistorical categories. Prior to the British colonial era, Dirks argues, caste was more unfixed, more socially and locally nuanced, and as such too various to be simply and manageably ordered. In the end, and despite some British objections, the official colonial view of Indian castes was indeed simplified and made universal and fixed rather than local and subject to change (5, 9–15, and 19–38). For supporting evidence, see Metcalfe, *Ideologies of the Raj*, 116–117, and Bayly, *Indian Society and the Making of the British Empire*, 155–159.

6. Chakrabarty, *Provincializing Europe*, 47–53.

7. Cohn, *Colonialism and Its Forms of Knowledge*, 57–75, and "The Recruitment and Training of British Civil Servants in India," 500–553.

8. Bayly, *Indian Society and the Making of the British Empire*, 37–39; Barnett, introduction to *Rethinking Early Modern India*, 11–32; de Almeida and Gilpin, *Indian Renaissance*, vii.

9. Grove, *Green Imperialism*, 372–76; Pratt reads Humboldt more critically in *Imperial Eyes*, 11–143.

10. Drayton, *Nature's Government*, xviii.

11. Majumdar, *Vanaspati*; Grove, *Green Imperialism*, 78–94.

12. Noltie, *Indian Botanical Drawings*; idem, *The Dapuri Drawings*; and idem, *Robert Wight and the Botanical Drawings of Rungiah & Govindoo*, 2: 13; Nair, "Native Collecting and Natural Knowledge," 284–288. Nair observes that Serfoji's network included company officials like the governors (John Huddleston and Lord Hobart), EIC surgeons (James Anderson, W. Ainslie, and W. S. Mitchell), residents (Benjamin Torin and William Blackburne, who later became an active council member of the Royal Asiatic Society) and other EIC officials (Colin MacKenzie of the Mysore Survey and the Madras collector and Tamil scholar Francis Whyte Ellis, and Tranquebar missionaries, who were also Banksian collectors and close to EIC naturalists Johann Gerhard Koenig, William Roxburgh, Anderson, Andrew Berry, Patrick Russell, Alexander Dalrymple, Andrew Ross, and Lady Clive.

13. Spivak, *A Critique of Postcolonial Reason*, 5–6, 238.

14. Derrida, *Archive Fever*, 3.

15. Suleri, *The Rhetoric of English India*, 3.

16. Foucault, *The Archaeology of Knowledge*, 126–131; Agamben, *Remnants of Auschwitz*, 143–145.

17. See Walkowitz, "Unimaginable Largeness," 216–239.

18. Cohn uses the first of these phrases ironically in *Colonialism and its Forms of Knowledge*, 134; Chow's caution appears in *Writing Diaspora*, 37.

19. In addition to unpublished materials in private collections and company records, the first publications on Indian natural history included Roxburgh's *Plants of the Coast of Coromandel*, idem, *Hortus Bengalensis*, the posthumously published idem, *Flora Indica*, and Wight's *Prodromus Florae Peninsulae Indiae*. Sir William Jones's *Asiatick Researches* regularly published essays on Indian natural history and other branches of knowledge that were widely quoted and reprinted in England. Modern scholarship about Indian natural history begins with Archer, *Natural History Drawings*, Desmond, *The European Discovery of the Indian Flora*, and idem, *Sir Joseph Dalton Hooker*.

20. Dalrymple, *White Mughals*, xliv–xlv. Bayly observes that in the decade 1770–1780, the "easy symbiosis between Europeans and Indians began to decline under the pressure of world war and commercial rivalry" (*Indian Society and the Making of the British Empire*, 69). Rocher also makes the case for a mix of political goals and orientalist knowledge gathering between 1771 and 1794 in "British Orientalism in the Eighteenth Century," 215–249.

21. Lelyveld, "The Fate of Hindustani," 195.

22. Bickerstaff, *The Sultan*.

23. Sabin, *Dissenters and Mavericks*, 23. Sabin notes that Walpole's "verbal play, surprising juxtapositions, and tonal disjunctions undercut justifications for British power in India a generation before the rhetorical models in defense of empire were fully set" (30).

24. Walpole, *Horace Walpole's Correspondence*, 22: 16, quoted by Sabin, *Dissenters and Mavericks*, 32. Sabin notes (29) that Walpole's self-presentation as the "consummate outsider" on parliamentary affairs masked his insider position as Robert Walpole's son.

25. "Letter Written by an Officer Who Lately Served in Bengal," *Gentleman's Magazine* (5 Mar. 1772), 69, quoted by Sabin, *Dissenters and Mavericks*, 31. Teltscher tracks discussion of India in British magazines in *India Inscribed*, 157–172.

26. Walpole, letter of 5 Mar. 1772, in *Correspondence*, 23: 387, quoted by Sabin, 40.

27. Walpole, *Correspondence*, 30 Apr. 1783, 25: 399–400, quoted by Sabin, 41.

28. Walpole, *Correspondence*, 30 Apr. 1786, 25: 642, quoted by Sabin, 42.

29. Mackenzie, *Man of Feeling*, 170; Cowper, *The Task*, 183, quoted by Rajan in *Under Western Eyes*, 83.

30. Moore, *Lalla Rookh*, 82, quoted in part by Rajan, *Under Western Eyes*, 232–333n.

31. Sharafuddin, *Islam and Romantic Orientalism*, 173.

32. Dow, *History of Hindostan*, 3: xxxv. Rajan assesses the impact of this and other British histories of India published between 1750 and 1820 in *Under Western Eyes*, 78–81. In *Castes of Mind* (20), Dirks notes that Dow's *History* is a translation of an earlier Indian text.

33. Goldsmith, *A History of the Earth, and Animated Nature*, 2: 225. Rajan, who

quotes both Dow and Goldsmith in *Under Western Eyes* (80), gives Goldsmith's work the more hopeful title *A History of Man*. Goldsmith's *History* tends rather to present humankind as curiosities of nature. Leask, in *British Romantic Writers and the East*, offers a virtual parade of rabidly anti-Indian writers that includes Southey and Beddoes (92–93). Rajan has his own parade, which includes this 1807 remark by General John Carnac: "The immense excavations cut out of the solid of Salsette, appear to me the operations of too great labour to have been executed by the hands of so feeble and effeminate a race as the aborigines of India have generally been held to be." Quoted by Rajan, *Under Western Eyes*, 95.

34. Shelley, *Philosophical View of Reform*, 7: 18.

35. Leask, *British Romantic Writers and the East*, 119–120 and 140.

36. Jones began publication of *Asiatic Researches* in Calcutta in 1784. The first issue included Hastings's congratulatory letter as the dedicatee and patron of the new publication. William Chambers and Jones, then others published successive editions of *The Asiatic Miscellany* (1785, 2 vols.; 1787, and 1818) and *The New Asiatick Miscellany* (1789). Along with romantic periodicals like the *Monthly Review*, which quoted large extracts of British writing on India, these publications quickly republished British writing about India for readers back in England. See Drew, *India and the Romantic Imagination* on Jones and the periodical press (43–82). As Rajan also notes in *Under Western Eyes* (85), Jones's writing created the taste and provided resources for the popular Oriental tales that followed, among them Beckford's *Vathek*, Southey's *The Curse of Kehama*, Thomas Moore's *Lalla Rookh*, Owenson's *The Missionary*, and Byron's *Lara* and *The Giaour*. Charles Robert Maturin's *Melmoth the Wanderer* stops in India.

37. Rajan, *Under Western Eyes*, 168.

38. Leask, *British Romantic Writers and the East*, 85 and 76. In a stunningly apt reading of the imperial gestures of Shelley's radical idealism, Leask notes (79) that Shelley, like other liberal reformers of his era, joined the project of reform at home with that of reform in the empire, a point that Patrick Brantlinger develops in *Rule of Darkness*, 27.

39. Mill, *History of British India* (1817) was first published in three volumes before British India existed as such. The 1820 edition, used here (9: 567), reached 9 vols. Leask and Javed Majeed each observe that Mill's rationale for British colonial rule was at once liberal and utilitarian, imagining a course of commercial and political improvement for India as an extension of progress at home. Majeed, *Ungoverned Imaginings*, 126–150. Leask, in *British Romantic Writers and the East* (87–88), is acerbic about the political doubleness of Scottish involvement in colonial India, from those Scottish advocates of liberty who became colonial profiteers working for Indian rulers (James Macpherson for the nabob of Arcot) to Mill's citation of Dr. Johnson on the fecklessness and superstition of the Scottish highlander as an analogue for the Hindu character and Mill's further insistence that Sir William Jones's orientalism aligns him with the despotism of the ancien régime.

40. Macaulay, *Thomas Babington Macaulay, Selected Writings*, 242–243. Young, editor of *Speeches by Lord Macaulay*, notes (351) the curious history of this essay, which appears in an extremely rare copy of a Calcutta edition, originally published in 1962 and found only in the Cambridge University Library; Rajan quotes the second Macaulay remark in *Under Western Eyes*, 180, citing as its source Clive, *Macaulay*, 372–373.

41. Raynal, *L'Histoire des deux Indes*, 1: 430–462. Although Raynal, like his British contemporaries, frequently describes the debilitating effects of Indian (especially

Hindu) religion on Indians, here he discusses the British control of the wealth of Bengal, the submissiveness of the Bengalese, British rapacity, and, just at the end, his hope that the British will grant the spirit of liberty to India, by which Raynal apparently means that the British will rule India liberally, not rapaciously. Sabin notes in *Dissenters and Mavericks* (37–38) that Arthur Wilson argues that Diderot silently co-authored Raynal's *L'Histoire des deux Indes*. Diderot is credited with having written at least one-third of the work. See Wilson, *Diderot*, 682–686.

42. Hegel repeats the argument that Indians have "achieved no foreign conquests, but have on every occasional been themselves vanquished": *Philosophy of History*, 140, quoted and partly retranslated by Rajan, *Under Western Eyes*, 100 and 103; and Hegel, *Werke*, 12: 178. The English translation used here is primarily Sibree's, as noted, but the last sentence is Rajan's more literal translation, *Under Western Eyes*, 103. See my chapter 7.

43. Kelley, *Reinventing Allegory*, 135–142, and Rajan, *Under Western Eyes*, 50–66, citing Milton, *Paradise Lost*, 9: 1105–06. Leask, *British Romantic Writers and the East* (105–107) reviews Schlegel's and others' admiration for India and Hegel's reaction. Mitter defends Indian art against Hegel's reading in *Much Maligned Monsters* (202–220).

44. Moore, *Lalla Rookh*, in *Poetical Works*, 55, 68, and 81. Here and in other quoted extracts in this chapter, I preserve original spellings of botanical names and Indian words. I argue for a more variable relation to Indian culture and flora in Moore and Jones than Fulford, Lee, and Kitson claim in "Indian Flowers and British Orientalism," 80–89.

45. Rajan, *Under Western Eyes*, 161–162.

46. Shelley, *Prometheus Unbound*, 2.4.63–65, in *Shelley's Poetry and Prose*, hereafter *SPP*, 248.

47. Kelley, "Reading Justice," 267–288.

48. Leask, *British Romantic Writers and the East*, 102.

49. Metcalf, *Ideologies of the Raj* (102) discusses the British inability to grasp the sexual/religious doubleness of what they believed to be "temple prostitution."

50. Wright makes this point in her introduction to Owenson, *The Missionary*, 36.

51. Jones, "Hymn to Lacshmi," in *Sir William Jones: Selected Poetical and Prose Works*, 153–167; hereafter *SPPW*.

52. Noltie's plant identification.

53. "Moore's Lalla Rookh," 562, quoted by Majeed, *Ungoverned Imaginings*, 102n.

54. Shelley, *Alastor*, in *SPP*, 157 and 453, pp. 78 and 84.

55. See Wright's introduction to Owenson, *The Missionary* for a discussion of Owenson's changes in the novel (53–77).

56. Picart, *Ceremonies and Religious Customs*, 4: 7–8 and plate on 9.

57. Graham, *Journal of a Residence in India*.

58. O'Quinn, "Haunted Trees," np.

59. Forbes, *Oriental Memoirs* 1: 28, quoted by de Almeida and Gilpin, *Indian Renaissance*, 44 and 306n.

60. Mukherjee, *Sir William Jones*, 77–79, 86–88; Raj, *Relocating Modern Science*, 95–97, 104–107; Bewell, "William Jones and Cosmopolitan Natural History," 167–180.

61. Jones, "Tenth Anniversary Discourse" (delivered 28 Feb. 1793), in *AR*, 4: xxx.

62. William Jones to Lady Spencer, 24 Oct. 1791, in *The Letters of Sir William Jones*, 2: 902.

63. Jones, 13 Mar. 1794, *Letters*, 2: 931. Jones died in April, 1794.

64. Mukherjee, *Sir William Jones*, 51–61; Jenny Uglow, *Lunar Men*, 338.

65. Mukherjee, *Sir William Jones*, 48.

66. Mukherjee, *Sir William Jones*, 86.

67. Jones to Lady Spencer, 24 Oct. 1791, *Letters*, 2: 902.

68. Robert Kyd to Bengal government, 1 June 1786. D.T.C. Vol. 7, folios 57–67, BMNH; quoted by Desmond, *European Discovery*, 57.

69. Roxburgh, *Flora Indica*, 2: 236, quoted by Desmond, *European Discovery*, 47.

70. Jones's letters on the spikenard show an accelerated learning curve, beginning with one to Patrick Russell in which Jones asks, "What is spikenard? I mean botanically, what is the natural order, class, genus &c. of the plant" (1787) and ending with another to Banks two years later in which Jones indicates he is certain that it is the Indian plant Jatamansi: Jones, *Letters*, 2: 777 and 843–844.

71. Jones, "Second Anniversary Discourse," in *AR*, 1: 336–337.

72 Rocher, "The Career of Rādhākānta," 627–633 and idem, "Weaving Knowledge," 51–79; Mukherji, "European Jones and Asiatic Pandits," 43–58.

73. Jones, *Letters*, 2: 815.

74. Jones, *Letters*, 2: 892.

75. Jones, *Letters*, 2: 707–708.

76. Jones, *Letters*, 2: 892.

77. Jones, "Design of a Treatise on the Plants of India," in *AR*, 2:272.

78. Jones, "Botanical Observations on Select Indian Plants," in *AR*, 4: 241.

79. Jones, *Letters*, 2: 776.

80. Jones, "Tenth Anniversary Discourse," in *AR*, 4: xxxi–xxxii.

81. Grove, *Green Imperialism*, 88–89.

82. Jones, "Botanical Observations," *AR*, 4: 233.

83. See Hanelt, *Mansfeld's Encyclopedia*—electronic database http://mansfeld.ipk-gatersleben.de/Mansfeld/Taxonomy derived from *Mansfeld's Encyclopedia of Agricultural and Horticultural Crops*, ed. P. Hanelt & IPK (New York: Springer, 2001).

84. Jones, "Botanical Observations," *AR*, 4: 236–238. In current taxonomic understanding, the term Scitamineæ designates a natural order of plants that includes mostly tropical herbs like ginger, banana, and plants that produce turmeric, arrowroot, and galangal. The term was first used in 1883 to refer to a specific and taxonomically stable group of flowering plants. Since 2003, it has been identified with Zingiberales, but assigned, in cladistic terms, to the commelinids in the monocots (Angiosperm Phylogeny Group 2 system of 2003).

85. Jones, "Botanical Observations," *AR*, 4: 237–238.

86. Jones, "Botanical Observations," *AR*, 4: 240.

87. Jones, "Botanical Observations," *AR*, 4: 262–263.

88. Foucault, *Discipline and Punish*, 190.

89. Jones, "Botanical Observations,"*AR*, 4: 266–267.

90. At least three English transcriptions of the title of this play exist: Jones's *Sacontalá, Shakuntala,* and *Śakuntalā.* For a modern translation, see Kālidasā's, *Śakuntalā* trans. Peter Khoroche (London: Folio Society, 1992). I use Jones's spelling to refer to his translation.

91. Jones, "Botanical Observations," *AR*, 4: 275–276.

92. Jones, *Sacontalá; or, the Fatal Ring*, in Jones, *Works*, 6: 217–218.

93. Jones, "Botanical Observations,"*AR*, 4: 275.

94. Jones, *Sacontalá*, Acts 2–4, in Jones, *Works*, 6: 232–233, 238, 245, 254, 260.

95. Jones, *Letters*, 2: 894; Figueira assesses Jones's influence in *Translating the Orient*.

96. The editor, Franklin, summarizes evidence for Jones's influence on the romantic oriental tale in Jones, *SPPW*, 80.

97. Jones, "Enchanted Fruit," in *SPPW*, ll.65–72, p. 83.

98. Jones, "Enchanted Fruit," in *SPPW*, ll.474–486, p. 95.

99. Franklin summarizes Jones's plot for his projected epic in a note: Jones, *SPPW*, 97n.

100. Jones, *SPPW*, 132.

101. Raj reviews Jones's legal work in *Relocating Modern Science*, 95–138.

102. Desmond summarizes the relationship between Koenig, the EIC, and Roxburgh's India career in *European Discovery*, 39–40 and 47–49; for bibliographic histories of the Roxburgh drawings, see Sanjappa, Thothathri, and Das, "Roxburgh's Flora Indica Drawings at Calcutta,"1; the Kew website for Roxburgh's *Flora Indica* is www.kew.org/FloraIndica; Sealey, "The Roxburgh Flora Indica Drawings at Kew," 296–399.

103. Desmond, *European Discovery*, 68–69.

104. Bayly, *Indian Society and the Making of the British Empire*, 45, 53.

105. Schwartz, "The Roxburgh Account of Indian Cotton Painting," 47–56.

106. Roxburgh, *Plants of the Coast of Coromandel*, 1: 4.

107. *Flora Indica*, hereafter *FI*, Kew MSS 1: 790, Royal Botanic Gardens and Herbarium Library, Kew. The printed text of this work, published after Roxburgh's death, introduced a different phonetic transcription for many of Roxburgh's Indian plant names (*FI*, 2: 577, 647 and elsewhere).

108. *FI*, Kew MSS 1: 795. Roxburgh nos. 663 and 664: "Tamara. Rheed. Mal 11.t.30. Padma. Asiat.: Res. 4. 286. Lall-pud[crossed out]ho of the Bengalese and Hind. Padma. Racta-ssaroruha, Racta-pala, Sanscrit names. See Asiat. Res.4. 286. Yerra-Tamara of the Telingas. Niluser. Pers." The last of these names is an addition in Roxburgh's handwriting.

109. Roxburgh, *FI*, Kew MSS 1: 130; *FL*, 1: 641–644.

110. Roxburgh, *FI*, Kew MSS 1: 1056. The printed edition reports slightly different spelling for these Indian plant names (*FI*, 1: 653). Singh and Singh, "On the Identity and Economico-medicinal Uses of Hastikarnapalasa."

111. *O. sanctum* is probably Thai basil; *O. basilicum* is the basil used in Western cooking.

112. Roxburgh, *FI*, 2:1604, Kew MSS Roxburgh #1190; Roxburgh #1345, Kew MSS 2: 1849 and Roxburgh #1851, Kew MSS 2: 1849; Sealey, "The Roxburgh Flora Indica Drawings," 363.

113. Roxburgh, *FI*, Kew MSS 1: 447.

114. In the printed edition of *Flora Indica*, Roxburgh's remark about arbitrary plant names corrects the first of these homonymic misspellings, but deletes the second altogether (*FI*, 1: 333). It is not clear whether Roxburgh made this correction or whether it was introduced years after his death as the work was prepared for publication.

115. In *The Body of the Artisan*, Pamela H. Smith presents images and what might be called "signatures" of European artists during the scientific revolution.

116. See works by Archer, Desmond, Noltie, and Nair cited earlier in this chapter; Llewellyn-Jones, *A Very Ingenious Gentleman*, idem, *A Man of the Enlightenment*, and, idem, essays in her edited collection, *Lucknow Then and Now* discuss Martin's commissioned drawings, especially those by Lucien Harris and J. P. Losty.

117. Archer, *Natural History Drawings*, 56; Desmond, *European Discovery*, x, 149, and 183.

118. Many of the drawings made for Robert Wight ended up in the herbarium of the Royal Botanic Garden at Edinburgh, where the plant taxonomist Henry Noltie found 711 of the drawings among 250,000 miscellaneous items in the "Cuttings Collection" of the garden's library, filed to illustrate the taxonomic sequence, but stripped of historical information about their provenance: Noltie, correspondence, 2004.

119. Noltie, *Dapuri Drawings*, 79, and Nair, "Native Collecting and Natural Knowledge."

120. More needs to be understood about the division of labor by jati or caste among Indian artists who worked outside the Mughal court workshops. These studies of artist and artisan communities indicate the direction of further research: Bearce, "Intellectual and Cultural Characteristics," 3–17; Carré, *The Travels of the Abbé Carré*, 2: 595; Patrick Russell's list of castes and crafts in Dalrymple's *Oriental Repertory*, 1: 49–52; Varadarajan, *South Indian Traditions of Kalamkari*, 12–13; Goswamy and Fischer, *Pahari Masters*, 7–10; and Hadaway, *Cotton Painting and Printing*, 17–19.

121. The list includes Col. and Lady Impey, the English landscape artist William Daniell (1769–1837), Roxburgh, Marquis Wellesley (1760–1842), the second Lord Edward Clive (1754–1839) and his wife, J. F. Cathcart, Benjamin Heyne, Francis Buchanan–Hamilton, Patrick Russell, Nathaniel Wallich, Robert Wight, Thomas Hardwicke, Claude Martin, a Frenchman attached to the court at Lucknow, the Raja Serfoji of Tanjore, and Alexander Gibson. Archer, *Natural History Drawings*, 55, notes that some Indian artists signed their work as "artists of Patna."

122. Poovey makes a similar argument about reading material evidence gleaned from ship logs for voyages between India, Europe, and the Far East against abstract imperial claims of the same epoch, as well as abstract claims modern scholars make without appealing to such evidence: "The Limits of the Universal Knowledge Project," 183–202.

123. Desmond, *European Discovery*, 50.

124. Desmond notes that in the 1790s Roxburgh paid artists about three rupees per drawing, as had Lady Anne Monson in 1775; Colonel Kyd calculated in 1785 that an artist could earn between 50 and 60 rupees per month, depending on his output: Desmond, *European Discovery*, 40–51, 149; Archer, *Natural History Drawings*, 2–62, and Noltie, *Dapuri Drawings*, 79. In some instances, artists were on salary and not paid by the piece. Writing in 1803 to a third party about Roxburgh's payment to an Indian painter of insects, Thomas Hardwicke reports that "the pay of this Painter is cleared up to the End of Febr. And as often as he wishes to draw any part of his allowed 30 Rupees per month, be so kind as to furnish him with a chit saying Painter Goordial [name added below line], on acct. of wages, Pay Capt Hardwicke." Roxburgh MSS, BMNH, Botany Library, London.

125. Roxburgh to Banks, 17 Aug. 1792; Banks to E.I.C., 4 July 1794; and Banks to Roxburgh, 29 May 1796. Desmond quotes these manuscript sources in *European Discovery*, 48 and 51.

126. Desmond, *European Discovery*, 148; Archer, *Natural History Drawings*, 56; Graham, *Journal of a Residence in India*, 146.

127. Archer remarks that this practice is typical of Mughal miniature painting techniques (*Natural History Drawings*, 58), but Goswamy and Fischer in *Pahari Masters*

and Bellinger and Hobhouse, *Fifty-one Flowers*, 15, suggest that the practice of overlaying paint was and still is common across India during this period and occurs in some Indian botanicals of this period.

128. Noltie, "Robert Wight and the Illustration of Indian Botany," 7; Archer, *Natural History Drawings*, 57.

129. Noltie. *Indian Botanical Drawings*, 29; In *Natural History Drawings*, 58, Archer notes Roxburgh's use of British botanical engravings to guide Indian artists.

130. Tobin, *Picturing Imperial Power*, 190.

131. Sowerby Correspondence, Section A, Boxes 19 and 20, A56–60, Smith to Sowerby, May 2, 1802, Mar. 14, 1803, Apr. 10, 1824, and Nov. 8, 1824, Natural History Museum (BMNH), London.

132. The curators of the Roxburgh drawings at Kew provide an overview of the Indian artists and their techniques and materials on the Royal Botanic Garden, Kew website: www.kew.org/data/floraIndicastatic/htm/artists.htm; Gettens observes that whereas Mughal artists used Indian yellow, orpiment was the source of the yellow used by Rajput artists, "Review of Devkar," 56.

133. Mackay and Sarkar, "Lalighat Pats," 135–142.

134. Archer, *Natural History Drawings*, 58.

135. Vajracharya, *Watson Collection of Indian Miniatures*, 103.

136. Fischer, "Foreword," Desmond Peter Lazaro, *Materials, Methods & Symbolism*, 8.

137. Sealey's catalogue of the Roxburgh drawings at Kew lists several breadfruit species, but none of their species names exactly corresponds to the drawing Sealey identifies as #250, in the Coromandel work. Roxburgh #474; *Plants of the Coromandel*, 65; *FI*, 3: 838–842; Fleming collection, BMNH, Botany Library, London.

138. Roxburgh #478 and #479, *Coromandel*, 71 and 72; *FI*, 3: 790.

139. Roxburgh #1345; *Coromandel*, 234; *FI*, 1: 428–430.

140. Archer, *Natural History Drawings*, 57–58.

141. Porter, *Painters, Paintings, and Books*, 75.

142. Roxburgh #658; *FI*, 2: 577–578.

143. Roxburgh #663 and 664; *FI*, 2: 647–50; Fleming Collection, Drawings in the Fleming collection of the Natural History Museum, British Museum, were collected by John Fleming, MD, FRS, who joined the Indian Medical services in 1768 and was in charge of the Calcutta Botanic Garden after Robert Kyd's death and during one of Roxburgh's leaves of absence, when Fleming would have also supervised the Indian botanical artists who worked there. Fleming returned to England in 1813. His *Catalogue of Indian Medical Plants and Drugs*, published in that year, echoes Roxburgh's medical interest in Indian plants.

144. In *Indian Renaissance* (93), De Almeida and Gilpin quote Jones's translation: "O my Sacontala! It well becomes thee, who are soft as the fresh blown Mallica, to water the tender shrubs of our father Canna; Who would attempt to cleave the hard wood Sami, with a leaf of the Blue Lotus?"

145. Roxburgh #925; *FI*, 1: 653–637.

146. Roxburgh #1483; *Coromandel*, 247; *FI*, 3: 167; Fleming # 21, Fleming Collection, Drawings in the Fleming collection of the Natural History Museum, British Museum.

147. Roxburgh #774; *FI*, 1: 245, RBG, Edinburgh collection.

148. Roxburgh #1190; *FI*, 3: 777–779; Piddington, *English Index*; s.v. "Calamus."

149. Roxburgh #553; *FI*, 1: 459–460.

150. Roxburgh #1270, Kew MSS 2: 1722, *FI*, 3: 131–132: "Tala indica. Lourier. Cochinch. 494. Cumbhica. The Sanscrit name [this phrase crossed out]. Kodda pail. Rheed ma. 11.t.32. Plantago aquatica. Rhump. Amb. 6.t.74. Neeroo boodooky of the Terlingas [the phrase is crossed out]. Tacca panna of the Hindoos, & Bengaleses [this provenance is crossed out]."

151. Roxburgh # 2399, Kew MSS 3: 325; *FI*, 3: 705.

152. Bosch, *The Golden Germ*, 55, 65–78, 127, 164. On color and flower symbolism in Indian textiles, see Jaykar, "Traditional Textiles in Indian Art," 7 and Krishna, "Flowers in Indian Textile Designs," 1–44.

153. Blunt "The Mughal Painters of Natural History," 48. On Mughal art within and beyond the Gangetic plane, see Crill, Stronge, and Topsfield, *Arts of Mughal India*, 139, 143, 212–222, 250.

154. The debate about whether Italian rather than Persian artists directed or created these Indian instances of *pietra dure* continues (at least one other was created during this period for the floor of a Hindu temple in Ahmedabad). See Nicholson, *The Red Fort*, 73. For roughly contemporary Italian *pietra dure* presentations of botanical motifs, see Tomasi and Hirschauer, *The Flowering of Florence*, 65.

155. Irwin, "Indian Textiles," 5.

156. Desmond, *European Discovery*, 145.

157. Welch, "Reflections," 11–13.

158. Ehnbom, "'Passionate Delineation and the Mainstream of Indian Painting,'" 184.

159. Vajracharya, "Atmospheric Gestation," 40–51.

160. Vajracharya, *Watson Collection of Indian Miniatures*, 155.

161. Rossi et al., *From the Ocean of Painting*, 40.

162. Varadarajan, *South Indian Traditions of Kalamkari*, 28 and 89.

163. Rossi et al., *From the Ocean of Painting*, 198; conversation with Lotika Varadarajan on astrological manuscripts, Nov. 2007.

164. Goswamy and Fischer, *Pahari Masters*, 64, 125, 141, 143, 162. In the case of the master known as Mahesh of Chamba (c. 1730–1777), the authors suggest that claims that he was trained in the Mughal style fail to recognize how slow the Pahari artists were to incorporate Mughal influence from the Gangetic plain (268).

165. Desmond, *European Discovery*, 50.

166. Noltie, *Robert Wight*, 2:23.

167. Archives, Botanical Survey of India, Botanic Garden, Calcutta, Howrah.

168. Desmond, *European Discovery*, 79.

169. Spivak, *A Critique of Postcolonial Reason*, 5–10, 205–208, 238, and 246.

170. Chakrabarty, *Provincializing Europe*, 112.

INTERLUDE 2: A ROMANTIC GARDEN

1. Shelley, *The Sensitive-Plant* and *The Triumph of Life*, in *Shelley's Poetry and Prose* (*SPP*), citations appear parenthetically in the text.

2. Goldstein, "Growing Old Together," in *Sweet Science: Romantic Materialism and the New Sciences of Life*."

3. La Mettrie, *L'Homme Plante*, 1: 72, 74, and 253–269.

4. Boury, "Irritability and Sensibility," 521–535; Guido Giglioni, "What Ever Hap-

pened to Francis Glisson?" 465–493; Wolfe and Terada, "The Animal Economy as Object and Program," 537–79; Bordeu, *Dissertatio Physiologica*, 1: 13.

5. Erasmus Darwin, *BG*, pt. 2, Canto 1, 31–32.

CHAPTER 7: RESTLESS ROMANTIC PLANTS AND PHILOSOPHERS

1. Round 1, Hegel: *Philosophy of Nature*, in Hegel, *Werke*, 9: 28–29. Unless otherwise noted, the German edition cited is Hegel's *Werke*, ed. Moldenhauer and Michel. Miller's translation, taken from the 1830 text, published as part 2 of Hegel's *Encyclopaedia of the Philosophical Sciences*, *Philosophy of Nature*, is slightly different (no innocent plant crushed at the end), 17. But see 276, where Hegel allies the "innocence of plants" with the "impotence of self-relation to inorganic being"; hereafter cited as Hegel, *PN*, trans. Miller. See also Petry's translation (1: 209–210), which includes the flower-crushing vignette and collates earlier lecture versions from 1805–1806 of material that Hegel recast for publication as *The Philosophy of Nature*; hereafter cited as Hegel, *PN*, trans. Petry. This translation is based on the text prepared by K. L. Michelet in 1842 and reprinted for modern readers by Hermann Glockner as vol. 9 in Hegel, *Sämtliche Werke*. See Hegel, *PN*, trans. Petry, 1: 122–123. Elaine Miller's *The Vegetative Soul* cites Hegel's remark in her critique of his view of vegetable, that is, plant nature (122).

Round 2, Goethe: The itinerary and origin of Goethe's remark, made to Thomas Johann Seebeck in 1812, are in part anecdotal. The version I have quoted appears in Hoffmeister, *Goethe und der Deutsche Idealismus*, 82. Schubert quotes the anecdote without the sentence about Hegel's monstrous reading of plant nature in *Goethe und Hegel*, 23–24. Goethe's remark about Hegel on plant nature appears in a letter to the botanist Thomas Seebeck. See Goethe, *Die Schriften zur Naturwissenschaft*, 9B: 354–55; hereafter, this title cited as Goethe, *LA*. Goethe appears here to paraphrase Hegel's *Phenomenology of Spirit* (1807) or the *Logic* (1812). Hoffmeister pitches for the *Phenomenology*, Schubert for the *Logic* (1: 46). Neither text quite matches the sentence Goethe quotes. Miller's and Petry's translations include similar remarks. See Hegel, *PN*, trans. Miller, 336, 350–351, where plant ruin becomes a stepping-stone to animal organism's greater animate powers, and *PN*, trans. Petry, 3: 101.

Round 3, Hegel: *Philosophy of History*, 32–33, in Hegel, *Werke*, 12: 49.

2. Boury, "Irritability and Sensibility," 521–535; Giglioni, "What Ever Happened to Francis Glisson?" 465–493; Wolfe and Terada, "The Animal Economy," 537–579; Bordeu, *Dissertatio Physiologica*, 1: 13; Miller, *The Vegetative Soul*, 45–78 and 99–148; Richards, *The Romantic Conception of Life*, 65–90, 114–145; Krell, *Contagion: Sexuality, Disease, and Death*, esp. 117–160; Steigerwald, "Instruments of Judgment," 79–131; Beiser, *The Romantic Imperative*, 156; Reill, *Vitalizing Nature*, 199–220; Rajan, "Excitability," 309–325; idem, "First Outline of a System of Theory," 311–335; and idem, "Toward a Cultural Idealism," 51–71.

3. Goethe published *Versuch die Metamorphose der Pflanzen zu Erkläre*, in English *Metamorphosis of Plants*, in 1790, then republished it with other essays on morphology in 1817–1824 and again in 1830. Arber translates the 1790 essay in "Goethe's Botany," 67–124; hereafter Goethe, *MP* and, when Arber's translation is at issue, trans. Arber. Two more recent translations appear in *Goethe's Botanical Writings*, trans. Mueller (hereafter Goethe, *Botanical Writings*) and *Goethe: Scientific Studies*, trans. Douglas Miller

(hereafter Goethe, *Scientific Studies*). Except when otherwise noted, the German edition of Goethe's other writings used here is vol. 12, *Sämtliche Werke*. Hegel does not mention plants in the early Jena manuscripts for his *Philosophy of Nature*, which discusses mineralogy and chemistry, topics of enormous interest to his generation of German intellectuals, many of whom had studied mineralogy and silver mining. His first discussion of plants or vegetable nature turns up in manuscripts from 1805–1806, which he presumably used or referred to in his Jena lectures of that academic year. In *Philosophy of Spirit* (1807), he presents plants and animals (except man) as lesser than spirit. Petry's English translation of Hegel's *Philosophy of Nature*, part 2 of the *Encyclopedia of the Philosophical Sciences* (1830), identifies passages that Hegel either used first in the Jena lectures of 1805–1806 or in the later Berlin lectures, beginning in 1819–1820, where he first mentions Goethe. Hegel's Berlin lectures on the philosophy of nature include sections and wording dating from 1817 and others introduced in either 1827 or 1830. Hegel's Jena lectures devote extensive analysis to nature and spirit, although his discussion of nature considerably exceeds that on spirit (185 to 100 pages). Hegel, "Jenaer Systementwürfe III," Rauch, "Hegel and the Emerging World," 175–181. The three later versions of Hegel's *Philosophy of Nature* are published as *Enzyklopädie der Philosophischen Wissenschaften im Grundrisse*. Schlegel, *Von der Weltseele*, 349.

4. Gillispie, *The Edge of Objectivity*, 192–199; Sachs, *History of Botany*, 157–158. Neubauer catalogues Goethe's warnings about the problem of hypothesis in "Organic Form in Romantic Theory," 212, 219, 220.

5. Hegel, *PN*, trans. Miller, lemma 345, 311; Hegel, *Werke*, 9: 380.

6. For further consideration of Kant's remarkable observations in the third *Critique*, see Stiegerwald, "Kant's Concept of Natural Purpose," 712–734; and Huneman, "From the *Critique of Judgment* to the Hermeneutics of Nature," 9.

7. In preserving Kant's reticence about attributing life to organized beings, I differ from Huneman (1), who uses the term to show how Kant's *Naturphilosophie* successors took their cue from the third *Critique*. Bowie notes in *Schelling and Modern European Philosophy*, 35, that on one occasion Kant use the word *life* in discussing Casper Friedrich Blumenbach's analysis of the *Bildungstrieb* or formative drive. But even here, Kant limits his use of the term to paraphrasing Blumenbach: "he rightly declares it to be contrary to reason that raw matter should originally have formed itself in accordance with mechanical laws, that life should have arisen from the nature of the lifeless." In *Strategy of Life*, Lenoir reviews philosophical as well as biological responses to Blumenbach and arguments about "vital materialism," the study of the presence of apparently vital forces alongside mechanical processes in biology (17–53).

8. Kant, *Critique of the Power of Judgment*, lemma 65, 244, hereafter *CJ*; and Kant, *Immanuel Kant: Schriften zur Ästhetik und Naturphilosophie*, vol. 3, 735; hereafter, this volume cited as *KU*.

9. Bowie, *Schelling*, 31–32.

10. Kant, *CJ*, lemma 63, 240–241; *KU*, 730–731.

11. Kant, *CJ*, lemma 64, 243, and lemma 66, 248; *KU*, 733 and 740–741.

12. Huneman, "From the *Critique of Judgment*," 10.

13. Hacking, *The Emergence of Probability*, and idem, *The Taming of Chance*; Daston, *Classical Probability in the Enlightenment*.

14. Huneman, "From the *Critique of Judgment*," 7.

15. Schelling, *Ideas for a Philosophy of Nature*, 41 and 53. Critical of vitalist argu-

ments in this essay, Schelling is only willing to grant the notion of a life force if it is understood as embodied in and by a polarity of forces that is modeled on principles of physics and chemistry (37). Although his account of plant development echoes Kant's in the third *Critique,* Schelling is here primarily concerned with the chemical processes suggested by how they take in oxygen and respond to their environment.

16. Jardine, *Scenes of Inquiry* (45–46) cites the third edition of Oken's *Lehrbuch* and notes that it was translated into English by Tulk as *Elements of Physiophilosophy*; see Grant, *Philosophies of Nature after Schelling*, and Beiser, "Hegel and *Naturphilosophie*," 135–147.

17. Nancy, *Hegel*, 8–19; in *Tarrying with the Negative* (153–55), Žižek suggests that the problem of contingency for Hegel turns on whether it is within or in the world.

18. Rajan, "Philosophy as Encyclopedia: Hegel, Schelling and the Organization of Knowledge," *Wordsworth Circle* 25, no. 1 (2004): 6–11, and idem, "System and Singularity," 137–149.

19. Kain, *Hegel and the Other*, 72.

20. Rauch, "Hegel and the Emerging World," 176.

21. Sachs, *History of Botany*, 79–80.

22. Holland, *German Romanticism*, 26, quoting Goethe, *LA*, II 9A: 97.

23. Holland, *German Romanticism*, 27, quoting Goethe, *LA*, II 9A: 97.

24. Holland, *German Romanticism*, 27, quoting Goethe, *LA*, I, 10: 65.

25. Goethe, *Botanical Writings*, 33; and Arber, "Goethe's Botany," *Metamorphosis of Plants*, 8. I have slightly retranslated the English so that it more closely echoes the 1790 German text that Arber reprints.

26. Goethe, *LA*, 2.9A: 58; Richards emphasizes the significance and mystery of Goethe's Sicilian interlude and *Urpflanze* intuition in *The Romantic Conception of Life*, 382–400.

27. Miller, *Vegetative Soul*, 74–75, and Tantillo, "Polarity and Productivity in Goethe's *Wahlverwandtschaften*," 310–25, and idem, *The Will to Create*, 16–26.

28. Goethe, *Metamorphosis*, in Goethe, *Scientific Studies*, lemma 73, 87; lemma 121, 97. Goethe may not have read Johann Friedrich Blumenbach's concept of *Bildungstrieb* by 1790: see Richards, *The Romantic Conception of Life*, 218–229; Lenoir, *The Strategy of Life*, 19–25; and Holland, *German Romanticism*, 66–68.

29. Miller, *Vegetative Soul*, 47.

30. Krell, *Contagion*, 4, argues that anastomosis is a digestive process, as Hegel had also claimed in *PN*, trans. Miller, 304; *Werke*, 9: 372.

31. Goethe, "*Die Absicht eingeleitet*," *LA* 17: 13–14; "The Purpose Set Forth" (dated Jena, 1807), Goethe, *Scientific Studies*, 63; "Our Objective is Stated" and the 1817–20 tables of contents for the *Morphology* volumes, Goethe, *Botanical Writings*, 23 and 255.

32. Tantillo, "Goethe's Botany and His Philosophy of Gender," 123–130 and *The Will to Create*, 48–57; and Koerner, "Goethe's Botany," 470–495.

33. Goethe's 1824 statement about intensification and his use of the term *polarity* in 1828 to mean all oppositions in nature, as well as his 1828 retrospective on Tobler's aphoristic essay on Nature, first published 1783 and often identified as Goethe's, are presented in Goethe, *Scientific Studies*, 98 and 3–5.

34. Goethe never completed "The Spiral Tendency." It was published in 1831 with the last reissue of his *Metamorphosis* essay. See the translators' notes in Goethe, *Scientific Studies*, 327, and *Botanical Writings*, 127.

35. Holland offers a nuanced reading of this poem in *German Romanticism*, 42–48.

36. My translation is slightly more literal than the English translation of Goethe's poem "The Metamorphosis of Plants" / "Die Metamorphose der Pflanzen," in Goethe, *Selected Poems*, 154–155. Nineteenth-century readers were familiar with Goethe's poem. Lindley uses its opening lines (untranslated) on the title page of *Ladies Botany*.

37. I have modified Middleton's translation slightly to preserve a closer approximation to the German.

38. These lines are quoted in succession from the poems "Parabasis" (c. 1820) and "Epirrhema" (c. 1819), in Goethe, *Poems and Epigrams*, 70–71.

39. Benveniste, "The Notion of 'Rhythm,'" 281–288; cited by Miller, *Vegetative Soul*, 47.

40. Lucas provides the English translation in "The 'Sovereign Ingratitude' of Spirit toward Nature," 131–150. In his analysis of Hegel's increasing attention to the superiority of spirit over nature from the early Jena writings through successive iterations of the *Philosophy of Nature* in the 1820s, Lucas quotes (137) Hegel's remarkable use of the phrase *sovereign ingratitude* in the *Philosophy of Spirit* to describe Spirit's view of its one-upmanship of nature (*Enzyklopädie: Sämtliche Werke*, 10: 29, lemma 381). Other analyses of Hegel on nature include Derrida, *Glas*, 245–251; Miller, *Vegetative Soul*, 119–147; Buchdahl, "Hegel's Philosophy of Nature," 257–667; Pinkard, *German Philosophy*, 266–79, and Pippin, "You Can't Get There from Here," 52–85.

41. Hegel to Paulus, 1814, *Briefe*, 1: 138; quoted in Findlay's foreword to *Hegel's Logic*, xxxv.

42. Hegel, *PN*, trans. Miller, 315; Hegel, *Werke*, 9: 386. Hegel similarly construes the Goethe's argument about the pistil, calling it "only a contracted leaf."

43 Hegel, *PN*, trans. Miller, 315; Hegel, *Werke*, 9: 386. Translation modified to echo the German more literally.

44. Adorno, "Aspects," in *Hegel: Three Studies*, 42, hereafter cited as *Hegel*; *Zur Metakritik der Erkenntnistheorie*, vol. 5, 286; hereafter this title cited as *Zur Metakritik*.

45. Koselleck, *Futures Past*, 126.

46. Adorno, *Minima Moralia*, 200; and idem, *Minima Moralia*, in *Gesammelte Schriften*, 4: 227–228.

47. Adorno, "Parataxis," and idem, "Parataxis: Zur späten Lyrik Hölderlin"; Kelley, "Adorno Nature Hegel," 161–198.

48. Adorno, "The Meaning of Working through the Past," 200.

CHAPTER 8: CONCLUSION

1. Mabey, *Nature Cure*, 146–147.
2. Bateman, *The Orchidaceae of Mexico and Guatemala*, 17.
3. Metcalf, *Ideologies of the Raj*, 17.
4. Bateman, *A Monograph of Odontoglossum*.
5. Deleuze and Guattari, *A Thousand Plateaus*, 10–11.
6. Wordsworth, *The Ruined Cottage and the Pedlar*, MS. D, l. 514, p. 73.
7. Solkin, *Art on the Line*; Thomas, *Romanticism and Visuality*.
8. http://en.wikipedia.org/wiki/Biddulph_Grange citing Stocks, "Biddulph Is a Horticultural Disneyland."

9. For a history of *phantasia* and its cognates, see Kelley, *Reinventing Allegory*, 109–28.

10. Gledhill, *The Names of Plants*, 185.

11. Jim Fulsom, director of the Henry E Huntington Gardens, noted this possibility (correspondence, Jan. 2004).

12. Milton, *Paradise Lost*, 2: 61–62, ll. 624–628.

13. Pijl and Dodson, *Orchid Flowers*, 129–130, 135–137.

14. Charles Darwin to *Gardener's Chronicle*, 4 or 5 June 1860, *Darwin Correspondence Project*, vol. 8, letter 2826, www.darwinproject.ac.uk.

15. Charles Darwin, *On various contrivances*, 68–69.

16. The orchid is the Peruvian *Trichoceros tupaipi*. See chapter 5.

Bibliography

"Account of the New Botanic Institution in Ireland," *Monthly Review* 1, no. 3 (1796): 208–210.

Adorno, Theodor W. "Aspects of Hegel's Philosophy." In *Hegel: Three Studies*. Translated by Shierry Weber Nicholsen, Cambridge, MA: MIT Press, 1999. 1–51.

——. *Zur Metakritik der Erkenntnistheorie. Drei Studien zu Hegel*. Edited by Rolf Tiedemann, Gretel Adorno, Susan Buck-Morss, and Klaus Schultz. Frankfurt am Main: Suhrkamp, 1995.

——. "The Meaning of Working through the Past." In *Can One Live after Auschwitz? A Philosophical Reader*. Translated by Henry W. Pickford. Stanford: Stanford University Press, 2003. 3–18.

——. *Minima Moralia: Reflections from a Damaged Life*. Translated by E. F. N. Jephcott. London: Verso, 1978.

——. *Minima Moralia: Reflexionen aus dem Beschädigten Leben*. In *Gesammelte Schriften*, vol. 4. Edited by Rolf Tiedemann. Frankfurt am Main: Suhrkamp, 1951.

——. *Negative Dialectics*. Translated by E. B. Ashton. New York: Continuum, 1973.

——. "Parataxis: On Hölderlin's Late Poetry." In *Notes to Literature*, vol. 2. Translated by Shierry Weber Nicolsen, New York: Columbia University Press, 1992. 109–149.

——. "Parataxis: Zur Späten Lyrik Hölderlin." In *Noten Zur Literatur*. Frankfurt: Suhrkamp, 1974. 447–491.

Agamben, Giorgio. *Remnants of Auschwitz: The Witness and the Archive*. New York: Zone Books, 2002. Repr. 1999.

Agrawal, Arun. "Dismantling the Divide between Indigenous and Scientific Knowledge." *Development and Change* 26 (1995): 413–439.

Aikin, John. *An Essay on the Application of Natural History to Poetry*. Warrington, Cheshire, UK: W. Eyres; London: J. Johnson. New York: Garland Publishing, 1970. Repr. 1777.

Alcock, Randal H. *Botanical Names for English Readers*. London: L. Reeve, 1876.

Allaby, Michael. *A Dictionary of the Plant Sciences*. 2nd ed. Oxford: Oxford University Press, 1998.

The Anti-Gallican. Jan. to Dec. 1–12 (1804).

Arber, Agnes, *The Natural Philosophy of Plant Form*. Cambridge: Cambridge University Press, 1950.

——, trans. "Goethe's Botany," with an introduction by the translator. *Chronica Botanica* 10, no. 2 (1946): 67–124.

Archer, Mildred. *Natural History Drawings in the India Office Library*. London: H. M. Stationery Office, 1962.

Aristotle. *Complete Works of Aristotle*. Edited by Jonathan Barnes. 2 vols. Princeton: Princeton University Press, 1984.

Armstrong, Isobel. "The Gush of the Feminine: How Can We Read Women's Poetry of the Romantic Period?" In *Romantic Women Writers: Voices and Countervoices*. Edited by Paula R. Feldman and Theresa M. Kelley. Hanover, NH: University Presses of New England, 1995. 13–32.

Arnold, David. *The Tropics and the Traveling Gaze: Indian, Landscape, and Science, 1800–1856*. Seattle: University of Washington Press, 2006.

The Asiatick Miscellany. Edited by William Chambers and Sir William Jones. 1785–1786 repr. London: John Murray, 1792.

Atran, Scott. *Cognitive Foundations of Natural History*. Cambridge: Cambridge University Press, 1990.

Baber, Zaheer. "Colonizing Nature: Scientific Knowledge, Colonial Power, and the Incorporation of Indian into the Modern World-System." *British Journal of Sociology* 52, no. 1 (2002): 37–58.

Baillon, Henri. *Étude générale du groupe des euphorbiacées*. Paris: Librairie de Victor Masson, 1858.

Baker, George, and Anne Baker. *Glossary of Northamptonshire Words and Phrases*. 2 vols. London: John Russell Smith; Northampton: Abel & Sons and Mark Dorman, 1854.

Banks, Joseph. Letter to E.I.C. July 4, 1794. D.T.C. vol. 9. Folios 52–56. BMNH.

——. Letter to William Roxburgh. 29 May 1796. Add. MSS 33980. Folios 65–66. British Library.

Barbauld, Anna Letitia. *The Poems of Anna Letitia Barbauld*. Edited by William McCarthy and Elizabeth Kraft. Athens: University of Georgia Press, 1994.

Barnett, Richard. Introduction to *Rethinking Early Modern India*. Edited by Richard Barnett. New Delhi: Manohar, 2002. 11–32.

Barrell, John. *The Idea of Landscape and the Sense of Place, 1730–1840: An Approach to the Poetry of John Clare*. Cambridge: Cambridge University Press, 1972.

Bartram, John. *Correspondence of John Bartram, 1734–1777*. Edited by Edmund Berkeley Jr. and Dorothy Smith Berkeley. Gainesville and Tallahassee: University of Florida Press, 1992.

Bartram, William. *The Travels of William Bartram*. Edited by Francis Harper. Athens: University of Georgia Press, 1998.

Bate, Jonathan. "Night Light on the Life of Clare." *John Clare Society Journal* 20 (July 2001): 141–154.

Bateman, James. A *Monograph of Odontoglossum*. London: L. Reeve, 1864–74.

——. *The Orchidaceae of Mexico and Guatemala*. London: J. Ridgway, 1837–43.

Bauer, Franz. *Illustrations of Orchidaceous Plants*. London, 1830–1838.

Bayly. C. A. *Indian Society and the Making of the British Empire*. In *The New Cambridge History of India*, part 1, vol. 2. Cambridge: Cambridge University Press, 1988.

Bearce, George D. "Intellectual and Cultural Characteristics of India in a Changing Era, 1740–1800." *Journal of Asian Studies* 25, no.1 (1965): 3–17.

Becker-Ho, Alice. *L'Essence du jargon*. Paris: Gallimard, 1994.

——. *Les Princes du jargon*. 1990; repr. Paris: Gallimard, 1993.

Beiser, Frederick C. *German Idealism: The Struggle against Subjectivism, 1781–1801*. Cambridge, MA: Harvard University Press, 2002.

——. "Hegel and Naturphilosophie." *Studies in the History and Philosophy of Science* 34, no.1 (2003): 135–147.

——. *The Romantic Imperative: The Concept of Early German Romanticism* Cambridge, MA: Harvard University Press, 2003.

Bellinger, Katrin, and Niall Hobhouse. *Fifty-one Flowers.* London: Colnaghi Gallery, Hobhouse, 2006.

Benedict, Barbara. *Curiosity: A Cultural History of Early Modern Inquiry.* Chicago: Chicago University Press, 2001.

Benjamin, Walter. *The Arcades Project.* Translated by Howard Eiland and Kevin McLaughlin. Cambridge, MA: Harvard University Press, 1999.

Benveniste, Émile. "The Notion of 'Rhythm' in Its Linguistic Expression." In *Problems in General Linguistics.* Coral Gables, FL: University of Miami Press, 1971. 281–288.

Bermingham, Ann. *Learning to Draw: Studies in the Cultural History of a Polite and Useful Art.* New Haven: Yale University Press, 2000.

Bewell, Alan J. "Erasmus Darwin's Cosmopolitan Nature." *ELH* 76, no. 1 (2009): 19–48.

——. "'Jacobin Plants': Botany as Social Theory in the 1790s." *Wordsworth Circle* 20, no. 3 (1989): 132–139.

——. "Keats's 'Realm of Flora.'" *Studies in Romanticism* 31, no. 1 (1992): 71–98. Repr. in *New Romanticisms: Theory and Critical Practice,* edited by David L. Clark and Donald C. Goellnicht. Toronto: University of Toronto Press, 1994. 71–100.

——. "'On the Banks of the South Sea': Botany and Sexual Controversy in the Late Eighteenth Century." In *Visions of Empire: Voyages, Botany, and Representations of Nature.* Edited by David Philip Miller and Peter Hanns Reill. Cambridge: Cambridge University Press, 1994. 173–196.

——. "William Jones and Cosmopolitan Natural History." *European Romantic Review* 16, no. 2 (2005): 167–180.

Bicheno, J. E. "On Systems and Methods in Natural History." *Transactions of the Linnean Society of London* 15, no. 2 (1826): 479–496.

Bickerstaff, Isaac. *The Sultan; or, A Peep into the Seraglio.* Corrected by Thomas Dibdin. London: Chiswick Press, 1817.

Blunt, Wilfrid [Jasper Walter], and William T. Stearn. *The Art of Botanical Illustration.* Rev. ed. Woodbridge, Suffolk: Antique Collectors' Club and Royal Botanic Gardens, Kew, 1995.

Blunt, Wilfrid Scawen. "The Mughal Painters of Natural History." *Burlington Magazine* 90, no. 539 (1948): 48–50.

Bordeu, Théophile de. *Dissertatio Physiologica.* In vol. 1, *Oeuvres complètes.* 2 vols., Ed. Richarand, Paris: Gaille et Ravier, 1818.

Borland, Francis. *Memoirs of Darien.* Glasgow: Hugh Brown, 1815.

Bosch, F. D. K. *The Golden Germ: An Introduction to Indian Symbolism.* 'S-Gravenhage, Holland: Mouton, 1960.

Bourdieu, Pierre. *The Logic of Practice.* Translated by Richard Nice. Stanford: Stanford University Press, 1990.

Boury, Dominique. "Irritability and Sensibility: Key Concepts in Assessing the Medical Doctrines of Haller and Bordeu." *Science in Context* 21, no. 4 (2008): 521–535.

Bowie, Andrew. *Schelling and Modern European Philosophy.* London: Routledge, 1993.

Bradbury, S. *The Evolution of the Microscope.* Oxford: Pergamon, 1967.

——. *The Microscope Past and Present.* Oxford: Pergamon, 1968.

Brantlinger, Patrick. *Rule of Darkness: British Literature and Imperialism, 1830–1914*. Ithaca, NY: Cornell University Press, 1988.

Broadway, Lucile. *Science and Colonial Expansion*. New York: Academic Press, 1979.

Brontë, Charlotte. *Jane Eyre*. Edited by Thomas Crawford. New York: Dover, 2002.

Brown, Bill. *The Material Unconscious*. Cambridge, MA: Harvard University Press, 1996.

——. "The Secret Life of Things (Virginia Woolf and the Matter of Modernity)." *Modernism/modernity* 6, no. 2 (1999): 1–28.

——. "Thing Theory." In *Things*. Edited by Bill Brown. Chicago: University of Chicago Press, 2004. 1–16.

Brown, Robert. "An account of a new Genus of Plants, named Rafflesia." First published in the *Transactions of the Linnean Society*: "An Account of a new Genus," 13 (Apr. 1821): 201–234. In Brown, *Miscellaneous Botanical Works*, 2 vols. Vol. 1. 1821 and 1844. Repr.; London: Ray Society, 1866. 367–398.

——. "On the Female Flower and Fruit of Rafflesia Arnoldi and on Hydnora Africana." First published in the *Transactions of the Linnean Society* 19 (1844): 221–247. In Brown, *Miscellaneous Botanical Works*, 2 vols. Vol. 1. 1821 and 1844. Repr.; London: Ray Society, 1866. 399–431.

——. *Observations on the Structures and Affinities of the More Remarkable Plants*. London: Thomas Davison, 1826.

——. "Some Observations on the Natural Family of Plants Called Compositae." *Transactions of the Linnean Society* 12 (1818): 76–150.

——. *Prodromus Florae Novae Hollandiae*. London: Richard Taylor, 1810. Repr. in *Historiae naturalis classica*, vol. 6. Edited by William T. Stearn. Wienheim: H. R. Englemann, 1960.

Brown, Thomas. *Observations on the Zoonomia of Erasmus Darwin, M.D.* Edinburgh: Mundell & Son, 1798.

Browne, Janet. "Botany for Gentlemen: Erasmus Darwin and *The Loves of the Plants*." *Isis* 80, no. 4 (1989): 593–621.

Browne, Patrick. *The Civil and Natural History of Jamaica*. London: B. White & Son, 1789.

Buchdahl, Gerd. "Hegel's Philosophy of Nature." *The British Journal for the Philosophy of Science* 23, no. 3 (1972): 257–667.

Buffon, G. L. L. *Histoire naturelle, générale et particulière*. vol. 1. Paris: L'Imprimerie royale, 1749. Translated by William Smellie as *Natural History: General and Particular*. London, W. Strahan & T. Cadell, 1785.

[Canning, George, and J. H. Frere.] "Loves of the Triangles." In *Poetry of the Anti-Jacobin*. London: Wright, 1799. 108–129, 134–141.

Carlson, Julie A. *England's First Family of Writers: Mary Wollstonecraft, William Godwin, Mary Shelley*. Baltimore: Johns Hopkins University Press, 2007.

Carr, D. J., ed. *Sydney Parkinson: Artist of Cook's Endeavour Voyage*. London: British Museum (Natural History), 1983.

Carré, Abbé. *The Travels of the Abbé Carré in India and the Near East 1672 to 1674*. Translated by Lady Fawcett. Edited by Sir Charles Fawcett. 3 vols. London: Hayluyt Society, 1947.

Casid, Jill H. *Sowing Empire: Landscape and Colonization*. Minneapolis: University of Minnesota Press, 2005.

Cesalpino, Andrea. Dedication to *De plantis libri XVI*. Florence: Giorgio Marescotti, 1583.

Chabran, Rafael, and Simon Varey. "Entr'Acte." In *Searching for the Secrets of Nature*. Edited by Simon Varey, Rafael Chabran, and Dora B. Weiner. Stanford: Stanford University Press, 2000. 106–107.

Chakrabarti, Pratik. "Medical Marketplace beyond the West: Bazaar Medicine, Trade and the English Establishment in Eighteenth-Century India." In *Medicine and the Market in England and its Colonies, 1450–1850*. Edited by Patrick Wallis and Mark Jenner. London: Palgrave, 2007. 196–215.

Chakrabarty, Dipesh. *Provincializing Europe: Postcolonial Thought and Historical Difference*. Princeton: Princeton University Press, 2000.

Chambers, Ephraim, and Abraham Rees. *Cyclopaedia, or, Universal Dictionary of the Arts and Sciences*. 4 vols. London: Printed for J. F. & C. Rivington, 1786.

Chatterjee, Partha. *Nationalist Thought and the Colonial World: A Derivative Discourse?* Minneapolis: University of Minnesota Press, 1986.

Chaudhuri, K. N. *The Trading World of Asia and the East Indian Company*. Cambridge: Cambridge University Press, 2006.

Cheyron, Henry. "L'amour de la botanique: les annotations de Jean-Jacques Rousseau sur la botanique de Régnault." *Littératures* 4 (1981): 53–95.

———. "Ray Et Sauvages Annotés Par Jean-Jacques Rousseau." *Littératures* 15 (1986): 82–99.

Chow, Rey. *Writing Diaspora: Tactics of Intervention in Contemporary Cultural Studies*. Bloomington: Indiana University Press, 1993.

Clare, John. *Cottage Tales*. Edited by Eric Robinson, David Powell, and P. M. S. Dawson. Manchester, UK: Carcanet, 1993.

———. *The Early Poems of John Clare, 1804–1822*. Edited by Eric Robinson and David Powell. 2 vols. Oxford: Clarendon Press, 1989.

———. *John Clare by Himself*. Edited by Eric Robinson and David Powell. New York: Routledge, 2002.

———. *The Later Poems of John Clare*. Edited by Eric Robinson. 2 vols. Oxford: Clarendon Press, 1984.

———. *The Natural History Prose Writings of John Clare*. Edited by Margaret Grainger. Oxford: Clarendon Press, 1983.

———. *Poems of the Middle Period, 1822–37*. Edited by Eric Robinson, David Powell, and P. M. S. Dawson. 5 vols. Oxford: Clarendon Press, 1996–2003.

———. Pforzheimer MSS. Pforzheimer Library. New York.

———. *The Shepherd's Calendar*. Edited by Eric Robinson, Geoffrey Summerfield, and David Powell. 2nd ed. Oxford: Oxford University Press, 1993.

Cleevely, R. J. "The Sowerbys and Their Publications in the Light of Manuscript Material in the British Museum (Natural History)." *Journal of the Society for the Bibliography of Natural History* 7, no. 4 (1976): 343–368.

Clive, John. *Macaulay: The Shaping of the Historian*. Cambridge, MA: Harvard University Press, 1987.

Cohn, Bernard S. *Colonialism and Its Forms of Knowledge: The British in India*. Princeton: Princeton University Press, 1996.

———. "The Recruitment and Training of British Civil Servants in India, 1600–1800." In *The Bernard Cohn Omnibus*. New Delhi: Oxford University Press, 1989. 500–553.

Coleridge, Samuel Taylor. *Lay Sermons*. Edited by R. J. White. Princeton: Princeton University Press, 1972.

Collinson, Peter. *"Forget Not Mee and My Garden": Selected Letters, 1725–1768, of Peter Collinson, F.R.S.* Edited by Alan W. Armstrong. Philadelphia: American Philosophical Society, 2002.

———. "Papers of Peter Collinson." Edited by William Massey. Library of the Linnean Society, London.

Cook, Elizabeth Heckendorn. "'Perfect' Flowers, Monstrous Women: Eighteenth-Century Botany and the Modern Gendered Subject." In *'Defects': Engendering the Modern Body*, edited by Helen Deutsch and Felicity Nussbaum. Ann Arbor: University of Michigan Press, 1999. 259–279.

Cornell, Drucilla, Michael Rosenfeld, and David Gray Carlson, eds, *Deconstruction and the Possibility of Justice*. New York: Routledge, 1992. 3–67.

Cowper, William. *The Task*. In *Poetical Works*, edited by H. S. Milford. 4th ed. London: Oxford University Press, 1934.

Crary, Jonathan. *Techniques of the Observer*. Cambridge, MA: MIT Press, 1991.

Crawford, Rachel. "Lyrical Strategies, Didactic Intent: Reading the Kitchen Garden Manual." In *Romantic Science*, edited by Noah Heringman. Albany: State University of New York Press, 2003. 199–222.

Crill, Rosemary, Susan Stronge, and Andrew Topsfield, eds. *Arts of Mughal India*. Ahmedabad: Mapin, 2004.

Curtis, Neal. *Against Autonomy: Lyotard, Judgement and Action*. Aldershot, UK: Ashgate, 2001.

Curtis, William. *The Botanical Magazine*. 42 vols. London: W. Curtis, 1787–1817.

Dalrymple, Alexander. *Oriental Repository. Oriental Repertory*. 2 vols. London: George Bigg, P. Emsly, 1793.

Dalrymple, William. *White Mughals*. London: Penguin, 2004.

Darwin, Charles. *Charles Darwin's Natural Selection*. Edited by R. C. Stauffer. Cambridge: Cambridge University Press, 1975.

———. *Darwin Correspondence Project*. www.darwinproject.ac.uk.

———. *On the various contrivances by which British and foreign orchids are fertilized by insects*. London: John Murray, 1802.

Darwin, Erasmus. *The Botanic Garden; a Poem, in Two Parts*. London: J. Johnson, 1791.

———. *Phytologia*. London: J. Johnson, 1800.

———. *System of Vegetables*. 2 vols. Lichfield: Botanical Society; London: John Jackson, 1783.

———. *Temple of Nature*. London: J. Johnson, T. Bensley, 1803.

———. *Zoonomia, or, the Laws of Life*. 2nd ed., corr. London: J. Johnson, 1796.

Daston, Lorraine. *Classical Probability in the Enlightenment*. Princeton: Princeton University Press, 1988.

Daston, Lorraine, and Katharine Park. *Wonder and the Order of Nature, 1150–1750*. New York: Zone Boos, 1998.

Daston, Lorraine, and Peter Galison. *Objectivity*. New York: Zone Books, 2007.

Davis, Natalie Zemon. *Women on the Margins: Three Seventeenth-Century Lives*. Cambridge, MA: Harvard University Press, 1995.

De Almeida, Hermione, and George H. Gilpin. *Indian Renaissance: British Art and the Prospect of India*. Aldershot, UK: Ashgate, 2005.

De Candolle, Alphonse. *Introduction a l'étude de la botanique, ou traité élémentaire de cette science*. 2 vols. Paris: Librairie encyclopédique de Roret, 1835.

De Candolle, Augustin Pyramus [A. P.] *Traité élémentaire de la botanique*. 2nd ed., rev. and aug. Paris: Deterville, 1819.

De Certeau, Michel. *The Practice of Everyday Life*. Translated by Steven Rendall. Berkeley and Los Angeles: University of California Press, 1984.

Deleuze, Gilles. *Difference and Repetition*. Translated by Paul Patton. New York: Columbia University Press, 1994.

———. *The Fold: Leibniz and the Baroque*. Translated by Tom Conley. Minneapolis: Minnesota University Press, 1993.

Deleuze, Gilles, and Félix Guattari. *A Thousand Plateaus*. Translated by Brian Massumi. Minneapolis: University of Minnesota Press, 1987.

Derrida, Jacques. *Archive Fever: A Freudian Impression*. Translated by Eric Prenowitz. Chicago: University of Chicago Press, 1996.

———. "*Différance*." In *Margins of Philosophy*, translated by Alan Bass. Chicago: Chicago University Press, 1982. 1–28.

———. "Force of Law: The 'Mystical Foundation of Authority.'" In *Deconstruction and the Possibility of Justice*, edited by Drucilla Cornell, Michael Rosenfeld, and David Gray Carlson, 3–67. New York: Routledge, 1992.

———. *Glas*. Translated by John P. Leavy and Richard Rand. Lincoln: University of Nebraska Press, 1986.

Desai, Chelna. *Ikat Textiles of India*. San Francisco: Chronicle Books, 1987.

Descourtilz, Michel Etienne. *Voyages d'un Naturaliste*. 3 vols. Paris: Dufar, 1809.

Desmond, Ray. *Dictionary of British and Irish Botanists and Horticulturalists*. London: Taylor & Francis, 1994.

———. *European Discovery of the Indian Flora*. Oxford: Oxford University Press and Royal Botanic Gardens, 1992.

———. *Sir Joseph Dalton Hooker: Traveller and Plant Collector*. Woodbridge, Suffolk: Antique Collector's Club, 1999.

De Zegher, Catherine. "Ocean Flowers and Their Drawings." In *Ocean Flowers: Impressions from Nature*, edited by Carol Armstrong and Catherine De Zegher. Princeton: Princeton University Press, 2004. 69–84.

Dirks, Nicholas. *Castes of Mind: Colonialism and the Making of Modern India*. Princeton: Princeton University Press, 2001.

Doherty, Meghan. "Robert Thornton's *A New Illustration:* Imaging and Imagining Nation and Empire." In *Visualising the Unseen, Imagining the Unknown, Perfecting the Natural: Art and Science in the 18th and 19th Centuries*, edited by Andrew Graciano. Cambridge: Cambridge Scholars Publishing, 2008. 49–82.

Dow, Alexander. *History of Hindostan*. 3 vols. London: S. A. Bechert & P. A. De Hontd, 1768–72.

Drayton, Richard. *Nature's Government: Science, Imperial Britain and the Improvement of the World*. New Haven: Yale University Press, 2000.

Drew, John. *India and the Romantic Imagination*. Delhi: Oxford University Press, 1987.

Druce, George C. *The Flora of Northamptonshire*. Arbroath: T. Buncle, 1930.

Dupré, John. *The Disorder of Things*. Cambridge, MA: Harvard University Press, 1993.

Dwyer, John D. "A Note on Plant Collectors and Localities in Panama." *Taxon* 17 (Feb. 1968): 107–108.

Edgerton MSS. British Library. London.

Edgeworth, Maria. *Belinda*. Edited by Kathryn Kirkpatrick. Oxford: Oxford University Press, 1994.

Edmondson, John. "Novelty in Nomenclature: The Botanical Horizons of Mary Delany." In Mark Laird and Alicia Weisberg-Roberts, eds. *Mrs. Delany & Her Circle*. New Haven: Yale University Press, 2009. 188–203.

Ehnbom, Daniel J. "'Passionate Delineation and the Mainstream of Indian Painting': The Mughal Style and the Schools of Rajasthan." In *Rethinking Early Modern India*, edited by Richard B. Barnett. New Delhi: Manohar Publishers, 2002. 179–192.

Encyclopaedia Britannica. 3 vols. Edinburgh: A. Bell & C. MacFarquhar or J. Balfour, 1771; *Encyclopaedia Britannica*. 10 vols. Edinburgh: A. Bell & C. MacFarquhar or J. Balfour, 1778; and *Encyclopaedia Britannica*. 18 vols. Edinburgh: A. Bell & C. MacFarquhar and J. Balfour, 1797. S.v. "Botany."

Engstrand, Iris H. W. "Mexico's Pioneer Naturalist and the Spanish Enlightenment." *Historian* 53: no.1 (1990): 17–32.

Ereshefsky, Marc. *The Poverty of the Linnaean Hierarchy*. Cambridge: Cambridge University Press, 2001.

Fara, Patricia. *Sex, Botany, and Empire*. New York: Columbia University Press, 2003.

Festa, Lynn. *Sentimental Figures of Empire in Eighteenth-Century Britain and France*. Baltimore: Johns Hopkins University Press, 2006.

Figueira, Dorothy Matilda. *Translating the Orient: The Reception of Śākuntala in Nineteenth-Century Europe*. Albany: State University Press of New York, 1991.

Figuier, Louis. *The Vegetable World*. New York: D. Appleton, 1867.

Fischer, Bernard. Foreword to Desmond Peter Lazaro. *Materials, Methods, and Symbolism in the Pichhvai Painting Tradition of Rajasthan*. Ahmedabad: Mapin Publishing, 2005.

Fletcher, Loraine. *Charlotte Smith: A Critical Biography*. Houndmills, Basingstoke, UK: Macmillan, 1998.

Forbes, James. *Oriental Memoirs*. 4 vols. London: T. Bensley, White, Cochrane, 1813.

Ford, Brian J. "Brownian Movement in Clarkia Pollen: A Reprise of the First Observations." *Microscope* 40, no. 4 (1992): 235–241.

Fothergill, John. *An Account of the life and travels in the work of the ministry, to which are added, divers epistles to Friends in Great Britain and America, on various occasions*. London: Mary Hinde, 1773.

———. *Some anecdotes of the late Peter Collinson, Fellow of the Royal Society, in a letter to a friend*. London: n.p., 1785.

Foucault, Michel. *The Archaeology of Knowledge*. Translated by A. M. Sheridan Smith. New York: Harper, 1972.

———. *Discipline and Punish: The Birth of the Prison*. Translated by Alan Sheridan. New York: Random House, 1977.

———. *Les Mots et les Choses*. Paris: Gallimard, 1966. Translated by Alan Sheridan as *The Order of Things*. New York: Random House, 1970.

François, Anne-Lise. *Open Secrets*. Stanford, CA: Stanford University Press, 2008.

Fulford, Tim. "Coleridge, Darwin, Linnaeus: The Sexual Politics of Botany." *Wordsworth Circle* 28, no.3 (1997): 124–130.

Fulford, Tim, Debbie Lee, and Peter J. Kitson. "Indian Flowers and Romantic Orientalism." In *Literature, Science and Exploration in the Romantic Era: Bodies of Knowl-*

edge. Edited by Fulford, Lee, and Kitson. Cambridge: Cambridge University Press, 2006. 71–89.

Garrison, Alysia E. "'Disdaining Bounds of Place and Time,' Staining Language with Furze and Burvine: John Clare's Nomadic Poetics." *Literature Compass* 3, no. 3 (2006): 376–387.

Gasché, Rodolphe. *Inventions of Difference: On Jacques Derrida*. Cambridge, MA: Harvard University Press, 1994.

Gascoigne, John. *Joseph Banks and the English Enlightenment*. Cambridge: Cambridge University Press, 1994.

———. "The Ordering of Nature and the Ordering of Empire: A Commentary." In *Visions of Empire*. Edited by David Philip Miller and Peter Hanns Reill. Cambridge: Cambridge University Press, 1994. 107–116.

Gaskell, Elizabeth. *Mary Barton*. Edited by Macdonald Daly. New York: Penguin, 1996.

Génlis, Stephanie de. *La Botanique historique et littéraire*. Paris: Chez Maradan, 1790.

Gentleman's Magazine. London: Edward Cave, 1731–1922.

Gettens, Rutherford J. "Review." *Bulletin of the Museum and Picture Gallery of Baroda*." *Studies in Conservation* 19 (1974): 56–57.

Gidal, Eric. *Poetic Exhibitions: Romantic Aesthetics and the Pleasures of the British Museum*. Lewisburg: Bucknell University Press, 2001.

Gigante, Denise. *Life: Organic Form and Romanticism*. New Haven: Yale University Press, 2009.

Giglioni, Guido. "What Ever Happened to Francis Glisson? Albrecht Haller and the Fate of Eighteenth-Century Irritability." *Science in Context* 21 (2008): 465–493.

Gillispie, Charles C. *The Edge of Objectivity: An Essay in the History of Scientific Ideas*. Princeton: Princeton University Press, 1960.

Gillray, James. "New Morality." *The Anti-Jacobin* (Aug. 1, 1798).

Gilmartin, Kevin. *Writing against Revolution: Literary Conservatism in Britain, 1790–1832*. Cambridge: Cambridge University Press, 2007.

Gladstone, Jo. "'New World of English Words': John Ray." In *Language, Self, and Society: A Social History of Language*, ed. Peter Brooks and Roy Porter. Cambridge: Cambridge University Press, 1991. 115–153.

Gledhill, D. G. *The Names of Plants*. 3rd ed. Cambridge: Cambridge University Press, 2002.

Godwin, William. "Fleetwood: Or, the New Man of Feeling." In *Collected Novels and Memoirs of William Godwin*, vol. 5. Edited by Pamela Clemit. London: William Pickering, 1992.

———. *Life of Chaucer*. London: Richard Phillips, 1803.

Goethe, Johann Wolfgang von. *Goethe: Scientific Studies*. Translated by Douglas Miller. New York: Suhrkamp, 1988.

———. *Goethe's Botanical Writings*. Translated by Bertha Mueller. Woodbridge, CT: Ox Bow Press, 1952.

———. *Poems and Epigrams*. Translated by Michael Hamburger. Cambridge: Cambridge University Press, 1980.

———. *Sämtliche Werke*. Vol 12, *Zur Naturwissenschaft überhaupt, besonders zur Morphologie*. Edited by Hans J. Becker, Gerhard H. Müller, John Neubauer, and Peter Schmidt. Munich: Carl Hanser Verlag, 1989.

——. *Die Schriften zur Naturwissenschaft*. Edited by Dorothea Kuhn, Wolf von Engelhardt, and Irmgard Müller. 23 vols. Weimar: Hermann Böhlaus Nachfolger, 1947.

——. *Selected Poems*. Translated by Christopher Middleton. Boston, MA: Suhrkamp & Insel, 1983.

——. *Versuch die Metamorphose der Pflanzen zu Erklären*. Gotha, Germany: C. W. Ettinger, 1790. Repr. in Arber, trans., "Goethe's Botany," *Chronica Botanica* 10, no. 2 (1946), 67–124.

Goldsmith, Oliver. *A History of the Earth, and Animated Nature*. 8 vols. 2nd ed, London: J. Nourse, 1779, 2:225.

Goldstein, Amanda Jo. "Growing Old Together: Figural Materialism in Shelley's *Triumph of Life*," in *Sweet Science: Romantic Materialism and the New Sciences of Life*," University of California, Berkeley, PhD dissertation, 2011.

Goodman, Kevis. "Romantic Poetry and the Science of Nostalgia." In *Cambridge Companion to British Romantic Poetry*, ed. James K. Chandler and Maureen N. McLane. Cambridge: Cambridge University Press, 2008. 195–216.

Goodridge, John, and Kelsey Thornton. "John Clare: The Trespasser." In *John Clare in Context*. Edited by Hugh Haughton, Adam Phillip, and Geoffrey Summerfield. Cambridge: Cambridge University Press, 1994. 87–129.

Goswamy, B. N., and Eberhard Fischer. *Pahari Masters: Court Painters of Northern India*. Delhi: Oxford University Press, 1997.

Gould, Stephen Jay. *Dinosaur in a Haystack*. New York: Random House, 1995.

Graham, Linda E., James M. Graham, and Lee W. Wilcox, eds. *Plant Biology*. 2nd ed. Upper Saddle River, NJ: Pearson / Prentice Hall, 2006.

Graham, Maria. *Journal of a Residence in India*. Edinburgh: George Ramsay, 1812.

Grandville, J. J., and Ch. Geoffroy. *Les Fleurs animées*. 2 vols. Paris: Garnier Frères, 1867.

Grant, Ian Hamilton. *Philosophies of Nature after Schelling*. London: Continuum, 2006.

Gray, Samuel F. *A Natural Arrangement of British Plants*. vol. 1. London: Baldwin, Cradock, & Joy, 1821.

Grosz, Elizabeth A. *Nick of Time: Politics, Evolution, and the Untimely*. Durham: Duke University Press, 2004.

Grove, Richard. *Green Imperialism: Colonial Expansion, Tropical Island Edens, and the Origins of Environmentalism, 1600–1860*. Cambridge: Cambridge University, 1995.

Guest, Harriet. *Small Change: Women, Patriotism, and Learning, 1750–1810*. Chicago: Chicago University Press, 2000.

Guignon, Charles. "Moods in Heidegger's Being and Time." In *What Is an Emotion?* Edited by Cheshire Calhoun and Robert C. Solomon. Oxford: Oxford University Press, 1984. 230–243.

Hacking, Ian. *The Emergence of Probability*. Cambridge: Cambridge University Press, 1975.

——. *The Taming of Chance*. Cambridge: Cambridge University Press, 1990.

Hadaway, W. S. *Cotton Painting and Printing in the Madras Presidency*. Madras: Government Press, 1917.

Hanelt, Peter, ed. *Mansfeld's Encyclopedia of Agricultural and Horticultural Crops*. New York: Springer, 2001. http://mansfeld.ipk–gatersleben.de/Mansfeld/Taxonomy.

Harley, R. D. *Artists' Pigments, c. 1600–1835*. 2nd ed. London: Archetype Publications, 2001.

Harris, Steven J. "Long-Distance Corporations, Big Sciences, and the Geography of Knowledge." *Configurations* 6, no. 2 (1998): 269–304.

Hayden, Ruth. *Mrs Delany: Her Life and Works*. 1980. Repr.; London: British Museum, 2000.

Heads, Michael. "Principia Botanica: Corizat's Contribution to Botany." *Tuatara* 27, no. 1 (1984): 26–45.

Heaney, Seamus. "John Clare's 'Prog.'" In *The Redress of Poetry*, 63–82. New York: Farrar, Strauss, & Giroux, 1995.

Hegel, Georg Wilhelm Friedrich. *Aesthetics: Lectures on Fine Art*. Translated by T. M. Knox. 2 vols. Oxford: Clarendon Press, 1975.

——. *Encyclopaedia of the Philosophical Sciences, Philosophy of Nature*. Translated by A. V. Miller. Oxford: Clarendon Press, 1970.

——. *Enzyklopädie: Sämtliche Werke*. Vol. 10. Jubiläum ed. Stuttgart: Fromann, 1958.

——. *Enzyklopädie Der Philosophischen Wissenschaften Im Grundrisse*. In *Gesammelte Werke*. Vols. 17, 19, and 20 (1817, 1827, and 1830). Edited by Hans-Christian Lucas, Udo Remeil, Wolfgang Bonsiepen, and Klaus Grotsch. Hamburg: Felix Meiner Verlag, 1965.

——. *Hegel's Logic, Being Part 1 of the Encyclopedia of the Philosophical Sciences (1830)*. Translated by William Wallace. Foreword by J. N. Findlay. 3rd ed. 1873. Repr., Oxford: Clarendon Press, 1975.

——. "Jenaer Systementwürfe III." In *Gesammelte Werke*, vol. 8. Edited by Johann Heinrich Trede and Rolf-Peter Horstmann. Hamburg: Felix Meiner Verlag, 1976.

——. *Phänomenologie Des Geistes*. Edited by Hans-Friedrich Wessels and Heinrich Clairmont. Hamburg: Felix Meiner Verlag, 1988.

——. *Phenomenology of Spirit*. Translated by A. V. Miller. Oxford: Oxford University Press, 1977.

——. *Philosophy of History*. Translated by John Sibree. Amherst, NY: Prometheus Books, 1991.

——. *The Philosophy of Nature*. Translated by Michael Petry. 3 vols. London: George & Allen, 1970.

——. *Sämmtliche Werke*. Edited by Hermann Glockner. 26 vols. Stuttgart: Fromann, 1927–40.

——. *Werke*. Edited by Eva Moldenhauer and Karl Markus Michel. 20 vols. Frankfurt am Main: Suhrkamp Verlag, 1970.

Heidegger, Martin. *Being and Time*. Translated by John Macquarrie and Edward Robinson. New York: Harper & Row, 1962. Originally published as *Sein und Zeit*. Tübingen: Max Niemeyer Verlag, 1967.

Henrey, Blanche. *British Botanical and Horticultural Literature before 1800*. 3 vols. London: Oxford University Press, 1975.

Hensen, Nilas Thode. "The Medical Skills of the Malabar Doctors in Tranquebar, India, as Recorded by Surgeon T. L. F. Folly, 1798." *Medical History* 49 (2005): 489–515.

Heringman, Noah. "'Stones So Wondrous Cheap.'" *Studies in Romanticism* 37, no. 1 (1998): 43–62.

Heyes, Robert. "'Looking to Futurity': John Clare and Provincial Culture." University of London PhD diss., 1999.

——. "Writing Clare's Poems." In *John Clare: New Approaches*. Edited by John Goodridge and Simon Kövesi. Peterborough: John Clare Society, 2000. 33–45.

Hill, John. Preface to *The Vegetable System.* London: J. Hill, 1759–75.
Hoffmeister, Johannes. *Goethe und der Deutsche Idealismus: Eine Einführung zu Hegel's Realphilosophie.* Leipzig: Felix Meiner Verlag, 1932.
Holland, Jocelyn. *German Romanticism and Science: The Procreative Poetics of Goethe, Novalis, and Ritter.* New York: Routledge, 2009.
Holmes, Richard. *The Age of Wonder: How the Romantic Generation Discovered the Beauty and Terror of Science.* New York: Pantheon, 2008.
Hong, Grace. "Colonialism, Transnationalism, and Sexuality in Shani Mootoo's *Cereus Blooms at Night.*" *Meridians* 7, no. 1 (2006): 73–103.
Hull, David. "Biological Species: An Inductivist's Nightmare." In *How Classification Works: Nelson Goodman among the Social Sciences.* Edited by Mary Douglas and David Hull. Edinburgh: Edinburgh University Press, 1992. 42–68.
——. *Metaphysics of Evolution.* Albany: State University of New York Press, 1989.
——. *Science and Selection.* Cambridge: Cambridge University Press, 2001.
Hulton, Paul. *Jacques Le Moyne De Morgues.* 2 vols. London: Trustees of the British Museum, 1977.
Humboldt, Alexander von. *Essai sur la géographie des plantes.* Paris: Chez Levrault, Schoell, 1806; and *Ideen zu einer Geographie der Pflanzen.* Tübingen: F. G. Cotta, 1807.
Huneman, Philippe. "From the *Critique of Judgment* to the Hermeneutics of Nature: Sketching the Fate of Philosophy of Nature after Kant." *Continental Philosophy Review* 39, no. 1 (2006): 1–34.
Ingrouille, Martin, and Bill Eddie. *Plants: Diversity and Evolution.* Cambridge: Cambridge University Press, 2006.
Irwin, John. "Indian Textiles in Historical Perspective." MARG 15, no. 4 (1962): 4–6.
Jackson, Benjamin Daydon. *A Glossary of Botanic Terms.* 4th ed. London: Gerald Duckworth, 1949.
Jackson, Maria E. *Botanical Dialogues between Hortensia and Her Four Children.* London: J. Johnson, 1797. Repr. *Botanical Lectures by a Lady.* London J. Johnson, 1804.
——. *The Florist's Manual or, Hints for the Construction of a Gay Flower Garden, with Observations of the Best Methods of Preventing the Depredations of Insects, by the Author of Bot. Dial. and Sketches Etc.* London: Henry Colburn, 1816.
——. *Sketches of the Physiology of Vegetable Life.* London: John Hatcherd, 1811.
"James Lee." *The Gentleman's Magazine* 65 (1795): 1052.
Jameson, Fredric. *Postmodernism, or, The Cultural Logic of Late Capitalism.* Durham: Duke University Press, 1991.
——. *A Singular Modernity: Essay on the Ontology of the Present.* London: Verso, 2002.
Janowitz, Anne F. "The Transit of Gypsies in Romantic Period Poetry." In *The Country and the City Revisited.* Edited by Gerald MacLean, Donna Landry, and Joseph P. Ward. Cambridge: Cambridge University Press, 1999). 213–230.
——. *Women Romantic Poets: Anna Barbauld and Mary Robinson.* Horndon, Tavistock, Devon: Northcote House Publishers, 2004.
Jardine, Nicholas. *Scenes of Inquiry.* 2nd ed. Oxford: Clarendon Press, 2000.
Jardine, Nicholas, and Andrew Cunningham, eds. *Romanticism and the Sciences.* Cambridge: Cambridge University Press, 1990.
Jaykar, Pupul. "Traditional Textiles in Indian Art." MARG 15, no. 4 (1962): 6–34.
Johnson, Samuel. *Dictionary.* 5th ed., corrected. 1773.
Jones, Colin, and Emily Richardson. "How Not to Laugh in the French Enlighten-

ment: The Saint-Aubin Livre De Caricatures." France Center Chicago, University of Chicago. http://fcc.uchicago.edu/pdf/enlightenment.pdf.

Jones, William. "Botanical Observations on Select Indian Plants." *Asiatic Researches*. 24 vols. 1799. Repr., New Delhi: Cosmo Publications, 1979, 4: 231–303; "A Catalogue of Indian Plants."*Asiatic Researches*, 4: 225–229; "The Design of a Treatise on the Plants of India." *Asiatic Researches*, 2:270–276; "On the Spikenard of the Ancients." *Asiatic Researches*, 2:315–325; "Second Anniversary Discourse." *Asiatic Researches*, 1: 335–342; "Tenth Anniversary Discourse." *Asiatic Researches*, 4: xii–xxxiv.

——. *The Letters of Sir William Jones*. Edited by Garland Cannon, 2 vols. Oxford: Clarendon Press, 1970.

——. *Sir William Jones: Selected Poetical and Prose Works*. Edited by Michael J. Franklin. Cardiff: University of Wales Press, 1995.

——. *Sacontalá. Vol. 6, in The Works of Sir William Jones*. 13 vols. 1807 repr. New Delhi: Agam Prahashan, 1976–80.

Jussieu, Antoine Laurent de. *Genera Plantarum, secundum ordines naturales disposita juxta methodum in Horto Regio Parisiensi exaratam*. Paris: Charles Herissant, 1789. Repr. Frans A. Stafleu. Weinheim, J. Cramer, 1964.

Kain, Philip J. *Hegel and the Other: A Study of the Phenomenology of the Spirit*. Albany: State University of New York Press, 2005.

Kain, Roger J. P., John Chapman, and Richard R. Oliver. *The Enclosure Maps of England and Wales, 1595–1918*. Cambridge: Cambridge University Press, 2004.

Kant, Immanuel. *Critique of the Power of Judgment*. Translated by Paul Guyer and Eric Matthews. Edited by Paul Guyer. Cambridge: Cambridge University Press, 2000.

——. *Immanuel Kant: Schriften zur Ästhetik und Naturphilosophie*. Edited by Manfred Frank and Véronique Zanetti. 3 vols. Frankfurt am Main: Deutscher Klassiker Verlag, 1996.

Keach, William. "A Regency Prophecy and the End of Anna Barbauld's Career." *Studies in Romanticism* 33, no. 4 (1994): 569–577.

Keats, John. *Complete Poems*. edited by Jack Stillinger. Cambridge, MA: Harvard University Press, 1978.

——. "La Belle Dame sans Merci." *The Indicator* no. 31 (May 10, 1820): 246–247.

Kejariwal, O. P. *The Asiatic Society of Bengal*. Delhi: Oxford University Press, 1988.

Kelley, Theresa M. "Adorno Nature Hegel." In *Language without Soil: Adorno and Late Philosophical Modernity*. Edited by Gerhard Richter. New York: Fordham University Press, 2009. 161–198.

——. "Poetics and the Politics of Reception: Keats's 'La Belle Dame sans Merci.'" *ELH* 54, no. 2 (1987): 333–362.

——. "Reading Justice: From Derrida to Shelley and Back." In "Romanticism and the Legacies of Jacques Derrida," edited by David Clark. Festschrift for Jacques Derrida, *Studies in Romanticism* 46, no. 3 (2007): 267–288.

——. *Reinventing Allegory*. Cambridge: Cambridge University Press, 1997.

Kent, Elizabeth. *Flora Domestica*. London: Taylor & Hessey, 1823.

King, Amy Mae. *Bloom: Botanical Vernacular in the English Novel*. Oxford: Oxford University Press, 2003.

King-Hele, Desmond. *Erasmus Darwin*. London: Giles de La Mare Publishers, 1999.

Klonk, Charlotte. *Science and the Perception of Nature*. New Haven: Yale University Press, 1996.

Koerner, Lisbet. "Goethe's Botany: Lessons of Feminine Science." *Isis* 84, no. 3 (1993): 470–495.

——. *Linnaeus: Nature and Nation*. Cambridge, MA: Harvard University Press, 1999.

Kopytoff, Igor. "The Cultural Biography of Things: Commoditization as Process." In *The Social Life of Things*. Edited by Arjun Appadurai, Cambridge: Cambridge University Press, 1986. 64–91.

Koselleck, Reinhart. *Futures Past: On the Semantics of Historical Time*. Translated by Keith Tribe. New York: Columbia University Press, 1979.

Krell, David Farrell. *Contagion: Sexuality, Disease, and Death in German Idealism and Romanticism*. Bloomington and Indianapolis: Indiana University Press, 1998.

Krishna, Vijay. "Flowers in Indian Textile Designs." *Journal of Indian Textile History* 7 (1967): 1–44.

Kriz, Kay Dian. "Curiosities, Commodities, and Transplanted Bodies in Hans Sloane's *Natural History of Jamaica*." In *An Economy of Colour*. Edited by K. Dian Kriz and Geoff Quilley. Manchester: Manchester University Press, 2003. 85–105.

Kuehni, Rolf G., and Andreas Schwartz. *Color Ordered: A Survey of Color Order Systems from Antiquity to the Present*. Oxford: Oxford University Press, 2008.

Kuhn, Bernhard. "'A Chain of Marvels': Botany and Autobiography in Rousseau." *European Romantic Review* 17, no. 1 (2006): 1–20.

Kumar, Deepak. *Science and the Raj*. Bombay: Oxford University Press, 1995.

Kumar, Deepak, ed. *Science and Empire*. Delhi: Anamika Prakashan, 1991.

Kyd, Robert. Letter to Bengal Government. 1 June 1786. D.T.C. vol. 7. Folios 57–67. BMNH.

Laird, Mark. *The Flowering of the Landscape Garden: English Pleasure Grounds, 1820–1800*. Philadelphia: University of Pennsylvania Press, 1999.

Laird, Mark, and Alicia Weiberg-Roberts, eds. *Mrs. Delany and Her Circle*. New Haven: Yale University Press, 2009.

La Mettrie, Julien Offray de. *L'Homme plante*. vol. 1, *Oeuvres philosophiques*. 3 vols. 1st ed. 1751; Berlin: Charles Tutot, 1796.

Langan, Celeste. *Romantic Vagrancy*. Cambridge: Cambridge University Press, 1995.

Larson, James L. *Interpreting Nature: The Science of Living Form from Linnaeus to Kant*. Baltimore: Johns Hopkins University Press, 1994.

——. *Reason and Experience: The Representation of the Natural Order in the Work of Carl von Linné*. Berkeley: University of California Press, 1971.

Latour, Bruno. *Reassembling the Social: An Introduction to Actor-Network-Theory*. Oxford: Oxford University Press, 2005.

Leask, Nigel. *British Romantic Writers and the East*. Cambridge: Cambridge University Press, 1992.

Lee, James. *An Introduction to Botany*. 3rd ed. London: J. F. and C. Rivington, 1776.

Lelyveld, David. "The Fate of Hindustani: Colonial Knowledge and the Project of a National Language." In *Orientalism and the Postcolonial Predicament*. Carol Breckenridge and Peter van der Veer, eds. Philadelphia: University of Pennsylvania Press, 1993. 189–214.

Lenoir, Timothy. *The Strategy of Life: Teleology and Mechanics in Nineteenth-Century German Biology*. Chicago: Chicago University Press, 1992.

"Letter Written by an Officer Who Lately Served in Bengal." *Gentleman's Magazine* 42 (Feb. 1772): 69.

Levine, Caroline. "The Criticism of Purpose." *The Valve*. Monday, Apr. 16, 2007. http://thevalve.org/go/valve/go/valve/the_criticism_of_purpose.

Lindley, John. *An Introductory Lecture Delivered at the Opening of the University of London*. London: Richard Taylor, 1829.

——. *Ladies Botany: Or, a Familiar Introduction to the Study of the Natural System of Botany*. 2 vols. 6th ed. London: Ridgway, 1862.

Lindley, John, and Thomas Moore. *The Treasury of Botany*. 2 vols. London: Longmans, Green, 1889.

Lindsey, Jack. Introduction to *Autobiography of Joseph Priestley*. Cranbury: Associated University Presses, 1970. 1–66.

Linnaeus, Carl von. *Genera Plantarum*. 5th ed. Stockholm: Lars Salvius, 1754.

——. *Philosophia Botanica*. Translated by Stephen Freer. Oxford: Oxford University Press, 2003.

——. *Species Plantarum*. Edited by William T. Stearn. 2 vols. 1753. Repr., London: Ray Society, 1957–59.

Llewellyn-Jones, Rosie. Introduction to *A Man of the Enlightenment in Eighteenth-Century India: The Letters of Claude Martin, 1766–1800*. New Delhi: Orient Blackswan, 2003.

——. *Lucknow Then and Now*. Mumbai: MARG, 2003.

——. *A Very Ingenious Gentleman: Claude Martin in Early Colonial India*. Delhi: Oxford University Press, 1992.

Locke, John. *Essay Concerning Human Understanding*. Edited by Peter H. Nidditch. Oxford: Clarendon Press, 1975.

Loewenstein, David. "Digger Writing and Rural Dissent in the English Revolution: Representing England as a Common Treasury." In *The Country and the City Revisited*, ed. Gerald MacLean, Donna Landry, and Joseph P. Ward. Cambridge: Cambridge University Press, 1999. 74–88.

Lucas, Hans-Christian. "The 'Sovereign Ingratitude' of Spirit toward Nature: Logical Qualities, Corporeity, Animal Magnetism, and Madness in Hegel's 'Anthropology.'" *Owl of Minerva* 23, no. 2 (1992): 131–150.

Lynch, Deidre Shauna. "'Young ladies are delicate plants': Jane Austen and Greenhouse Romanticism," *ELH* 77, no. 3 (2010): 689–729.

Lyotard, Jean-Francois. *Peregrinations: Law, Form, Event*. New York: Columbia University Press, 1988.

Mabberley, David. *Jupiter Botanicus: Robert Brown and the British Museum*. London: British Museum, 1985.

Mabey, Richard. *Flora Britannica*. London: Chatto & Windus, 1996.

——. *Nature Cure*. London: Pimlico, 2006.

——. "Pastoral Suite." *Guardian Unlimited*. Saturday, 29 July 2006. www.guardian.co.uk/books/2006/jul/29/featuresreviews.guardianreview2.

Macaulay, Thomas Babington. *Speeches by Lord Macauley with His Minute on Education*. Edited by G. M. Young. 1935. Repr., New York: AMS Press, 1979.

——. *Thomas Babington Macaulay, Selected Writings*. Edited by John Clive and Thomas Pinney. Chicago: University of Chicago Press, 1972.

Mackay, Christine, and Aditi Nath Sarkar. "Lalighat Pats: An Examination of Techniques and Materials." In *Scientific Research on the Pictorial Arts of Asia: Proceedings of the Second Forbes Symposium at the Freer Gallery of Art*, edited by Paul Jett,

John Winter, and Blythe McCarthy, 135–142. Washington, DC: Freer Gallery of Art, n.d.

Mackay, David. "Agents of Empire: The Banksian Collectors and Evaluation of New Lands." In *Visions of Empire*. Edited by David Philip Miller and Peter Hanns Reill. Cambridge: Cambridge University Press, 1996. 38–57.

Mackenzie, Henry. *Man of Feeling*. Edited by Hamish Miles. London: Scholartis Press, 1928.

Magee, Judith. *The Art and Science of William Bartram*. University Park, PA: Pennsylvania State University Press; London: Natural History Museum, 2007.

Mahood, M. M. *The Poet as Botanist*. Cambridge: Cambridge University Press, 2008.

Majeed, Javed. *Ungoverned Imaginings: James Mill's History of British India*. Oxford: Clarendon Press, 1992.

Majumdar, Girija Prasanna. *Vanaspati: Plants and Plant-Life as in Indian Treatises and Traditions*. Calcutta: University of Calcutta Press, 1925.

Marshall, P. G., ed. *The Oxford History of the British Empire*. vol. 2: *The Eighteenth Century*. Oxford: Oxford University Press, 2001.

Martin, Philip W. "John Clare's Gypsies: Problems of Placement and Displacement in Romantic Critical Practice." In *Placing and Displacing Romanticism*. Edited by Peter J. Kitson. Aldershot: Ashgate, 2001. 48–59.

Mayall, David. *English Gypsies and State Policies*. Hatfield, UK: University of Hertfordshire Press; Centre de recherches et siganes. Université René Descartes, Paris, 1995.

Mayr, Ernst. "Another Look at the Species Problem." In *What Makes Biology Unique?* Cambridge, MA: Harvard University Press, 2004. 171–193.

——"The Species Category." In *Toward a New Philosophy of Biology*. Cambridge, MA: Harvard University Press, 1988. 315–334.

——. "Species Concepts and Definitions." In *The Species Problem: A Symposium*. Edited by Ernst Mayr. Washington, DC: American Association for the Advancement of Science, 1957. 1–20.

McKusick, James. "'A Language That Is Ever Green': The Ecological Vision of John Clare." *University of Toronto Quarterly* 61, no. 2 (1991–92): 226–249.

McLachlan, Herbert. *Warrington Academy: Its History and Influence*. Manchester: Chetham Society, 1943.

McLean, R. C., and W. R. Ivimey Cook, eds. *Textbook of Theoretical Botany*. 2 vols. London: Longmans, 1951.

McQuat, Gordon R. "The Origin of 'Natural Kinds': Keeping Essentialism at Bay in the Age of Reform." *Intellectual History Review* 19, no. 2 (2009): 211–230.

——. "Species, Rules, and Meaning: The Politics of Language and the Ends of Definitions in 19th Century Natural History." *British History of the Philosophy of Science* 27, no. 4 (1996): 473–519.

Metcalf, Thomas R. *Ideologies of the Raj*. Cambridge: Cambridge University Press, 1994.

Mill, James. *History of British India*. 9 vols. 2nd ed. 1820. Repr., New Delhi: Associated Publishing House, 1972.

Miller, David Philip, and Peter Hans Reill, eds. *Visions of Empire*. Cambridge: Cambridge University Press, 1996.

Miller, Elaine P. *The Vegetative Soul: From Philosophy of Nature to Subjectivity in the Feminine*. Albany: State University of New York Press, 2002.

Miller, Eric. “Enclosure and Taxonomy in John Clare.” *Studies in English Literature, 1500–1900* 40, no. 4 (2000): 635–657.
Miller, Philip. *Gardeners Dictionary*. 16th ed. London: John & Francis Rivington, J. Hinton, Hawes, Clarke, & Collins, R. Horsfield, 1775.
Milton, John. *Paradise Lost*. Edited by Merritt Y. Hughes. New York: Odyssey Press, 1935.
Mingay, G. E. *Parliamentary Enclosure in England: An Introduction to its Causes, Incidence and Impact, 1750–1850*. London: Longmans, 1997.
Mitchell, Robert. “Cryptogamia.” *European Romantic Review* 21, no. 5 (2010): 631–651.
Mitter, Partha. *Much Maligned Monsters: History of European Reactions to Indian Art*. Oxford: Clarendon Press, 1977.
Monthly Review. London, 1749–1845. Vols. 1–81, May 1749–Dec. 1789 and 2nd series. vols. 1–108, Jan. 1790–Nov. 1825.
Moore, Lisa. *Sister Arts: The Erotics of Lesbian Landscapes*. Minneapolis: University of Minnesota Press, 2011.
Moore, Thomas. *Lalla Rookh*. In *Poetical Works*. Edited by David Herbert. Edinburgh: William P. Nimmo, 1873. 49–125.
“Moore’s Lalla Rookh.” *Critical Review* 5 (1817): 562.
Mootoo, Shani. *Cereus Blooms at Night*. New York: Avon, 1996.
Morton, Allen G. *History of Botanical Science*. New York: Academic Press, 1981.
Morton, Timothy. *Ecology without Nature: Rethinking Environmental Aesthetics*. Cambridge, MA: Harvard University Press, 2007.
Mukherjee, S. N. *Sir William Jones: A Study in Eighteenth-Century British Attitudes to India*. Cambridge University Press, 1968.
Mukherji, Abhijit. “European Jones and Asiatic Pandits.” *Journal of the Asiatic Society of Calcutta* 27, no. 1 (1985): 43–58.
Nair, Savithri Preetha. “Native Collecting and Natural Knowledge, 1798–1832; Raja Serfoji II of Tanjore as a ‘Centre of Calculation.’” *Journal of the Royal Asiatic Society*. Series 3. 15, no. 3 (2005): 279–302.
Nancy, Jean-Luc. *Hegel: The Restlessness of the Negative*. Translated by Jason Smith and Steven Miller. Minneapolis: Minnesota University Press, 2002.
———. *The Inoperable Community*. Translated by Peter Connor, Lisa Garbus, Michael Holland, and Simona Sawhney. Minneapolis: University of Minnesota Press, 1991.
Neeson, J. M. *Commoners: Common Right, Enclosure and Social Change in England, 1700–1820*. Cambridge: Cambridge University Press, 1993.
Neubauer, John. “Organic Form in Romantic Theory: The Case of Goethe’s Morphology.” In *Romanticism across the Disciplines*. Edited by Larry H. Peer. Lanham, MD: University Press of America, 1998. 207–230.
The New Asiatic Miscellany. Calcutta: Joseph Cooper, 1789.
Newton, Michael. “Review of *Echolalias: On the Forgetting of Language*, by Daniel Heller-Roazen.” *London Review of Books* 28, no. 4 (2006): 25–26.
Ngai, Sianne. *Ugly Feelings*. Cambridge, MA: Harvard University Press, 2005.
Nicholson, Louise. *The Red Fort, Delhi*. London: Tauris Parke Books, 1989.
Nickelson, Kärin. *Draughstmen, Botanists, and Nature: The Construction of Eighteenth-Century Botanical Illustrations*. Dordrecht: Springer, 2006.
Noltie, Henry J. *The Dapuri Drawings: Alexander Gibson and the Bombay Botanic Gar-*

dens. London and Edinburgh: Antique Collectors' Club and Royal Botanic Garden Edinburgh, 2002.

——. *Indian Botanical Drawings, 1793–1868*. Edinburgh: Royal Botanic Garden, 1999.

——. *Robert Wight and the Botanical Drawings of Rungiah and Govindoo*. 3 vols. Edinburgh: Royal Botanical Garden, Edinburgh, 2007.

——. "Robert Wight and the Illustration of Indian Botany: The Hooker Lecture." *Linnean Society of London*. Special Issue no. 6 (2005).

Oerlemans, Onno. *Romanticism and the Materiality of Nature*. Toronto: University of Toronto Press, 2002.

Oken, Lorenz. *Lehrbuch*. 3rd ed. Zurich: F. Schulthess, 1843; Translated by A. Tulk, *Elements of Physiophilosophy*. London: Ray Society, 1847.

O'Quinn, Daniel. "Haunted Trees circa 1793." In "Romantic Frictions," edited by Theresa Kelley. *Praxis* 2011. www.rc.umd.edu/praxis/.

Osborn MSS Files, Beinecke Library, Yale University.

Owenson, Sydney, Lady Morgan. *The Missionary*. Edited by Julia M. Wright. Peterborough, Ontario: Broadview Press, 2002.

Packham, Catherine. "The Science and Poetry of Animation: Personification, Analogy, and Erasmus Darwin's Loves of the Plants." *Romanticism* 10, no.2 (2004): 191–208.

Page, Michael. "The Darwin before Darwin: Erasmus Darwin, Visionary Science, and Romantic Poetry." *Papers on Language and Literature* 41, no. 2 (2005): 146–169.

Paine, Thomas. *Rights of Man*. 2 vols. London: J. Parsons, 1792.

Pascoe, Judith. "Female Botanists and the Poetry of Charlotte Smith," in *Re-Visioning British Women Writers, 1776–1837*. Edited by Carol Shiner Wilson and Joel Haefner. Philadelphia: University of Pennsylvania Press, 1994. 193–209.

——. "'Unsex'd Females': Barbauld, Robinson, and Smith." In *The Cambridge Companion to English Literature, 1740–1830*. Edited by Tom Keymer and Jon Mee. Cambridge: Cambridge University Press, 2004. 211–226.

Pavord, Anna. *The Naming of Names: The Search for Order in the World of Plants*. London: Bloomsbury, 2005.

Phillips, Henry. *Flora Historica or the Three Seasons of the British Parterre*. 2nd. rev. ed. 2 vols. London: E. Lloyd & Son, 1829.

Picart, Bernard. *The Ceremonies and Religious Customs of the Various Nations of the World*. Vol. 4, London: William Jackson, 1733.

Piddington, H. *English Index to the Plants of India*. 1832. Repr., New Delhi: Today & Tomorrow Printers, 1980.

Pijl, Leendert van der, and Calaway H. Dodson. *Orchid Flowers: Their Pollination and Evolution*. Coral Gables, FL: University of Miami Press, 1966.

Pinkard, Terry. *German Philosophy, 1760–1860: The Legacy of Idealism*. Cambridge: Cambridge University Press, 2002.

Pippin, Robert B. "You Can't Get There from Here; Transition Problems in Hegel's *Phenomenology of Spirit*." In *The Cambridge Companion to Hegel*. Edited by Frederick C. Beiser. Cambridge: Cambridge University Press, 1993, 52–85.

Pollan, Michael. *The Botany of Desire: A Plant's-Eye View of the World*. New York: Random House, 2001.

Poovey, Mary. "The Limits of the Universal Knowledge Project: British India and the East Indiamen." *Critical Inquiry* 31, no. 1 (2004): 183–202.

Porter, Dahlia. "Scientific Analogy and Literary Taxonomy in Darwin's *Loves of the Plants*." *European Romantic Review* 18, no. 2 (2007): 213–221.

Porter, Yves. *Painters, Paintings and Books*. Delhi: Manohar, 1994.

Pratt, Mary Louise. *Imperial Eyes*. 2nd ed. New York: Routledge, 2007.

Priestley, Joseph, *Memoirs of Dr. Joseph Priestley to the Year 1975*. 2 vols. Northumberland: n.p., 1804.

Raina, Dhruv. "Betwixt Jesuit Enlightenment Historiography: Jean-Sylvain Bailly's *History of Indian Astronomy*," *Revue d'histoire des mathématiques* 9 (2003): 253–306.

Raistrick, Arthur. *Quakers in Science and Industry*. London: Bannisdale Press, 1950.

Raj, Kapil. *Relocating Modern Science: Circulation and the Construction of Scientific Knowledge in South Asia and Europe*. Delhi: Permanent Black, 2006.

Rajan, Balachandra. *Under Western Eyes: India from Milton to Macaulay*. Durham, NC: Duke University Press, 1999.

Rajan, Tilottama. "Dis-Figuring Reproduction: Natural History, Community, and the 1790s Novel." *CR: The New Centennial Review* 2, no. 3 (2002): 211–252.

——. "Excitability: The (Dis)Organization of Knowledge from Schelling's *First Outline* (1799) to *Ages of the World* (1815), *European Romantic Review* 21, no. 3 (2010): 309–325.

——. "First Outline of a System of Theory: Schelling and the Margins of Philosophy, 1799–1815." *Studies in Romanticism* 46, no. 3 (2007): 311–335.

——. "System and Singularity From Herder to Hegel." *European Romantic Review* 11, no. 2 (2000): 137–149.

——. "Toward a Cultural Idealism: Negativity and Freedom in Hegel and Kant." In *Idealism without Absolutes: Philosophy and Romantic Culture*, edited by Tilottama Rajan and Arkady Plotnitsky, 51–71. Albany: State University of New York Press, 2004.

Rauch, Leo. "Hegel and the Emerging World: The Jena Lectures on *Naturphilosophie*, 1805–06." *The Owl of Minerva* 16, no. 2 (1985): 175–181.

Raven, Peter H., Ray F. Evert, and Susan E. Eichhorn, eds. *Biology of Plants*. 7th ed. New York: W. H. Freeman, 2005.

Ray, John. *A Collection of English Words*. Edited by R. C. Alston, English Linguistics, 1500–1800. Menston, England: Scolar Press, 1969. Repr. of 1691 ed.

——. *Dictionariolum Trilinguae*. Edited by William T. Stearn. London: Ray Society, 1981. Repr. of 1675 ed.

——. Preface to *Methodus plantarum nova*. London: Henry Faithorne & John Kersey, 1682.

Raynal, Guillaume Thomas François, Abbé. *L'Histoire philosophique et politique des établissements et du commerce des Européens dans les deux Indes*. [*L'Histoire des deux Indes*.] 5th ed. 7 vols. Maestricht: Jean-Edme Dufour & P. Roux, 1777.

Reeder, Kohleen. "The 'Paper Mosiack' Practice of Mrs. Delany and Her Circle." In Mark Laird and Alicia Weisberg-Roberts, eds., *Mrs. Delany and Her Circle*. New Haven: Yale University Press, 2009. 224–235.

Reill, Peter Hanns. *Vitalizing Nature in the Enlightenment*. Berkeley: University of California Press, 2002.

"Review of *Ladies' Botany*, by John Lindley, and *Edwards's Botanical Register*, edited by John Lindley." *Analyst* 1 (1834): 139–141.

Richards, David. *Masks of Difference*. Cambridge: Cambridge University Press, 1994.

Richards, Robert J. *The Romantic Conception of Life: Science and Philosophy in the Age of Goethe*. Chicago: Chicago University Press, 2002.

Richardson, Alan. *Literature, Education, Romanticism: Reading as Social Practice, 1780–1832*. Cambridge: Cambridge University Press, 1994.

Ritvo, Harriet. *The Platypus and the Mermaid and other Figments of the Classifying Imagination*. Cambridge, MA: Harvard University Press, 1997.

Robinson, Eric. "John Clare: 'Searching for Blackberries.'" *Wordsworth Circle* 38, no. 4 (2007): 208–212.

Rocher, Rosane. "British Orientalism in the Eighteenth Century: The Dialectics of Knowledge and Government." In *Orientalism and the Postcolonial Predicament*, edited by Carol Breckenridge and Peter van der Veer, 215–249. Philadelphia: University of Pennsylvania Press, 1993.

——. "The Career of Rādhākānta Tarkavāgīśa, an Eighteenth-Century Pandit in British Employ." *Journal of the American Oriental Society* 109, no. 4 (1989): 627–633.

——. "Weaving Knowledge: Sir William Jones and Indian Pandits." In *Objects of Enquiry*, edited by Garland Cannon and Kevin R. Brine. New York: New York University Press, 1991. 51–79.

Roper, Derek. *Reviewing before the Edinburgh, 1788–1802*. London: Methuen, 1978.

Rossi, Barbara, Roy C. Craven, and Stuart Welch. *From the Ocean of Painting*. New York: Oxford University Press, 1998.

Rousseau, Jean-Jacques. *The Collected Writings of Rousseau*. Translated by Alexandra Cook. Edited by Christopher Kelly. Vol. 8. Hanover: University Press of New England, 2000.

Roxburgh, William. *Flora Indica*. Kew MSS. Royal Botanic Gardens and Herbarium Library, Kew.

——. *Flora Indica*. 3 vols. Serampore: W. Thacker; London: Parbury, Allen, 1832.

——. *Hortus Bengalensis*. Serampore: Mission Press, 1814.

——. Letter to Joseph Banks. 17 Aug. 1792. Add. MSS 33979. Folios 1761–63. Botany Library. BMNH.

——. Papers. Fleming Collection. BMNH.

——. *Plants of the Coast of Coromandel*. 3 vols. London: W. Bulmer and Co., Shakspeare Printing Office, for George Nicol, 1795–1819.

Ruiz, Hipólito. *The Journals of Hipólito Ruiz: Spanish Botanist in Peru and Chile, 1777–1788*. Translated by Richard Evans Schultes and María José Nemry von Thenen de Jaramillo-Arango. Portland, OR: Timber Press, 1998.

Ruse, Michael. Introduction to *What the Philosophy of Biology Is*. Edited by Michael Ruse. Dordrecht and Boston: Kluwer Academic, 1989: 1–15.

Russell, Patrick. "Castes and Crafts." In Alexander Dalrymple, *Oriental Repertory*, vol. 1. London: George Biggs, P. Emsly, 1793. 49–52.

Ruwe, Donelle R. "Charlotte Smith's Sublime: Feminine Poetics, Botany, and Beachy Head." *Prism(s)—Essays in Romanticism* 7 (1999): 117–132.

Sabin, Margery. *Dissenters and Mavericks: Writings about India in English, 1765–2000*. New York: Oxford University Press, 2002.

Sachs, Aaron. "The Ultimate 'Other': Post-Colonialism and Alexander von Humboldt's Ecological Relationship with Nature." *History and Theory* 42 (Dec. 2003): 111–135.

Sachs, Julius von. *History of Botany, 1530–1860*. Translated by Henry E. F. Garney, rev. Isaac Bayley. Oxford: Clarendon Press, 1890.

Sanjappa, M., K. Thothathri, and A. R. Das. "Roxburgh's Flora Indica Drawings at Calcutta." *Bulletin of the Botanical Survey of India* 33, nos. 1–4 (1991): 1–232.

Saunders, Gil. *Picturing Plants: An Analytical History of Botanical Illustration*. Berkeley and Los Angeles: California University Press, 1995.

Schelling, F. W. J. *Ideas for a Philosophy of Nature*. Translated by Errol E. Harris and Peter Heath. Cambridge: Cambridge University Press, 1988.

Schiebinger, Londa. *Nature's Body*. Boston: Beacon Press, 1993.

——. *Plants and Empire*. Cambridge: Cambridge University Press, 2004.

Schiebinger, Londa, and Claudia Swan, eds. *Colonial Botany*. Philadelphia: University of Pennsylvania Press, 2005.

Schiebinger, Londa, and Sara Stidstone Gronim. "What Jane Knew: A Woman Botanist in the Eighteenth Century." *Women's History* 19, no. 3 (2007): 33–59.

Schiller, Friedrich. *Naïve and Sentimental Poetry and On the Sublime*. Translated by Julius A. Elias. New York: Frederick Ungar, 1980.

Schlegel, Friedrich. *Von der Weltseele* [On the world soul]. Vol. 2 in *Friedrich Schlegel Kritische Ausgabe*. Edited by Ernst Behler, Jean-Jacques Anstett, and H. Eichner. Paderhorn: Ferdinand Schöningh, 1958.

Schubert, Johannes. *Goethe Und Hegel*. Leipzig: Felix Meiner Verlag, 1932.

Schwartz, Janelle. *Wormwork*. Minneapolis: University of Minnesota Press, 2012.

Schwartz, P. R. "The Roxburgh Account of Indian Cotton Painting: 1795." *Journal of Indian Textile History* 4 (1959): 47–56.

Schwarz, Henry. *Writing Cultural History in Colonial and Postcolonial India*. Philadelphia: University of Pennsylvania Press, 1997.

Scott, David. "Rousseau and Flowers: The Poetry of Botany." *Studies on Voltaire and the Eighteenth Century* 182 (1979): 73–86.

Sealy, J. R. "The Roxburgh Flora Indica Drawings at Kew." *Kew Bulletin* nos. 2 and 3 (1956): 296–399.

Secord, Anne. "Botany on a Plate: Pleasure and the Power of Pictures in Promoting Early Nineteenth-Century Scientific Knowledge." *ISIS* 93, no. 1 (2002): 28–57.

——. "Corresponding Interests: Artisans and Gentlemen in Nineteenth-Century Natural History." *British Journal for the History of Science* 27 (1994): 383–403.

——. "Science in the Pub: Artisan Botanists in Early Nineteenth-Century Lancashire." *History of Science* 32 (1994): 269–315.

Seward, Anna. *Letters of Anna Seward*. 6 vols. Edinburgh: George Ramsay, 1811.

Sharafuddin, Mohammed. *Islam and Romantic Orientalism*. London: I. B. Tauris, 1994.

Shelley, Mary. *Frankenstein*. Edited by D. L. Macdonald and Kathleen Scherf. Peterborough, Ontario: Broadview Press, 1994.

Shelley, Percy. *Philosophical View of Reform*. Vol. 7 in *The Complete Works of Percy Bysshe Shelley*. 10 vols. Edited by Roger Ingpen and Walter E. Peck. New York: Gordian Press, 1965.

——. *Shelley's Poetry and Prose*. edited by Donald H. Reiman and Neil Fraistat. 2nd ed. New York: W. W. Norton, 2002.

Shteir, Ann. *Cultivating Women, Cultivating Science: Flora's Daughters and Botany in England, 1760–1860*. Baltimore: Johns Hopkins University Press, 1996.

——. "Gender and 'Modern' Botany in Victorian England." In *Women, Gender, and Science: New Directions*, edited by Sally Gregory Kohlstedt and Helen E. Longino, *Osiris* 12 (1997): 29–38.

Simpson, David. *Situatedness, or, Why We Keep Saying Where We're Coming From*. Durham: Duke University Press, 2002.

——. "Speaking Place: The Matter of Genre in 'The Lament of Swordy Well.'" *Wordsworth Circle* 34, no. 3 (2003): 131–134.

——. *Wordsworth: Commodification, and Social Concern: The Poetics of Modernity*. Cambridge: Cambridge University Press, 2009.

——. *Wordsworth's Historical Imagination*. New York: Methuen, 1987.

Singh, R. S., and A. N. Singh. "On the Identity and Economico-medicinal Uses of hastikarnapalasa (Leea macrophylla Roxb., family: Ampelidaceae) as Evinced in the Ancient (Sanskrit) Texts and Traditions." *Indian Journal of the History of Science* 16, no. 2 (1981): 219–222.

Sivarajan, V. V., and N. K. P. Robson. *Introduction to the Principles of Plant Taxonomy*. Cambridge: Cambridge University Press, 1991.

Slaughter, Thomas P. *The Natures of John and William Bartram*. New York: Alfred A. Knopf, 1996.

Sloan, Phillip R. "The Buffon-Linnaeus Controversy." *Isis* 67 (1976): 356–375.

——. "Natural History, 1670–1802." In *Companion to the History of Modern Science*. Edited by G. N. Cantor, R. C. Olby, J. R. R. Christie, and M. J. S. Hodge. New York: Routledge, 1990. 295–313.

Smith, Adam. *Essays on Philosophical Subjects*. Edited by W. P. D. Wightman and J. C. Bryce. Oxford: Clarendon Press, 1980.

Smith, Charlotte. *The Collected Letters of Charlotte Smith*. Edited by Judith Phillips Stanton. Bloomington: Indiana University Press, 2003.

——. *Minor Morals*. 2 vols. London: Sampson Low, 1799.

——. *The Poems of Charlotte Smith*. Edited by Stuart Curran. New York: Oxford University Press, 1993.

——. *The Young Philosopher*. Edited by Elizabeth Kraft. Lexington: University Press of Kentucky, 1999. Repr. of 1798 ed.

Smith, J. E. Correspondence MSS. Library of the Linnean Society, London.

——. *Introduction to Physiological and Systematical Botany*. London: Longman, Hurst, Reese, & Orme, and J. White, 1807.

——. "Review of *Flora Britannica*." *Monthly Review* 47 (1805): 362.

Smith, J. E., and James Sowerby. *English Botany, or, Coloured Pictures of British Plants*. 36 vols. London: J. Davis, 1790–1814.

Smith, Pamela H. *The Body of the Artisan*. Chicago: Chicago University Press, 2004.

Snyder, Gary. *The Practice of the Wild*. San Francisco: North Point Press, 1990.

Solitarius. "Remarks upon Zoological Nomenclature and Systems of Classification." *Field Naturalist* 1 (1833): 521–528.

Solkin, David H., ed. *Art on the Line: The Royal Academy Exhibitions at Somerset House, 1780–1836*. New Haven: Paul Mellon Centre for Studies in British Art and the Courtald Institute Gallery, Yale University Press, 2001.

Soltis, Douglas, Pamela S. Soltis, Peter K. Endress, and Mark W. Chase. *Phylogeny and Evolution of Angiosperms*. Sunderland, MA: Sinaur Associates, 2005.

Sorenson, Janet. "Vulgar Tongues: Canting Dictionaries and the Language of the People in Eighteenth-Century Britain." *Eighteenth-Century Studies* 37, no. 3 (2004): 435–454.

Sowerby, James. "Notes and Correspondence about Fungi." MSS SOW. British Museum (Natural History).

Spary, E. C. *Utopia's Garden: French Natural History from the Old Regime to the Revolution*. Chicago: Chicago University Press, 2000.

Spivak, Gayatri. *A Critique of Postcolonial Reason: Toward a History of the Vanishing Present*. Cambridge, MA: Harvard University Press, 1999.

Srivastava, Ashok Kumar. *Mughal Painting: An Interplay of Indigenous and Foreign Traditions*. New Delhi: Munshiram Manoharlal Publishers, 2000.

Stafford, Barbara Maria. *Voyage into Substance*. Cambridge, MA: MIT Press, 1984.

Stafleu, Franz A. Introduction to A. K. de Jussieu. *Genera Plantarum*. Vol. 35 in *Historiae Naturalis Classica*. edited by J. Cramer and H. K. Swann. New York: Cramer & Weinheim, 1964.

——. *Linnaeus and the Linnaeans*. Utrecht: A. Oosthoek, 1971.

Stearn, William S. *Botanical Latin*. 4th ed. Portland: Timber Press, 2004.

——, trans. and ed. *Historiae naturalis classica* by Robert Brown. Vol. 6. Wienheim: H. R. Englemann, 1960.

——. "Linneaus's System of Sexual Classification," in Linnaeus, *Species Plantarum*, 1:25–34.

Stearn, William T. "Alexander von Humboldt and Plant Geography." In William T. Stearn, ed., *Humboldt, Bonpland, Kunth*. 116–120.

——. ed. *Humboldt,Bonpland, Kunth, and Tropical Botany*. Lehre: Verlag von J. Cramer, 1968.

——. "Humboldt's *Essai*," in William T. Stearn, ed., *Humboldt, Bonpland, Kunth*. 351–357.

—— "An Introduction to the *Species Plantarum* and Cognate Botanical Works of Carl Linnaeus." In Carl von Linnaeus, *Species Plantarum*. 2 vols. 1753. Repr., London: Ray Society, 1957–59. 1:1–167.

Steigerwald, Joan. "Instruments of Judgment: Inscribing Organic Processes in Late Eighteenth-Century Germany." *Studies in History and Philosophy of Biological and Biomedical Sciences* 33 (2002): 79–131.

——. "Kant's Concept of Natural Purpose and the Reflecting Power of Judgement." *Studies in History and Philosophy of Biological and Biomedical Sciences* 37, no. 4 (2006): 712–724.

Stevens, Peter F. *The Development of Biological Systematics: Antoine-Laurent de Jussieu, Nature, and the Natural System*. New York: Columbia University Press, 1994.

Stewart, Susan. *On Longing: Narratives of the Miniature, the Gigantic, the Souvenir, the Collection*. Baltimore: Johns Hopkins University Press, 1984.

Stocks, Christopher. "Biddulph Is a Horticultural Disneyland." *Independent*. Sunday, Sept. 24, 2006. Features, 54.

Suleri, Sara. *The Rhetoric of English India*. Chicago: Chicago University Press, 1992.

Summerhayes, V. S. *Wild Orchids of Britain*. London: Collins, 1951.

Tantillo, Astrida Orle. "Goethe's Botany and His Philosophy of Gender." *Eighteenth-Century Life* 22, no. 2 (1998): 123–130.

——. "Polarity and Productivity in Goethe's *Wahlverwandtschaften*." *Seminar: A Journal of Germanic Studies* 36, no. 3 (2000): 310–325.

——. *The Will to Create: Goethe's Philosophy of Nature*. Pittsburgh: University of Pittsburgh Press, 2002.

Taussig, Michael. *Mimesis and Alterity: A Particular History of the Senses*. New York: Routledge, 1993.

Teltscher, Kate. *India Inscribed: European and British Writing on India, 1600–1800*. Delhi: Oxford University Press, 1995.

Terada, Rei. *Looking Away: Phenomenality and Dissatisfaction, Kant to Adorno*. Cambridge, MA: Harvard University Press, 2009.

Thomas, Sophie. *Romanticism and Visuality: Fragments, History, Spectacle*. New York: Routledge, 2008.

Thornton, Robert. "Memoirs of the late James Lee." Preface to James Lee, *Introduction to the Study of Botany*. 4th ed. London: F. C. & J. Rivington, 1810.

——. *Temple of Flora*. London: R. Thornton, 1799–1807.

——. *Temple of Flora*. Bibliographic notes by Handasyde Buchanan. Edited by G. Grigson. London: William Collins, 1951.

Tilkin, François, ed. *Le groupe de Coppet et le monde moderne: conceptions, images, débats*. Geneve: Droz, 1998.

Tobin, Beth Fowkes. *Colonizing Nature: The Tropics in British Arts and Letters, 1760–1820*. Philadelphia: University of Pennsylvania Press, 2005.

——. *Picturing Imperial Power: Colonial Subjects in Eighteenth-Century Painting*. Durham: Duke University Press, 1999.

Tomasi, Lucia Tongiorgi, and Gretchen A. Hirschauer. *The Flowering of Florence: Botanical Art for the Medici*. Washington, DC: National Gallery of Art, 2002.

Trattinnick, Leopold. *Archiv der Gewächskunde*. Vienna: Trattinnick, 1812–14.

Tulk, Alfred. *Elements of Physiophilosophy*. London: Ray Society, 1847.

Turner, Michael. *English Parliamentary Enclosure: Its Historical Geography and Economic History*. Hamden, CT.: Archon Books, 1980.

Uglow, Jenny. *The Lunar Men: Five Friends Whose Curiosity Changed the World*. New York: Farrar, Straus & Giroux, 2002.

Vajracharya, Gautama V. "Atmospheric Gestation: Deciphering Ajantin Ceiling Paintings and Other Related Works." Pts. 1 and 2. *MARG* 55, no. 2 (2003): 41–57; 55, no. 3 (2004): 40–51.

——. *Watson Collection of Indian Miniatures at the Elvehjem Museum of Art*. Madison: University of Wisconsin Press, 2002.

Varadarajan, Lotika. *South Indian Traditions of Kalamkari*. Ahmedabad: National Institute of Design, 1982.

Vardy, Alan. *John Clare: Politics and Poetry*. New York: Palgrave Macmillan, 2003.

Wakefield, Priscilla. *A Brief Memoir of the Life of William Penn*. Philadelphia: Isaac Prince, 1818.

——. *An Introduction to Botany, in a Series of Familiar Letters*. Dublin: Thomas Burnside, 1796.

Walkowitz, Rebecca. "Unimaginable Largeness: Kazuo Ishiguro, Translation, and the New World Literature." *Novel* 40, no. 3 (2007): 216–239.

Wallace, James. "Part of a Journal Kept from Scotland to New Caledonia in Darien." *Philosophical Transactions of the Royal Society* 22 (1700–1701): 536–543.

Walpole, Horace. *Horace Walpole's Correspondence*. Edited by W. S. Lewis. 48 vols. New Haven: Yale University Press, 1937–1983.

Ward, John. "Objections to the appointing of Public days for admitting all Persons to the Museum without distinction." BL Add MS 6179, f.61, British Library.

Watts, Michael. *The Dissenters*. 2 vols. Oxford: Clarendon Press, 1995.

Welch, Stuart Cary. "Reflections upon Barbara Rossi's 'Ocean.'" In *From the Ocean of*

Painting: India's Popular Paintings, 1589 to the Present. Edited by Barbara Rossi, Roy C. Craven, and Stuart Cary Welch. New York: Oxford University Press, 1998, 11–13.

White, Daniel F. *English Romanticism and Religious Dissent*. Cambridge: Cambridge University Press, 2006.

Wight, Robert and G. A. Walker Arnott. *Prodromus Florae Peninsulae Indiae Orientalis*. London: Parbury, Allen & Co., 1834.

Wilkins, John. *Essay towards a Real Character, and a Philosophical Language*. London: Royal Society, 1668.

Williams, Raymond. *Keywords*. Rev. ed. New York: Oxford University Press, 1983.

Williams, William Carlos. *Paterson*. Edited by Christopher MacGowan. Rev. ed. New York: New Directions, 1992.

Wilson, Arthur. *Diderot*. New York: Oxford University Press, 1972.

Withering, William. *An Arrangement of British Plants*. 4 vols. 3rd ed. Birmingham: M. Swinney, 1796.

Wolfe, Charles T., and Motoichi Terada. "The Animal Economy as Object and Program in Montpellier Vitalism." *Science in Context* 21, no. 4 (2008): 537–579.

Wollstonecraft, Mary. *Vindications of the Rights of Woman*. In *The Vindications*. Edited by D. L. Macdonald and Kathleen Scherf. Peterborough, Ontario: Broadview Press, 1997. 99–344.

Wood, Neville. "On Making the English Generic Names of Birds Correspond to the Latin Ones." *Analyst* 2 (1835): 238–239.

——. "On the Study of Latin, More Especially as Regards the Medical Profession." *Analyst* 3 (1835): 46–53.

Wordsworth, William. *The Fenwick Notes of William Wordsworth*. Edited by Jared Curtis. London: Bristol Classical Press, 1993. 2nd rev. and corr. edn. Humanities e-book, 2011.

——. *The Prelude, 1799, 1805, 1850*. Edited by Jonathan Wordsworth, Stephen Gill, and Stephen Parrish. New York: W. W. Norton, 1979.

——. *The Ruined Cottage and the Pedlar*. Edited by James Butler. Ithaca, NY: Cornell University Press, 1979.

——. "The Thorn." In *The Lyrical Ballads, and Other Poems, 1797–1800*. Edited by Jared Curtis, James Butler, and Karen Green. Ithaca, NY: Cornell University Press, 1992.

Yahav-Brown, Amit. "Gypsies, Nomadism, and the Limits of Realism." *Modern Language Notes* 21 (2006): 1124–1147.

Yelling, J. A. *Common Field and Enclosure in England, 1450–1850*. Hamden, CT: Archon Press, 1977.

Zander Handwörterbuch dem Pflanzennamen. Stuttgart: Eugen Ulmer, 2000. Repr. of 1927 ed.

Žižek, Slavoj. *Tarrying with the Negative: Kant, Hegel, and the Critique of Ideology*. Durham, NC: Duke University Press, 1993.

Index

Adanson, Michel, 20, 36, 37, 39
Addison, Joseph, 26
Adelaide, Queen Dowager, 248
Adorno, Theodor, 2, 11, 222, 241–45; "Aspects of Hegel's Philosophy," 242–43, 244; *Minima Moralia*, 222, 242–45; *Negative Dialectics*, 242
aesthetics, 5, 7, 11, 14, 16, 45, 55, 56–57; and W. Bartram, 77; and Franz Bauer, 75; and Brown, 43; and color, 73, 74; and Hegel, 223; and illustrations, 6, 70–72, 73, 163, 200, 209; and Jones, 181; Kant on, 219; and Roxburgh, 193, 196–97
affines, 46, 51
affinities, 37, 40, 46, 51, 118, 123, 153, 254; and Bateman, 248–49; and E. Darwin, 88, 89; and Goethe, 21; and Hegel, 224; and Linneaus, 20, 21; and morphology, 20–21; natural, 18, 37; and Natural System, 37; and Rheede, 185
Agamben, Giorgio, 167
Agrawal, Arun, 284n2
Aikin, John, 59, 66, 101, 123, 124; *An Essay on the Application of Natural History to Poetry*, 123, 124
Alcock, Randal, 48
Alexander the Great, 182
algae, 5, 32, 41, 66, 113, 267n52
allegory, 12, 78, 170, 174, 221, 241; and Clare, 145, 155, 156; and Jones, 184, 189, 191. *See also* figures/figuration
aloe, 160
American cowslip, 32–33
Amra'taca, 188
analogy, 218; to animals, 40, 84, 87, 102, 104, 106, 123, 260; and E. Darwin, 78, 79, 80, 81, 82, 83, 84, 87, 88, 89; and Goethe, 227–28, 232–33; and Hegel, 173, 218, 222–23, 226; to humans, 210–11; and Jackson, 104, 106; and Kant, 219; and La Mettrie, 210–11; and Oken, 220; and P. Shelley, 211–14; and C. Smith, 119, 121, 122; and Wollstonecraft, 102
Anastatica (Rose of Jericho), 65
Anemone nemorosa, 66
angiosperms, 32, 37, 43
Anglican communion, 59
animals, 4, 13, 109, 110, 211, 221, 241; analogy to, 40, 84, 87, 102, 104, 106, 120, 123, 260; circulatory systems of, 8, 40, 82, 95, 231; and E. Darwin, 87, 260; and Grey, 111; and Hegel, 215, 221, 241; Linnaeus's classification of, 10; and middle kingdom of plants, 4, 7, 216–17; and plant-animals, 217; similarities to, 40–42, 81, 82, 246–47; and C. Smith, 123
animate life, 11, 34, 82, 215
animation, 13, 34, 41, 82, 83, 91, 92, 147; and Goethe, 216, 217, 218, 227; and Hegel, 216, 217, 218, 227, 241; and Shelley, 211, 215. *See also* life
Annual Register, 64
Anti-Gallican, 52
Anti-Jacobin, 184
Antirrhinum majus L. (snapdragon, calf's snout), 49
Archer, Mildred, 197, 201, 202–3
Archilochus colubris, 77
archives, 56, 70, 167, 208, 209
aristocracy, 53, 58, 111, 114, 154, 248–49
Aristotle, 7, 9, 18, 19, 26, 44, 73, 217; and essential kinds, 23, 39; and species, 24–25, 27, 30, 51, 72
Arnold, David, 284n2

Arnott, George Arnott Walker, *Prodromus Florae Peninsulae India Orientalis*, 268n66
arrowroot, 288n84
Artis, Edmund Tyrell, 145, 146, 279–80n12
artists, 6, 57, 58, 60, 69, 70, 77; Indo-Persian, 203. *See also* drawings; Indian artists; watercolor
Arum maculatum (cuckoo pint, priest's pintle, wake robin, lords and ladies), 49, 86, 107, 109, 117, 121, 126, 150, 154
Ascending Jussieua, 186–87
Asiatic Miscellany, The, 171, 181, 286n36
Asiatic Researches, 171, 179, 181, 182, 286n36
Asóca, 187–88
Atimucta (Bengal Banisteria), 188
Atran, Scott, 23, 24, 26, 30, 38, 44, 266n10
Aurangzeb, 207

Baber, Zaheer, 284n2
Baie-a-ondes (bayaonde), 161
Baillon, Henri, 17, 18
Baker, Anne, 279n12; *Glossary of Northamptonshire Words and Phrases*, 279n12, 282–83n60
Baker, George, *Glossary of Northamptonshire Words and Phrases*, 279n12, 282–83n60
banana, 35, 288n84
Banks, Joseph, 42, 49, 53, 58, 59, 60, 63; and Francis Bauer, 70, 74, 198–99, 272n21
Barbauld, Anna Letitia Aikin, 14, 59, 66, 99–103, 123, 277n33; "Inscription for an Ice-House," 277n33; "The Rights of Woman," 101; "To a Lady, with some painted flowers," 99–103; "To Mrs. P, with some Drawings of Birds and Insects," 99
Barbauld, Rochemont, 101
Barbauld family, 278n65
Barrell, John, 138–39
Bartram, John, 24, 33–34, 45, 53, 58, 60, 61, 268n58
Bartram, William, 59, 60, 62–63, 70, 76–77; *Travels*, 62–63
Bassia, 183–84
Bateman, James, 15–16, 116, 247–60; and aristocracy, 248–49; audiences of, 249; *The Orchidaceae of Mexico and Guatemala*, 15–16, 116, 247–60; and P. Shelley, 253–56
Bauer, Ferdinand, 43, 74
Bauer, Franz (Francis), 43, 58, 70, 74–75, 259
Bauhin, Casper, 48
Bauhin, John, 48
Bauhinia, 48
Bayly, C. A., 165
Beaufort, Daniel August, 112
Beaufort, Harriet, *Dialogues on Botany*, 112
Beckford, William, 286n36; *Vathek*, 286n36
Bee Ophrys, 76
Behn, Aphra, *Oroonoko*, 91
Beiser, Frederick, 293n2, 295n16
Benedict, Barbara, 62
Benjamin, Walter, 3
Benveniste, Emile, 235
Bermingham, Ann, 93
Bewell, Alan J., 4, 79–80, 95
Bewick, Thomas, 45; *History of British Birds*, 110–11
Bhawani Das, 197
Bhúchampaca, 185
Bible, 61, 65, 100, 156
Bicheno, J. E., 22–23, 36
Bickerstaff, Isaac, *The Sultan*, 169
Billings, John, 279n12
birds, 99, 111, 116, 247; and Bateman, 250, 258, 259; and Clare, 8, 127, 132, 134, 146, 148, 152, 153, 154, 155; and names, 8, 41, 46; and C. Smith, 117, 118, 121, 123, 124
bird's foot trefoil, 8, 123
Blackburne, Anna, 66
Blake, William: *Milton*, 242; *Songs of Experience*, 87
bluebell woods, 126, 127
bluecap, 154
blue tit, 154
Blumenbach, Casper Friedrich, 294n7
Bonpland, Aime, 10, 11
Borassus flabelliformis, 202
Bordeu, Theophile de, 217
Borges, Jorge Luis, 4
Borland, Francis, 263n4
Bosch, F. D. K., 206
Botanical Magazine, 60, 71, 200, 260

Botticelli, Alessandro, *Primavera*, 90
Bourdieu, Pierre, 93
breadfruit, 192
British Museum, 42, 44–45
Brontë, Charlotte, *Jane Eyre*, 110–11
Brown, Bill, 2, 3, 54
Brown, Robert, 17–18, 20, 41, 42, 58, 75, 109, 261; and Brownian motion, 40; career and achievements of, 42–44; and Linnaean system, 35–36, 37; and monsters, 45, 260
Browne, Patrick, 160–61; *Civil and Natural History of Jamaica*, 160–61
Bry, Theodore de, *America*, 91
Buffon, Georges Louis Leclerc, comte de, 10, 22, 36, 48, 64, 68, 268n63; "Botanique," 35; *Histoire naturelle, générale et particulière*, 34–35
Buffonia tenifolia, 48
Burke, Edmund, 64, 67, 180; *Reflections on the Revolution in France*, 99
Burkean contract, 138
Burning Bush, 61
Burns, Robert, 131
Bute, John Stuart, Earl of, 47, 59
buttercup, 154
Byron, Lord, 131; *Don Juan*, 124; *The Giaour*, 286n36; *Lara*, 286n36

cabbage rose, 53
Cactus Grandiflora (prickly pear cactus), 253
Calamariae, 35
Calcutta Botanic Garden, 180–81, 195, 198, 208–9
Caleandra baueri, 259
Caltha palustris (horse-blobs, marsh marigold, kingcup, May blobs), 49, 148
Camalata, 174
Campbell, John, *Pleasures of Hope*, 256
Canchrà, 186–87
Candolle, Alphonse de, 36, 37, 39, 40
Candolle, Augustin Pyramus de, 20, 37–38, 39, 48, 266n10, 268n66, 269n79; and gaps in nature, 34, 38; and Natural System, 36, 37; *Theorie elementaire de la botanique*, 47
Canna flaccida (Bandana of the Everglades, Golden Canna), 77
Canning, George, "Loves of the Triangles," 274n72
Cardamine pratensis (cuckoo flower), 49, 150, 155
Caribbean, 15, 67, 159–61, 160
Carlson, Julie, 96
Carolina grass, 65
Casid, Jill, 160, 283n3
cassava, 160
Cassia, 87
Catasetum maculatum orchid, 256
Catasetum orchids, 251–52
categories, 8, 11, 19, 22, 56, 118, 160, 163; and Buffon, 64; and Cesalpino, 5–6; and Clare, 155, 156, 158; and E. Darwin, 81, 89; and Hegel, 223–24; subdivision of, 22
Cats and Kittens, 149
celandine, 148
cereus plant, 159–60
Certeau, Michel de, 133
Cesalpino, Andrea, 9, 17, 18, 21, 25; *De Plantis*, 5–6
Chakrabarti, Pratik, 284n2
Chakrabarty, Dipesh, 164, 166, 209
Chambers, William, 286n36
Charlotte, Queen, 49
Chatterjee, Partha, 284n2
chay, 194
Chiasson, Dan, 151
children, 69, 93, 94, 106, 118; and Clare, 128, 131, 132, 133; and Jackson, 104–8
Chow, Rey, 167
cinnamon, 192
cladistics, 18, 30, 55
Clare, John, 33, 49, 116, 126–58, 248; autobiographical fragments of, 156; and beating of bounds, 127–28, 129; and botanical poetics, 151–58; and commonability, 14, 127, 128, 129, 131, 132, 133, 134–42, 152; and cottage in Northborough, 138; and enclosure, 14, 15, 127, 128, 129, 131, 132, 134–35, 136, 137, 138, 140, 141, 144, 151, 154, 281n35; and Epping Forest asylum, 132, 144, 156; friendships of, 279–80n12; and gypsies, 129, 131, 133, 134, 136, 142–45; handlist of English orchises, 282n59; and Helpston, 128, 129, 132, 133, 135, 138, 139, 140, 141, 142, 148, 154; and Henderson, 281–82n53; and Kent, 282n55;

Clare, John *(cont.)*
and localism, 127, 128, 132, 135, 144, 149; and names, 8, 14–15, 49, 126–27, 128–29, 131, 135, 145, 147, 148–49, 150, 152, 153, 154, 155, 157, 158; and orchises, 121, 129, 146, 149–51, 157–58, 247, 261; and plant exchanges, 53; and plant grammars, 145–51; and Royce wood, 152–53; and situatedness, 131–34; and C. Smith, 120–21; and sociability, 128, 131, 133–34, 140–41; Works: "The Botanist's Walk," 152; "The Bounty of Providence," 135; "Emmonsales Heath," 140; "The Flitting," 138, 154; "The Gipsey Camp," 144; "The Gipsey's Song," 143; "Going to the Fair," 143; "Helpstone," 140; journal entry for 1824, 156; "Journey out of Essex," 131; "June," 154; "The Lament of Swordy Well," 141, 152, 155, 156, 158; letter on natural history of 1823, 150; "May," 130, 141, 154; *Midsummer Cushion*, 138; "The Mores," 139, 141; "October," 136, 141; "On Visiting a Favourite Place," 138; "The Opening of the Pasture," 127, 141; "Pleasures of Spring," 140, 152–53; "A Ploughmans Skill at Classification after the Lineian Arrangement," 147–48; "The Progress of Rhyme," 130; "Reccolections after a Ramble," 152; *The Rural Muse*, 152; *The Shepherd's Calendar*, 126, 130, 136, 141, 145, 154; "Swordy Well," 157–58; "To the Snipe," 154–55; *The Village Minstrel*, 152; "The Wild Flower Nosegay," 153–54
Clark, Stephen, *Hortus Anglicus*, 36
class, biological, 19, 22, 26, 29, 32, 87, 88, 104
class, social, 6, 59, 63, 135, 136, 139, 141, 251, 252. *See also* aristocracy; commoners; middle class
classification, 8, 9, 12, 24, 30, 162, 163, 220; and C. Smith, 117, 118, 121; and E. Darwin, 86, 89; debates about, 6–7; and Goethe, 220, 232, 236; and individuals, 28; and Jones, 183–85, 184; Kant on, 219; and Linnaeus, 4; and monsters, 260; and *Monthly Review*, 67. *See also* taxonomy
clematis, 117
Clive, Robert, 168
coffee, 160, 181, 192
Cohn, Bernard, 165
coir, 181
Coleridge, Samuel Taylor, 27, 29; *Aids to Reflection*, 250–51
Collinson, Peter, 24, 33, 43, 53, 58, 59, 60, 61; and botanical enigmas, 45; and Linnaeus's nomenclature, 47
Collinsonia, 82
Collinsonia Canadensis (horsemint), 24
colonialism, 15, 55, 56, 160, 187, 191, 253, 283n8; and Hastings impeachment, 169, 170; and India, 161, 164, 169–73; and Indian artists, 197, 208, 209. *See also* empires/imperialism
color, 69, 73–74, 93, 149, 201; dyes for, 162, 194, 196, 201; fugitive, 71; and Indian art, 207; and Indian artists, 199, 202–3, 204, 205, 206. *See also* illustrations
columbine, 149
commerce, 5, 59, 70, 79, 121, 124, 181, 191; and Roxburgh, 192, 193, 194, 196
commodities, 2, 7, 54, 55, 71
commoners, 134, 135, 136–37, 139, 141–42, 144, 154. *See also* class, social
Compositae, 35
Coniferae, 265n7
conifers, 32, 43
"Conserva aegagpropila," 83
Constable, John, 45
contingency, 14, 219, 220, 244, 245; and Goethe, 222, 234, 235, 240; and Hegel, 8, 11, 221–22, 225, 240, 241, 242, 295n17
Cook, James, 79, 88, 115
Coppet Castle, 58
cornflower, 91, 154
corn poppies, 146
correspondents, exchange of plants by, 2, 53, 58, 59, 65, 133
cotton, 160, 162, 181, 196, 204
Cowper, William, 170; *The Task*, 121
cowslip, 32–33, 126, 128, 148, 154, 279n10
cranesbills, 146
crowflower, 129, 152
crowfoot, 121, 282–83n60
Crown Imperial, 52–53
Cruciferae, 265n7
Cruikshank, George, 77–78, 248; *Exhibition Extraordinaire in the Horticultural Room*, 77–78

Cruikshank, Isaac, 256
Cryptogamia, 19–20, 32, 33, 34, 67, 104; and Linnaeus, 5, 6, 25, 228
cryptogams / cryptogamic plants, 35, 40–41, 66, 113, 114, 193, 267–68n52; and Jackson, 106, 109–10; and Linnaeus, 109–10
cuckoo bud, 8, 49, 129, 150, 152, 282–83n60
cuckoos, 146, 149, 150–51
Cucurbitaceae, 35
Curcuma (Turmeric), 83, 288n84
Curran, Stuart, 120
Curtis, William, 108, 109, 110; *Beauties of Flora*, 60; *Flora Londinensis*, 60, 72, 200
Cusa (Cusha), 196
cycads, 32, 43
Cycnoches, 259, 260
Cycnoches edgertonianum, 258
Cycnoches loddigesii, 258
Cycnoches ventricosum (swan orchid), 116, 258, 259
Cyperaceae, 35
Cypripedium, 259
Cyrtopodiums, 252

Dalrymple, Alexander, 192, 284n12; *Oriental Repertory*, 192
Dalrymple, William, 168
Daniell, Thomas, 178
Dante, *Divina Commedia*, 214
Darwin, Charles, 37, 42, 50, 57, 81, 93, 260, 265n7; and agency, 260–62; *The Effects of Cross- and Self Fertilisation*, 260; and evolution, 10, 14, 19, 29, 34, 81; and gaps in nature, 38; and Linnaean system, 10, 18; and Natural System, 21; *On the Various Contrivances*, 260; *Origin of Species*, 10, 14, 18, 31, 37; and species, 19, 31; and twining plants, 147; and Venus fly trap, 40
Darwin, Erasmus, 59, 66, 67, 78–89, 93, 98, 116, 215, 260; and allegory, 78; and E. Darwin, 122; and females, 83, 85–87, 89; and Jackson, 104, 109, 110; and W. Jones, 190; and Linnaean system, 8, 11, 14, 21, 33, 51, 56, 57, 79, 80–81, 83, 84, 85, 88, 89; and Linnaeus, 148; and Polwhele, 118; and P. Shelley, 210, 211, 212; and C. Smith, 118; and Thornton, 76; and Wakefield, 104; and women, 95; and Wordsworth, 274n72; Works: *The Botanic Garden*, 8, 11, 14, 51, 67, 78–89, 93, 104, 122; *Families of Plants*, 80; *Phytologia*, 80, 84; *Systema Naturae*, 59; *System of Vegetables*, 80; *Temple of Nature*, 80, 89; *Zoonomia*, 80, 89
Daston, Lorraine, 3–4
Davies, H., 62
Davis, Samuel, 179
death, 210, 213, 214, 215, 217, 235
Decandria Pentagynia, 108–9
de Certeau, Michel, 112
Delany, Mary Granville, 14, 111, 112, 114–15, 116, 278n58; *Passiflora laurifolia*, 115; *Physalis*, 115
Delessert, Benjamin, 248
Delessert, Madeleine-Catherine, 94, 248
Deleuze, Gilles, 13, 93, 125, 131, 241, 254
Delhi: Qu-tab Minar, 207; Red Fort, 207
delphinium, 91
de Man, Paul, 214
Demerara, 251
Derrida, Jacques, 11, 28, 40, 109, 163, 167, 264n26, 269n87
Descourtilz, M. E., *Voyages d'un naturaliste*, 161
Desmond, Ray, 197
de Zegher, Catherine, 13
Dhal-samuds, 196, 196
Diandria, 32
Dictamina, 86
difference, 2, 11, 23, 56, 93, 97, 163, 164; and Clare, 128, 131, 133, 149; and Goethe, 228, 237; and taxonomy, 9; and Wallace, 1
differentiae, 9, 18, 25, 223–24, 225
differentiation, 18, 19, 21–22, 51, 125, 163, 226
Digynia, 32
Din, Shaikh Zain al-, 197
Dioecia, 32
Dionaea muscipula (Venus flytrap), 8, 40, 76, 77, 110, 217, 218
Dirks, Nicholas, 284n5
dissenters, 59, 65, 67, 101, 278n65
Dodecatheon meadia L., 32–33
Dodson, C. H., 260
dog rose, 53
Doherty, Meghan, 76

domesticity, 90, 97, 112, 116
Dorstenia, 48
double columbine, 91
Douglas, Gina, 277n20
Dow, Alexander, 285–86n33; *History of Hindostan*, 170–71
Dragon Arum, 56, 76
dragon trees, 126, 127
Drake, Sarah, 249
drawings, 57, 58, 60, 70, 111, 163, 192; of Frances Beaufort Edgeworth, 113–14; and Smith and Sowerby, 72–73. *See also* artists; illustrations; Indian artists
Drayton, Richard, 4, 165
Drosera (sundew), 110
Druce, George C., 153, 282–83n60; *The Flora of Northamptonshire*, 282–83n60
Dupre, John, 22

East India Company (EIC), 54, 58, 166, 178, 179, 181, 193; business of, 162; and Jones, 15; and Kyd, 180; and Roxburgh, 15, 168, 181, 192, 196, 199, 208; and H. Walpole, 169
East Indian companies, 161
Edgeworth, Frances Anne Beaufort, 14, 111, 112–14, 116
Edgeworth, Maria, 95; *Belinda*, 44
Edgeworth, Michael Pakenham, 113
Edgeworth, Richard Lovell, 95, 111
Edinburgh Review, 64
Edmundson, John, 115
Edwards, Sydenham, *Botanical Register*, 41
Ehret, George, 160
Emmerson, Mrs., 133
empires/imperialism, 1, 5, 14, 23, 51, 56, 62, 174; and botany, 7; and exchange of specimens and drawings, 58; and India, 15, 161, 162, 163, 164, 168; and Jones, 166, 180, 182, 191; and Mootoo, 160; and P. Shelley, 286n38; and taxonomy, 4; and Thornton, 76, 252. *See also* colonialism
Encyclopaedia Britannica, 63–64, 68, 69; "Botany," 41–42
Endeavour (ship), 42, 58, 60, 272n21
English rose, 53
Ephedra, 43
Epipendra, 259
Ereshefsky, Marc, 30, 266n11
Erica, 117
Erica pubescens, 116
Euphorbia, 17
Europe, 55, 161, 163, 164, 165, 198, 207
evolution, 18, 30, 84, 220, 265n7; and C. Darwin, 10, 14, 19, 29, 34, 81
experimentation, 36, 80, 109, 110, 147

families, 19, 20, 37, 38, 39, 80, 118, 146, 148
Fatepur Sikri, 207
fens, 134, 139, 142
Fern Chafers, 124
fern owls (goatsucker bird), 123, 124, 152
ferns, 5, 32, 34, 67, 113, 146, 147, 267n52
Fichte, Johann Gottlieb, 218
Ficus bengalensis (banyan tree), 173, 177–78, 179
Ficus indica, 86, 275n81
Ficus religiosa (bodhi tree), 177, 206
Figuier, Louis, *The Vegetable World*, 23
figures/figuration, 8, 13, 21, 48, 53–54, 56, 111, 220; and Barbauld and Wollstonecraft, 102; and W. Bartram, 63; and Bateman, 257; and Cesalpino, 5–6; and Clare, 131, 132, 135, 147, 149, 152, 153, 154, 155, 156–57, 158; and E. Darwin, 57, 79, 81, 87, 88, 89, 215; of Delany and Grey, 116; and Fothergill, 61; and Goethe, 227, 232, 233; and Jones, 187–88, 190; and Linnaeus, 5; and monstrosity, 3; and P. Shelley, 211, 255; and C. Smith, 121, 122, 124, 125. *See also* allegory; metaphors; personification; synecdoche
Firistah, 170–71
Fitzwilliam, Charles William Wentworth Fitzwilliam, Earl, 131, 248
Fitzwilliam family, 129, 139, 146, 279n21
flag iris, 121, 146, 155
Fleming, John, 291n143
fleur de lis, 53
Flinders, Matthew, 42
flor de corpus, 252
flor de los muertos, 252
flor de los santos, 252
flor de maio, 252
flowers, 32, 37, 94, 120, 123, 129, 222; Grandville's manual for care of, 91; pressed, 14, 70, 111, 115–16, 258, 259; Victorian

language of, 90, 275n1; women as, 90, 97, 101; women as painters of, 92
Fly Ophrys, 76
Forbes, James, 203–4; *A Banyan Tree*, 179; *Oriental Memoirs*, 178, 179; "A View of Cubber Burr," 178
forests, 134, 139, 142
Foster, Georg, 188
Fothergill, John, 77; *Some anecdotes of the late Peter Collinson*, 60
Foucault, Michel, 4, 23, 112, 167, 187, 208
foulroyce, 152–53
Fox, Charles James, 169
France, 6, 67, 118, 120
François, Anne-Lise, 5, 264n11
Franklin, Benjamin, 180
Franklin, Michael J., 191
French Revolution, 64, 65, 67, 142
Frere, J. H., "Loves of the Triangles," 274n72
Fucus peniculus, 41
fungi, 32, 67
furze, 127

galangal, 288n84
Galison, Peter, 3–4
Ganesh, 178
Garrison, Alysia, 145
genetics, 18, 43, 55, 254
Genista (Dyer's broom), 85
Gentleman's Magazine, 64–67, 169
genus/genera, 11, 18, 19, 21, 25, 27, 31, 38; and Bicheno, 22–23; and binomial nomenclature, 32, 46, 47, 48; Cesalpino on, 9; and E. Darwin, 89; and Goethe, 235–36; and Hegel, 224–25; and individuals, 45; and Jackson, 109; and A.-L. de Jussieu, 20, 37, 118; Kant on, 219; and Linnaeus, 20, 21, 22, 33–34; and names, 46, 48; and species, 19; and tribe, 118
George III, wife and daughters of, 58
George IV, 49, 52
Geranium (Pelargonium; cranes, cranebills), 49, 117, 119–20
Gestner, W. E., 208
Gibson, Alexander, 197
Gigante, Denise, 263n3, 264n17
Gilchrist, John, 168
Gillray, James, "New Morality," 274n72
ginger, 288n84
Glaucium phoenicium, 72
Gledhill, D. G., 49, 50, 266n18
Gnetum, 43
goatsbeard, 121
goatsucker bird, 124
Godwin, William, *Fleetwood*, 96–98
Goethe, Johann Wolfgang von, 17, 21, 74, 220, 230, 295n33; and anastomosis, 229–31, 234–35; and contingency, 222, 234, 235, 240; and Cryptogamia, 228; and *Gestalt*, 230; and Hegel, 15, 28, 216, 222, 227, 231, 234, 235, 236, 237–41; and Linnaeus, 229, 232, 236; and mechanism, 8–9; and metamorphosis, 15, 28, 40, 215, 221, 227–37, 238–39, 240; and nature, 227, 229, 230, 231, 232, 234, 235, 236–237, 238, 239, 240, 241, 295n33; and *Naturphilosophie*, 217, 218, 227, 228; and spiral vessels, 231–32; and *Urpflanze*, 227, 228; Works: *Elective Affinities*, 21, 228; *Farbenlehre*, 74; *Faust*, 228, 238; Italian journal of 1787, 228; *Metamorphosis of Plants*, 8, 15, 89, 215, 218, 222, 228–32, 234–35, 237, 238–39, 241; "The Metamorphosis of Plants," 222, 232–34; *On Morphology*, 238; "On the Cotyledons," 227; "Problems," 235; *Scientific Studies*, 236; "The Spiral Tendency," 295n34; *Die Wahlverwandtschaften*, 228
Goldsmith, Oliver, 171, 285–86n33
Goldstein, Amanda, 210, 214
Goodman, Kevis, 132
Goodridge, John, 144
Gould, Stephen Jay, 81
Govindoo (artist), 197
Graham, Maria, 177–78, 199
Graham, Robert, 199
Grainger, Margaret, 153
Graminaceae (grasses), 265n7
Gramineae (grasses), 35
Grammatophyllum speciosum, 252
Grandville, J. J., 90; *Les Fleurs animées*, 91–92
Granium columbium, 49
Gray, Samuel F., 268n57
Great Britain, 55, 67, 120, 160, 169, 182, 191; and India, 15, 161, 162, 163, 164
great flowered jasmin, 186

Greville, R. K., 54–55, 56
Grew, Nehemiah, 267n50
Grey, Katharine Charteris (Lady Jane Grey of Groby), 14, 111, 112, 114, 115–16, 248, 257–58, 259, 278n58
Grey family, 111
Griffith, William, 193
Griffiths, Ralph, 64
Grosz, Elizabeth, 93
Grove, Richard, 4, 185
Guatari, Félix, 13, 254
Gynandria, 32, 86, 107

habitus, 93
Haller, Albrecht von, 217
Hamilton, Elizabeth, 169
Hardwicke, Thomas, 290n124
Hare, James, 208
Harris, Steven J., 284n2
Hastings, Warren, 168, 169, 170, 178, 180, 181, 286n36
headache orchises, 146
Heaney, Seamus, 155
hearts-ease, 53, 91
heaths, 142
Hedwig, Johann, 108; *Species Muscorum Frondosorum*, 110
Hedysarum gyrans, 82, 110
Hegel, G. W. F., 27–28, 216, 220–27, 237–43, 287n42, 295n17; and Adorno, 2, 241–45; and animals, 215; and contingency, 11; and flowering plants, 236; and genera, 235; and Goethe, 15, 28, 216, 222, 227, 231, 234, 235, 236, 237–41; and India, 172–73; and Indian art, 163; and inner directedness, 220; and Linnaeus, 239; and material particulars, 11; and mechanism, 8–9; and metamorphosis, 238–39, 240; and Moore, 174; and nature, 8, 28, 215, 220–21, 223, 225–27, 237, 238, 240–41; and *Naturphilosophie*, 28, 217, 221, 225, 238, 241; and singularity, 8; and spiral vessels, 240; and Spirit/spirit, 28, 215, 218, 221, 237, 238, 239, 240–41, 243, 244, 293–94n3; Works: *Encyclopedia of the Philosophical Sciences*, 218, 227; *Lectures on the Philosophy of Spirit*, 237; *Logic*, 225–26, 227; *Phenomenology of Spirit*, 222–25; *Philosophy of History*, 172–73, 242; *Philosophy of Nature*, 217-18, 222, 227, 234, 238, 293–94n3; *Philosophy of Spirit*, 293–94n3
Heidegger, Martin, 133, 134, 144
Henderson, Joseph, 129, 145, 146, 147, 149–50, 279–80n12, 282n59
Henderson, Peter, 281–82n53
Henry VIII, 111
Hensen, Nilas Thode, 284n2
Herder, Johann Gottfried, 217; *Ideen zur Philosophie der Geschichte der Menschheit*, 227
hermaphrodites, 32, 35, 147, 190–91, 212, 230
hernshaw, 127
Hertford, Isabella Anne Seymour-Conway, Marchioness of, 52–54, 56, 57, 63
Hieracium (hawkweed), 123
Hill, John, 33
Hillia parasitica, 48
Himalayan plants, 165
Himalayas, 194
Hindus/Hinduism, 175, 177, 195, 223
Hodges, William, *Travels in India*, 178
Hofmeister, Wilhelm, 42
Hölderlin, Johann Christian Friedrich, 244
Holland, Jocelyn, 227
hollyhock, 91
hooded aron, 154
Hooker, Joseph Dalton, 194, 208
Hope, John, 70, 192, 193
horse blob, 148
Hudson, William, *Flora Anglica*, 115
Hull, David, 28, 29, 30
human beings, 13, 26, 40, 62, 121, 160, 217, 258; analogy with, 211–14; and Goethe, 227–28, 231, 232–33; and Hegel, 215, 221, 241; Kant on, 219; resemblance to, 82, 247
Humboldt, Alexander von, 10–11, 42, 52, 165, 237, 248, 283n8; *Cosmos*, 10; *Essai sur la géographie des plantes*, 11
Huneman, Philippe, 294n7
Huntington Botanical Gardens, 151
hybrids, 13, 30, 34, 149, 155, 164, 246
hylozoism, 219
Hypericum, 121

Ibbetson, Agnes, 95–96, 110, 277n20
illustrations, 1, 6, 14, 55, 57, 60, 69–78; and Brown, 43; cross sections in, 75;

development of trade in, 6; and engravings, 57, 69, 71, 73, 107; as essential, 70; as group portraits vs. still lifes, 76; magnification in, 75; and mimesis, 13; need for, 55; and picture plane, 199–200; and printing techniques, 6; public taste for, 69; and species type, 200–201; and watercolor, 74, 111, 112, 113–14, 199, 202; by women, 111, 112. *See also* artists; color; drawings; Indian artists; watercolor
Impatiens (touch me not), 86
Impatiens noli tangere (touch me not), 49
Impey, Lady, 197
India, 15, 53, 65, 114, 161, 165, 287n42; and anticolonialism, 169–70; as effeminate, 171, 175; as female, 172–73, 174, 176–77; informants in, 168, 197; and textiles, 206, 207
India Bill of 1783, 169
Indian arrowroot, 160–61
Indian art, 15, 165, 194
Indian artists, 15, 290n120, 290n124, 292n154; and aesthetics, 163; and color, 199, 202–3; and F. Edgeworth, 113–14; exchanges with, 164–65; paints of, 201; perspective of, 168; and picture plane, 199–200; and Roxburgh, 192, 197–209; and Serfoji II, 166; and species type, 200–201; style of, 201–7; techniques of, 290–91n127. *See also* artists
Indian gardeners, 166, 185, 195
Indian Mutiny of 1857, 168, 177
Indian pepper, 192, 196
indigo, 162, 181, 192, 196
individuals, 11, 23, 26, 55, 156, 240; and species, 19, 28–30. *See also* particulars/particularity; singularity
inner-directedness, 8, 218, 219, 220, 221
insects, 76, 84, 111, 157, 218, 247, 260, 261; and Bateman, 250, 251, 256–57, 258; and C. Smith, 120, 123
Irwin, John, 206

Jackson, Maria, 14, 94, 104–10, 121, 277n42, 277n44; *Botanical Dialogues between Hortensia and Her Four Children*, 104–8, 109, 110; *Botanical Lectures by a Lady*, 104, 108–10; *Sketches of the Physiology of Vegetable Life*, 104, 110
Jacquin, N. J., 48
Jama (rose-apple), 177
Jameson, Fredric, 133
Jardine, Nicholas, 39, 264n16
jasmine, 186
Játi, 186
Johnson, Joseph, 277n44
Johnson, Samuel, 26, 286n39; *Dictionary of the English Language*, 24
Jones, Anna Maria, 179
Jones, William, 15, 65, 165, 171, 179–91, 204, 286n36, 286n39; and imperialism, 162; and Indian knowledge, 186; and Indian literature and culture, 173; informants of, 168, 186–87, 191; and Linnaeus, 183–85, 186–87; and Moore, 174; and Owenson, 175, 177; and Priestley, 180; and religion, 65; and Roxburgh, 196; sources of, 181, 182, 185; and spikenard, 288n70; Works: "Anniversary Discourse," 182; "Botanical Observations on Select Indian Plants," 181, 185–87, 191; *Britain Discovered*, 191; "Catalogue of Indian Plants," 185; "Design of a Treatise on the Plants of India," 162, 179–80, 181, 184; *The Enchanted Fruit or the Hindu Wife*, 181, 184, 189–91; "Hymn to Ganga," 191; letters from India, 179; letters to Davis, 179; letter to Banks, 182–83, 189; letter to Russell, 183; "The Religious Use of Botanical Philosophy," 65; *Sacontalá*, 181, 188–89, 191, 204
Josephine, Empress, 52, 53, 56, 57, 63
Jussieu, Adrien de, 36
Jussieu, Antoine de, 36
Jussieu, Antoine-Laurent de, 34, 40, 42–43, 118, 269n79; *Genera Plantarum*, 20, 36–37, 89; and Natural System, 20, 21, 23, 36–37, 89, 264n19
Jussieu, Bernard de, 36, 37, 39
Jussieu, Joseph de, 36

Kaempferia rotunda, 185
Kālidasā, 181, 188–89
kamalam, 207
Kant, Immanuel, 27, 44; *Critique of Judgment*, 8, 217, 218–20, 294–95n15, 294n17; *Metaphysical Foundations of Natural Science*, 219
Kantianism, 13, 56

Keats, John, 64, 176; *Hyperion* poems, 124
Kent, Elizabeth, 282–83n60, 282n55; *Flora Domestica*, 148, 150
Kew Gardens, 58, 60, 70, 74, 192, 203
kingdoms, 7–8, 11, 13, 68, 109, 115, 122; and Bateman, 250, 251; and Clare, 124, 128–29, 151, 155; and E. Darwin, 82, 87, 88, 89; and Five Kingdoms scheme, 41; middle position among, 6, 7, 216–17, 241
King-Hele, Desmond, 274n72
Knowles, Mary, 98–99, 277n27
Koenig, Johann Gerhard, 181, 183, 184, 192, 284n12
Kopytoff, Igor, 71
Koselleck, Reinhard, 242
Krishna, 204, 206
Kriz, Kay Dian, 76
Kumar, Deepak, 284n2
Kyd, Robert, 180–81, 291n143

ladysmock, 150
Lakshman Singh, 197
La Mettrie, Julien Offray de, *L'Homme Machine*, 210–11
La'ngali, 186–87
Larson, 20, 265n7
Latin/Latin nomenclature, 63, 68, 113, 117, 195; and Clare, 127, 147, 149, 150, 155. *See also* nomenclature
Latour, Bruno, 59, 133
laurel, 52
Leask, Nigel, 171, 172, 175, 285–86n33, 286nn38, 39
Lee, Ann, 60
Lee, James, 57–58, 59–60, 63, 272n17, 272n21; *Introduction to the Science of Botany*, 59
Leea macrophylla, 196
Leea macrophylla (Roxburgh Indian artist), 204
Leibniz, Gottfried, 34, 38
Le Moyne de Morgues, Jacques, "A Young Daughter of the Picts," 90–91
Lepanthes calodictyon, 151
Levine, Caroline, 56–57
Lichen pyxidalus, 113
lichens, 5, 32, 34, 41, 42, 113
life, 7, 8, 215, 217, 218, 241, 294n7; animate, 11, 34, 84, 215; material forms of, 210; and materiality, 215; and P. Shelley, 210, 211, 214, 215. *See also* animation; vitality
Liliaceae (lilies), 100–101
Lindley, John, 82, 94, 259; *Ladies Botany*, 41, 94
Linnaean system, 10, 14, 21, 56, 57, 66, 67, 160; and artisans and amateur naturalists, 265n6; and Bicheno, 36; and Brown, 75; and Buffon, 22, 34–35; and Clare, 128, 147–48, 150; and C. Darwin, 19; and E. Darwin, 79, 80–81, 83, 84, 85, 86, 87, 88, 89; and Delany, 115; and F. Edgeworth, 113; and Goethe, 230–31; and K. Grey, 116; and Jackson, 14, 107; and Jones, 182–85, 187, 188, 189–90; and A.-L. de Jussieu, 89, 118; and Martyn, 94; political economy of, 21; and reproductive parts of flowers, 19–20; and Rousseau, 94; and Roxburgh, 195; and sexuality, 56, 57, 147–48; and C. Smith, 117, 121; subdivision of, 22; and Thornton, 75; and Wakefield, 103, 104
Linnaean systematic, 14, 19–20, 85, 147
Linnaeus, Carl, Jr., 107
Linnaeus, Carl von, 3, 30, 64, 65, 66, 269n79; and affinities, 37; and amnesiac, 45–46; and artificial vs. natural systematics, 20; and Bartram, 53; and Bicheno, 22–23; and binomial nomenclature, 32; and binomial system of Latin names, 5; and Blackburne, 66; and Buffon, 64; categories of, 22; and Clare, 15, 127; and classification, 4; and climate, 34; and color, 73, 149; correspondents of, 24; and creation of species, 61; and Cryptogamia, 5, 6, 25, 228; and cryptogams, 40, 109–10; and E. Darwin, 8, 11, 56, 57, 59, 80, 86, 89, 148; and *differentiae*, 18; and figuration, 5; and Foucault, 4; and gaps in nature, 38; *Genera Plantarum*, 80; and Glasnevin botanic institution, 68; and Goethe, 228, 229, 232, 236; and Hegel, 239; and images, 70; and India, 165–66; and Jackson, 104, 106, 108–10; and A.-L. de Jussieu, 20; and Jones, 181, 184, 186–87; and Latin, 9, 33, 46, 47, 63; and J. Lee, 59; and *Monthly Review*, 68; and Mootoo, 160; and names, 9, 10, 46; and natural plant families, 35; and

networks, 58, 59; and nomenclature, 5, 46, 47, 48; and orchids, 107; *Philosophia Botanica*, 18, 35, 70; and reproduction, 20, 32, 37, 221; and Roxburgh, 195; and Sanskrit, 182, 184; and sexuality, 5, 23, 31, 33, 98, 230–31; and P. Shelley, 212; and C. Smith, 123; and J. E. Smith, 36, 266n16; and species constancy, 266n10; species identity, 23; *Species Plantarum*, 6, 18, 48, 63, 166, 195; *Systema Naturae*, 33; and systematics, 4; and taxonomy, 18; and Thornton, 76; and tripartite division of nature, 241; and visibility, 5; and Wakefield, 103, 104; and W. Bartram, 62, 63; and women, 98
Linnean Society of London, 36, 42, 59, 64, 72, 110
Lobelia, 86
Locke, John, 25–26, 30, 34; *Essay concerning Human Understanding*, 9
Loddiges, Conrad, 116
London Medical Journal, 65
lotus, 77, 195, 207
Loudon, J. C., *Hortus Britannicus*, 36
Lucas, Hans-Christian, 237
Lunar Society, 80, 98, 101, 180
Lychnis, 108–9
Lynch, Deidre, 90
Lyotard, Jean-François, 50

Mabberley, David, 265n2
Mabey, Richard, 126, 127–28, 151, 154–55, 246–47, 251; *Nature Cure*, 128, 246–47
Macauley, Thomas Babington, *Minute on Education*, 172
MacKenzie, Colin, 284n12
Mackenzie, Henry, *Man of Feeling*, 170
Macpherson, James, 286n39
Mádhavi, 188–89
Madhuca, 183–84
Magnol, 39
Mahábhárata, 188, 190
Mahesh of Chamba, 292n164
maidenhair fern, 146, 147
Malabar coast, 165–66, 178, 187, 197
Málatí, 186
Mallicá, 188
Mancinella, 86
Mangifera indica (mango tree), 195
Mann, Horace, 169
Mansur, 206
Marat, Jean-Paul, 278n65
Marsh, Herbert, 133
Marsh, Marianne, 133
Martyn, Thomas, *Letters on the Elements of Botany*, 94
Masdevallias, 259
Massey, William, 61
matter/materiality, 2, 10, 14, 54, 55–56, 57, 113, 210; and Delany, 111, 114, 116–17; experience of, 3; and figural possibility, 53–54; and Goethe, 8, 215, 227, 238, 240; and K. Grey, 116–17; and Hegel, 221, 226–27, 238, 239, 241, 243; and life, 215; and mind, 217, 220; and P. Shelley, 214
Maturin, Charles Robert, *Melmoth the Wanderer*, 286n36
Mayr, Ernst, 30
mechanism / mechanical processes, 210, 215, 217, 218, 219, 220, 221, 240; and Goethe, 218, 227, 231–32; and Goethe vs. Hegel, 8–9, 218
Mendel, Gregor, 29
Mercurialis perennis (dog mercury), 152
Merian, Maria Sibylla, 76, 77
Mesipilus germanica (medlars), 137
metaphors, 9–10, 21, 47, 78, 81, 151, 157. *See also* figures/figuration
Metcalf, Thomas, 284nn4, 5, 287n49, 296n3
metonymy, 55, 257. *See also* figures/figuration
Meyer, Ernst, 235, 236
microscopes, 2, 25, 27, 35, 40, 43, 95, 111; and Jackson, 110; and Natural System, 20; and C. Smith, 119; and Smith and Sowerby, 73
middle class, 60, 69, 111, 112
Mill, James, *History of British India*, 172, 286n39
Miller, Elaine, 229
Miller, Eric, 128
Miller, Philip, 33–34, 64; *Gardeners Dictionary*, 33, 59, 272n17
Milton, John, 124, 259, 260; *Paradise Lost*, 173, 177
Milton Hall, 129, 146, 279–80n12
mimicry, 13, 16, 164, 250–51, 258, 262; and Clare, 151, 152; and C. Smith, 122, 123, 247

mimosa, 34, 40
Mimosa, 83
Mimosa pudica, 211, 212
Mimosa urens, 161
mind, 28, 215, 217, 220, 236
minerals, 4, 7–8, 216–17, 218, 241
Mirabilis jalapa (marvel of Peru, four o'clock), 91
Mjoberg, Bengt, 75
Monandria, 32
Monoecia, 32
Monogynia, 32
Monson, Anne, 290n124
monstrosity, 25, 29, 31, 34, 44–45, 53, 61–62, 260; and Aristotle, 44; and Bateman, 251, 259, 260; and Brown, 43–44, 45, 260; and E. Darwin, 78, 79, 81, 84–85, 86, 87, 88; and Goethe, 222, 229; and taxonomy, 3, 6, 27; and Wallace, 1, 2–3, 43, 44, 251
Monthly Magazine, 68
Monthly Review, 64, 67–68, 286n36
Moore, Lisa, 114
Moore, Thomas, 168, 174; *Lalla Rookh*, 92–93, 170, 172, 174, 177, 190, 286n36
Mootoo, Shani, *Cereus Blooms at Night*, 15, 159–60
morality, 83, 87–88, 117, 118, 119, 120
morphology, 13, 14, 20–21, 35, 37, 149, 222, 237
Morton, Allen, 265n2
Morton, Timothy, 131
mosses, 32, 34, 41, 267n52
mountain thyme, 123
mourning iris, 91
Mughals, 171, 182, 197, 201, 202, 206, 207, 290n120

Nair, Savithri Preetha, 166, 284n12
Nama of Sila'n, 186
names, 7, 11, 24, 31, 48–49, 67, 160, 194–96; and animation, 41; and Bateman, 249–50, 257; common, 32, 47, 48–49, 50, 113, 117, 123, 127, 128, 131, 147, 148–49, 150, 154, 155, 157; and Delany, 115; and discoverer or first describer, 47, 48, 56, 266n23; and F. Edgeworth, 113; and figures, 56; and genus, 46, 48; and Goethe, 232; and Greville, 55; and K. Grey, 116; and India, 161, 163, 165; and Jones, 182–83, 184, 185, 187, 188, 189; and letter of Marchioness of Hertford, 63; and Linnaeus, 9, 10, 46; local Mexican, for orchids, 252; and Mabey, 246–47; and *Monthly Magazine*, 68; rules about, 47–48; and P. Shelley, 256; and C. Smith, 117, 121, 123, 124, 125. *See also* Latin / Latin nomenclature; nomenclature; *under* Clare, John
Nancy, Jean-Luc, 142
Napoleon I, 52
Napoleonic Wars, 120
Natural History Museum, John Fleming collection, 198
"Naturalist's Ramble in the North, A," 66
Natural System, 10, 18, 19, 32, 89, 94, 149, 160; and Bicheno, 22–23; and Brown, 36, 42; development of, 5, 14, 36–45; and A.-L. de Jussieu, 20, 89; and morphology, 20–21, 37
Naturphilosophie, 8, 15, 40, 217, 218, 220, 294n7; and Goethe, 227, 228, 238; and Hegel, 28, 221, 225, 238, 241
Nelumbium speciosum (Tamara alba), 195, 203
Nelumbo lutea (American lotus), 77
Neoplatonism, 211, 213, 214, 215
Nepenthes (pitcher plants), 40
Nepenthes distillatoria, 259
New Asiatic Miscellany, The, 171, 286n36
Newtonianism, 215
Ngai, Sianne, 133, 144, 151, 263n3
Nickelsen, Karin, 71, 72
Noltie, Henry J., 58, 166, 197, 199, 200, 268n66, 269n116, 290n118
nomenclature, 14–15, 19, 66, 67, 68, 72, 80, 117, 160; and Bateman, 249–50; debates about, 6–7, 14, 56; development of, 45–50; and Ibbetson, 95; and Jones, 189–90; and Linnaeus, 5, 19; and Rousseau, 94. *See also* Latin / Latin nomenclature
no me olvides, 252
Northern Circars, 197
nyctanthes (sweet-night flower), 174
Nyctanthes tristis, 186
Nymphaea alba, 195
Nymphaea lotus (Roxburgh Indian artist), 203
Nymphea Lotos, Blue Water-Lily of Guzerat (Forbes), 203–4

Ocimum basilicum, 289n111
Ocimum caryophyllatum, 196
Ocimum gratissimum, 196
Ocimum sanctum, 196, 289n111
Ocymum Goolaltulasi, 196
Ocymum Ram-tulasi, 196
Odontoglossia, 249
Oerlemans, Onno, 263n3
Oken, Lorenz, 217; *Lehrbuch der Naturphilosophie*, 220
Oldenlandia umbellata (madder root), 194
Ondaatje, William, 35, 268n66
ophrys, 50, 76, 123, 157, 266n16
Ophrys Loefelli (dwarf Ophrys), 266n16
Ophrys muscifera, 123
opium, 181
Opulus, 84
Opuntia, 192
orange lily, 91
orchids, 13, 49, 116, 194, 218, 251–52, 258, 259, 260; and Bateman, 15–16, 247–60; and Brown, 42; *Catasetum*, 251–52; *Catasetum maculatum*, 256; Central American, 249; *Cycnoches ventricosum*, 116, 258, 259; and C. Darwin, 260–62; and F. Edgeworth, 114; and K. Grey, 115–16; hybridizing behavior of, 252; and Mabey, 246–47; and mimicry, 16, 258; popular craze for, 249, 252; rara avis (rare bird), 116; Venus's slipper, 259
orchises, 13, 20, 50, 107, 109, 246–47; bee, 121, 150, 157, 261; and Clare, 121, 129, 146, 149–51, 157–58, 247; fly, 121; headache, 146; and Jackson, 109; and C. Smith, 121, 123, 247
Orchis mascula (early purple orchis), 150
Orchis morio, 150
Orme, Robert, 180
Ovid, P. Naso, 82, 186
Owenson, Sydney, 168; *The Missionary*, 174, 175–77, 286n36
ox-eye daisy, 152
Oxeye Wood, 152

Paine, Thomas, *Rights of Man*, 102
Palmae, 35
Palmira, 86
Pandanus odoratissimus, 195, 202
Papilionaceae, 35
Parkinson, Stansfield, 60
Parkinson, Sydney, 58, 60, 64–65; *Journal of a Voyage to the South Seas*, 65
particulars/particularity, 2, 3, 7, 11, 13, 18, 19, 71; and Adorno, 242, 243, 245; and Clare, 128, 132, 133, 154, 155, 157; and Goethe, 237, 240; and Hegel, 28, 222, 241, 243. *See also* individuals; singularity
Passiflora (passion-flower), 107
Pavord, Anna, 46, 265n7
Peltandra virginica (arrow arum), 77
Penn, William, 103
Pennant, Thomas, 59, 272n17; *Genera of Birds*, 146
Pentandria Monogynia, 33
pepper, 160
personification, 12, 63, 76, 90, 120, 121, 159, 211; and Clare, 141, 145, 152, 154, 155, 156; and E. Darwin, 78, 79, 82, 83, 85, 89. *See also* figures/figuration
Philosophical Magazine, 64
Philosophical Transactions of the Royal Society, 64
phylogeny, 10, 19, 29, 30
pichwai (painted cloths), 204
pistils, 89, 103, 107, 115, 229, 230, 231; and Linnaeus, 5, 19, 20, 32, 107
Pitt, William (the younger), 169
Plantago media (hoary plantain), 46, 47
Plants of Righteousness, 61
Plato, 18, 25
Platonism, 38, 39
Poa Cynosuroides, 196
Pole, Elizabeth, 81
Pollan, Michael, 13
Polwhele, Richard, 99, 278n65; *The Unsex'd Females*, 95, 117–18
Polyandria Polygynia, 67
Polygamia, 32, 35
Poovey, Mary, 290n122
Pope, Alexander, 138; *The Rape of the Lock*, 191
Porter, Tom, 146, 149, 279n12
Portland, Margaret Cavendish Bentinck, Duchess of, 94
Powis, Earl, 248
Prakash, Gyan, 166–67
Pratt, Mary Louise, 283n8

Price, Richard, 180; A *Discourse on the Love of Our Country*, 65
Priestley, Joseph, 65, 66, 98, 101, 180
Priestley, Mary, 98, 99, 101
Priestley family, 278n65
print culture, 57, 69, 70, 93
Pterocaulon ycnostachyum (black root), 77
Pterostylis obtusa, 116
Pulteney, Richard, 64

Quakers, 14, 45, 59, 60–61, 62, 103, 272n21, 278n65

Radcliffe, Emma, 36
Radstock, Lord, 133, 135, 248, 279–80n12
Rafflesia arnoldii, 43, 75, 260
ragged robin, 152
ragwort, 154
Raina, Dhruv, 284n2
Raj, Kapil, 284n2
Rajan, Balachandra, 171, 173, 174, 285–86n33, 286n36
Ranunculus aquaticus, 229
Ranunculus repens (creeping crowfoot), 49
rattan palm, 205
Rauch, Leo, 226
Ray, John, 9, 10, 37, 38, 48, 127, 267n50
Raynal, Guillaume-Thomas, 172, 286–87n41; *L'Histoire des deux Indes*, 15, 159, 161
Red Fort, Diwan-i-Am (Delhi), 207
reed bird, 121
Reeder, Kohleen, 115
reproduction, 5, 18, 26, 30, 38, 109–10, 113; and Cryptogamia, 19–20, 25; and Goethe, 229, 230–31; and Linnaeus, 19–20, 32, 37, 221. *See also* sexuality
Rhapir rotang, 205
Rheede, Hendrik Adriaan van, 181, 187; *Hortus Indicus Malabaricus*, 166, 184–85, 186, 194
Richards, David, 91
Richardson, Samuel, *Pamela*, 90
Ritvo, Harriet, 266n23
Robinson, Eric, 136
rose campion, 91
Rousseau, Jean-Jacques, 65, 93–94, 95, 142, 161, 214–15, 248; and Barbauld, 101; *Émile*, 94, 108; *Julie, or La nouvelle Héloise*, 117; *Letters on the Elements of Botany*, 94; letters to Margaret Cavendish-Bentinck, 94; letters to Madeleine-Catherine Delessert, 94; *Lettres élémentaires sur la botanique à Madame de L*, 93–94, 101; *Reveries of the Solitary Walker*, 94; and P. Shelley, 210; and C. Smith, 117; and Wakefield, 103
Roxburgh, William, 15, 58, 207, 209, 291n143; and E. Darwin, 85; *Flora Indica*, 192–93, 194, 195, 196, 197, 203, 205, 289n114; and Indian artists, 168, 197, 198–208; and Indian flora, 181, 192–97; *Plants of the Coast of Coromandel*, 67, 85, 181, 192, 193–94, 195, 197, 199, 202, 204
Roxburghia gloriosoides, 181
Roxburgh Indian artists: *Bombax heptaphyllum*, 204; *Calamus rotang*, 205; *Menyanthes cristata*, 205; *Nelumbium speciosum*, 203; *Pistia stratiotes*, 205; *Trapa bisnosa*, 202–3; *Tricosanthes heteroclita*, 205–6
Royal Botanic Garden, Edinburgh, 198, 199, 204, 290n118
Royal Botanic Garden, Kew, 58, 60, 70, 74, 192, 203
Royal Society, 9, 64
ruby-throated hummingbird, 77
rue, 53
Rumphius, Georg Eberhard, 251, 252
Rungiah (artist), 197
Russell, Patrick, 182, 184, 284n12, 288n70
Rusticus, "Progress of Botany in England," 64

Saccharum exaltatum (Roxburgh Indian artist), 204
Sachs, Aaron, 283n8
Sachs, Julius von, 37, 38, 227, 228, 229, 232, 265n2, 266n10
Śakuntalā (Kālidasā), 181, 188–89
Said, Edward, 174
Saint-Aubin, Charles Germain de, 92
sallow palms, 149
Sanskrit, 182, 183, 184, 185, 195, 196
Sarracenia (pitcher plants), 40, 259
Sarracenia adunca (pitcher plants), 110
Schelling, F. W. J., 217, 218, 220, 294–95n15
Schiebinger, Londa, 160
Schiller, Friedrich, *Naïve and Sentimental Poetry*, 27

Schlegel, Friedrich, 173, 218, 225
Schomburghk, Robert, 257
Schomburgkia tibicinis, 257
Scitaminea, 288n84
Secord, Anne, 58, 63, 93, 265n6
Sempervivum (houseleek, welcome home husband however drunk you be), 49
sensitive plant, 34, 77, 83, 92, 110, 217, 218
Sép'halicá, 186
Serfoji II, 166, 197, 284n12
Seward, Thomas, 84
sexuality, 5, 6, 35, 88, 96–98, 177, 268n57; and Bateman, 250, 251–52; and Clare, 147; and E. Darwin, 21, 33, 51, 57, 78, 79, 80, 81, 82–89; and F. Edgeworth, 114; and Goethe, 229, 230–31, 233, 234; and Hegel, 239; and Jackson, 104, 106–7; and Jones, 184, 188–91; and Linnaeus, 5, 21, 33, 98, 230–31; and Mootoo, 159–60; and Polwhele, 118; and P. Shelley, 212; and Thornton, 76; and Wakefield, 103; and Withering, 80; and Wollstonecraft, 101–2. *See also* reproduction
Shakespeare, William, 26, 49, 250, 251; *A Midsummer Night's Dream*, 87; *The Tempest*, 214
Sharafuddin, Mohammed, 170
Shelley, Mary, *Frankenstein*, 43, 44, 62
Shelley, Percy Bysshe, 168, 177; *Alastor*, 176, 253–56; *The Assassins*, 171–72; and Bateman, 253–56; and imperialism, 171–72, 286n38; and love, 211, 212, 213; *Philosophical View of Reform*, 171; *Prometheus Unbound*, 174–75; *The Sensitive-Plant*, 15, 210, 211–14, 215; *Triumph of Life*, 15, 210, 214–15; vitality and decay in, 210–15
Shipley, Bishop, 180
Shteir, Ann, 93, 95, 276n17
Siegesbeck, Johann G., 268n57
signatures, doctrine of, 4, 6
similes, 78, 81, 87, 124, 255. *See also* figures/figuration
Simpson, David, 132, 133, 155
Simpson, Ned, 279n12
singularity, 8, 11, 19, 28, 29, 128–29, 133, 156. *See also* individuals; particulars/particularity
Skinner, George, 256
slavery, 11, 15, 56, 65, 66, 67, 139, 142; and Grandville, 92; and Hastings, 180; and Latin America, 160; and C. Smith, 119, 120
Sloane, Hans, 44
Smith, Adam, 44
Smith, Bernard, 200
Smith, Charlotte, 14, 77, 94, 110, 116, 117–25, 247; *Beachy Head, Fables, and Other Poems*, 117, 120, 121–25; *Conversations Introducing Poetry*, 117, 119, 120, 121; *Elegiac Sonnets*, 118; *The Emigrants*, 118, 120; "Flora," 120–21; *History of Birds*, 117; "The Horloge of the Fields," 121; "The Kalendar of Flora" ("Wildflowers"), 121; *Minor Morals*, 117, 119, 120; *Poems*, 118, 120–21, 122, 123, 124; *Sketches of Natural History*, 117; "To a Geranium," 119–20; "To the fire-fly of Jamaica," 119; "A Walk in the Shrubbery," 118; *The Young Philosopher*, 117
Smith, Pamela H., 197
Smith, James Edward, 17, 41, 42, 58, 73, 77, 150, 201; *English Botany*, 60, 66, 72–73, 121, 123, 150, 201; and Ibbetson, 95–96; *Introduction to Physiological and Systematical Botany*, 110; and Jackson, 110; and Linnaeus, 36, 266n16; and names, 49–50; and orchids, 261; *Plantarum Icones*, 68; and reviewers, 66, 67; review of *Flora Britannica*, 41; and C. Smith, 121, 123; and Sowerby, 60, 68, 72–73; and working class botanists, 63
Snyder, Gary, 134
Solander, Daniel Carlsson, 115
"Sorrowful Nyctanthes," 186
Southey, Robert, *The Curse of Kehama*, 286n36
Sowerby, James, 49, 58, 60, 66, 72–73, 74, 150; and documentary rigor, 201; *English Botany*, 60, 66, 72–73, 121, 123, 150, 201; on monstrosity, 61–62; and Wollstonecraft, 97
species, 9, 10, 20, 24–25, 32, 34, 56, 118; and appearance, 24, 25, 26, 27; and Aristotle, 30; and Clare, 129, 148, 149, 150, 156; and creation, 37, 61; and E. Darwin, 80, 89; debates about, 14; and *differentiae*, 9, 18; and *eidos*, 24, 26; emergence and death

species *(cont.)*
of, 38; as fixed, 21, 24, 25, 34, 80, 266n10; and Goethe, 235; and Hegel, 28, 224–25; and Indian illustrations, 200–201; and individuals, 19, 28–30; Kant on, 219; as kind and appearance, 35; and Le Moyne, 91; and Linnaeus, 19, 21; and Locke, 25–26; as manageable and coherent, 51; and names, 46, 48; newly discovered, 21–22; and nominal vs. real, 25–26; number and kind of as constant, 72; plant as representative of, 54; and Plato, 25; and reproductive parts, 5; and resemblance across kingdoms, 7–8; and singularity, 8; Sowerby on, 62; and taxonomy, 9, 31; as term, 24
specimens, 1, 55, 56, 72, 107–8, 113, 115
Spectator, 26
Spenser, Edmund, *The Faerie Queene*, 173
spikenard, 182, 185, 288n70
spiral vessels, 40, 231–32, 240
Spirit/spirit, 7, 8, 11, 172, 173, 174, 217, 245; and buds/flowers analogy, 222–23; as crushing innocent flowers, 220–21, 222; and differences and contradictions, 224; and Goethe, 218, 238, 239, 240, 241; and Hegel, 28, 215, 218, 221, 237, 238, 239, 240–41, 243, 244, 293–94n3; and nature, 217, 220–21, 225, 226, 227
Spivak, Gayatri, 166, 167, 208
Stael, Mme. de, 58
Stafford, Barbara, 28
Stafleu, Franz, 40
Stahl, Edmund, 217
stamen/stamina, 33, 35, 89, 103, 107, 115, 183, 184; and Goethe, 230, 231; and Linnaeus, 5, 19, 20, 32, 107; and Wollstonecraft, 99, 102
Stamford Champion, 157
Stearn, William T., 264n15, 264n19, 268nn56, 57
Stemona tuberosa, 181
Stewart, Frances Anne, 112
stinking Pothos, 76
Strelitzia reginae (crane flower, bird of paradise), 49, 74
subalterns, 166–67, 186, 187, 209
sugarcane, 160
Suleri, Sara, 167
sunflower, 83–84
sycamore trees, 35
symbols, 27, 29, 206
synecdoche, 75, 193–94, 233. *See also* figures/figuration
Synesthesia, 84
Syngenesia, 32, 83
Syngenesia Monogynia, 35

Tabernar, Ann, 138
Tahiti, 79, 88
tamarind, 184
taxa, higher, 19, 20; liliaceous, papilionaceous, and umbellate plants, 94
taxonomy, 1, 8, 23, 30, 57, 58, 71, 95; and affinities, 37; and Bateman, 249–50; and Clare, 128, 145, 146, 147, 148, 149, 150, 153, 154, 155, 156; comparison in, 37–38; and E. Darwin, 80, 84; debates about, 14; and difference, 9, 40; and Goethe, 230–31; and Greville, 55; and Hegel, 223–24, 225–26; and imperialism, 4; implications of, 17–18; and India, 164; and Indian artists, 209; intuitive judgments in, 38, 39; and Jackson, 104, 107; and Jones, 182–83, 190; and Linnaeus, 18; and Locke, 25; and monstrosity, 3, 6, 27, 44, 45; and Mootoo, 159, 160; rigor in, 56; and Roxburgh, 194–95; and singularity, 28; and C. Smith, 118–19; Sowerby on, 62; and species, 9, 31; specimens and illustrations in, 55; and *tâtonnement*, 39. *See also* classification
tazetta narcissus, 91
tea, 181
Tennant, William, 199
Theophrastus, 18, 19, 265n7
Thomson, James, 138
Thornton, Kelsey, 144
Thornton, Robert, 272n21; production of *Temple of Flora* by, 69, 73, 75–76; *The Temple of Flora*, 56, 93, 159–60, 193, 201, 248, 252
thyme, 127
tiggalu, 207
Titan arum, 43
tobacco, 181
Tobin, Beth, 200
Tournefort, Joseph Pitton de, 31, 37

Transactions of the Linnean Society, 36, 64
Trapa bisnosa (*Trapa natans*, water chestnut), 84–85, 202
Trattinnick, Leopold, 73
Trichoceros tupoipi, 151
Triodopsis albolabris (Whitelip Snail), 77
tulips, 91, 174
turmeric, 288n84
Turner, Dawson, 41, 67
Turner, J. M. W., 28, 45, 200; *Snow Storm*, 28; *Ulysses Deriding Polyphemus*, 28
Turner, William, *A New Herball*, 49, 50
tutsan-leaved dog's-bane, 76

Uglow, Jenny, 80, 98
Umbelliferae, 20, 38, 265n7
University of Edinburgh, 58, 70
Upas (hydra-tree of death, poison tree), 43, 86–87
Urtica, 86

Vaillant, Sebastien, 31, 267n50
Valeriana jatamansi, 181
Vallisneria spiralis, 110
Vanàdósini, 188, 189
Vardy, Alan, 131, 150, 281n53
variation, 30, 126, 128, 153
varieties, 26, 34, 62, 68
Vedas, 206
vegetable lamb, 218
Venus's slipper orchid, 259
Viola tricolor (meet her i'th'entry kiss her i'the'buttery, leap up and kiss me), 49
Virgil, 214; *Eclogue II*, 100
Vishnu Prasad (Vishnupersaud), 197
vitalism, 8, 15, 40
vitality, 88, 241; and Goethe, 8, 232, 239, 240; and Hegel, 215, 225, 241; and P. Shelley, 210, 211, 215. *See also* life

Wakefield, Priscilla, 94; *Introduction to Botany*, 103–4
Wales, James, *Cubeer Bur, the Great Banyan Tree*, 178
Wallace, James, 5, 7, 8, 11, 17, 263n4; and monstrosity, 1, 2–3, 43, 44, 251, 260; "Part of a Journal," 1
Waller, Mary, 112
Waller, Robert, 112
Wallich, Nathaniel, 192–93
Wallich, Robert, 54–55, 56, 58
Walpole, Horace, 99, 169–70, 263n4, 285n24
Walpole, Robert, 285n24
Warrington Academy, 101
wastes, 139
water blob, 148
watercolor, 74, 111, 112, 113–14, 199, 202
Watt, James, 98
Wedgwood, Josiah, 98
Welch, Stuart Cary, 206
Wellington, Marquis of, 52
Whatman paper, 198, 205
wheat, 70, 74, 75
wheat ear, 121, 154, 155
White, Gilbert, 56; *Natural History of Selbourne*, 148
Wight, Robert, 54–55, 58, 197, 290n118; *Prodromus Florae Peninsulae India Orientalis*, 268n66
Wilkins, John, *Essay towards a Real Character*, 10
Willdenow, Carl Ludwig, 270n116
Williams, Raymond, 127, 133
Williams, William Carlos, 157
Wilson, J., 66
Withering, William, 80, 98; *An Arrangement of British Plants*, 103, 104, 113; *A Botanical Arrangement of British Plants*, 67
Withers, Augusta, 248, 249
woe water, 126, 127
Wollstonecraft, Mary, 14, 58, 95, 97–98, 277n44; and Barbauld, 99–103; *A Vindication of the Rights of Woman*, 94, 100, 101, 106
women, 14, 58, 93, 103, 116, 268n57, 278n65; as amateurs, 111, 112; as artists, 69, 70, 93, 111, 125; and authority, 103; and Barbauld, 99–103; and beauty, 92, 100; as botanical ornamentation, 90; and botanical production, 93; and children, 93; and E. Darwin, 81, 85, 87, 89; and Delany, 114; and education, 93–95; as floral grotesques, 92; as flowers, 90, 91–92, 97, 101; and Godwin, 97; and Grandville, 91–92; handcoloring by, 69; Hegel on, 173; and illustrations, 111, 112;

women *(cont.)*
and Jackson, 104–7; and Jones, 190; and Linnaeus, 98; and Lunar Society, 98; male writing about, 100; and ornament, 93; and Owenson, 177; painting of flowers by, 92; as passive flowers, 92, 93; polite accomplishments for, 102–3; as public figures and authorities, 95; and Rousseau, 94, 95; and Saint-Aubin, 92; and sexuality, 98; and Shelley, 177; as virginal, 90; and Wollstonecraft, 99–103, 106
Wordsworth, William, 52, 130–31, 156, 157, 172, 213, 254, 255; and Bateman, 256; and Brown, 42; and individuals, 28, 71; vantage point of, 12–13; Works: advertisement to *Lyrical Ballads*, 274n72; *Intimations Ode*, 28, 71, 156; "Nutting," 157; "Preface to *Lyrical Ballads*," 157; *The Prelude*, 12–13; *Ruined Cottage*, 254; "The Solitary Reaper," 28; "The Thorn," 131
working class, 58–59, 63
Wright, William, *Account of the Medicinal Plants growing in Jamaica*, 65

yellow horned poppy, 91
yellow sidesaddle flower, 76
yoni and lingam, 177, 178

Zingiberales, 288n84